A GUIDE

Coastal Ecosystem Processes

 *Marine Science Series*

The CRC Marine Science Series is dedicated to providing state-of-the-art coverage of important topics in marine biology, marine chemistry, marine geology, and physical oceanography. The Series includes volumes that focus on the synthesis of recent advances in marine science.

CRC MARINE SCIENCE SERIES

SERIES EDITORS

Michael J. Kennish, Ph.D.
Peter L. Lutz, Ph.D.

PUBLISHED TITLES

The Biology of Sea Turtles, Peter L. Lutz and John A. Musick
Chemical Oceanography, Second Edition, Frank J. Millero
Coastal Ecosystem Processes, Daniel M. Alongi
Ecology of Estuaries: Anthropogenic Effects, Michael J. Kennish
Ecology of Marine Bivalves: An Ecosystem Approach, Richard F. Dame
Ecology of Marine Invertebrate Larvae, Larry McEdward
Morphodynamics of Inner Continental Shelves, L. Donelson Wright
Ocean Pollution: Effects on Living Resources and Humans, Carl J. Sindermann
Physical Oceanographic Processes of the Great Barrier Reef, Eric Wolanski
The Physiology of Fishes, David H. Evans
Pollution Impacts on Marine Biotic Communities, Michael J. Kennish
Practical Handbook of Estuarine and Marine Pollution, Michael J. Kennish
Practical Handbook of Marine Science, Second Edition, Michael J. Kennish

FORTHCOMING TITLES

Chemosynthetic Communities, James M. Brooks and Charles R. Fisher
Environmental Oceanography, Second Edition, Tom Beer
Handbook of Marine Mineral Deposits, D.S. Cronan
Intertidal Deposits: River Mouths, Tidal Flats, and Coastal Lagoons,
   Doeke Eisma

Coastal Ecosystem Processes

Daniel M. Alongi, Ph. D.
Australian Institute of Marine Science
Townsville, Queensland, Australia

CRC Press
Boca Raton Boston New York Washington London

Acquiring Editor:	Paul Petralia
Project Editor:	Sarah Fortener
Marketing Manager:	Becky McEldowney
Cover designer:	Dawn Boyd
Manufacturing:	Carol Royal

Library of Congress Cataloging-in-Publication Data

Alongi, D. M. (Daniel M.)
    Coastal ecosystem processes / by Daniel M. Alongi
        p. cm. – (Marine science series)
    Includes bibliographical references and index.
    ISBN 0-8493-8426-5 (alk. paper)
    1. Coastal ecology. 2. Food chains (Ecology). I. Title. II. Series.
QH541.5.C65A58 1997
577.5'1—dc21   97-36262
                                                          CIP

No claim to original U.S. Government works
International Standard Book Number 0-8493-8426-5
Library of Congress Card Number 97-36262
Printed in the United States of America 1 2 3 4 5 6 7 8 9 0
Printed on acid-free paper

PREFACE

This book is about how food webs process energy and nutrients in the coastal ocean. I have taken a process-functional approach to show not only how coastal ecosystems rely on exchanges among biota, but also how they are influenced by physical, chemical, and geological forces.

In keeping with the purpose of the Marine Science Series, this volume focuses on recent advances on how food webs and nutrient cycles are closely intertwined with the dynamics of coastal seas, land, and atmosphere — a recurring theme in this book. I weave this information into present marine ecological concepts not only for the sake of completeness, but also to provide some feel for progress and where major gaps lie. This book may serve as an advanced graduate text for marine ecology courses; a very extensive bibliography is provided partly for this reason.

It is becoming clearer from the burgeoning literature that the energetics of coastal organisms are not divorced from the surrounding environment, but are indeed subtly and beautifully interwoven into the machinery of a very dynamic, if increasingly fragile, coastal ocean. This book has a strong microbial flavor — a reflection of the dominant role microorganisms are now understood to play in food webs and nutrient cycles.

It was my original intention to write a book exclusively about tropical marine ecology, but it became painfully obvious that tropical marine science is still in its infancy. Nevertheless, it became equally clear that many of our dearest-held concepts and paradigms concerning the structure and function of marine ecosystems — nearly all developed from the study of temperate and boreal seas — are not necessarily applicable to the tropics. There are, of course, natural constraints on the processing of energy and matter within all ecosystems, but the magnitude of such flows in the tropical coastal ocean are often different from those in seas at higher latitudes. I have therefore made some attempt to redress the imbalance of previous books by treating tropical and temperate ecosystems with equal emphasis, where possible, and to show, for instance, that mangrove forests are as different from temperate salt marshes as temperate deciduous forests are from tropical rainforests.

The uniqueness and intricacies of food webs in mangroves, sandy beaches, and other coastal habitats are why major habitats are dealt with separately; trophic complexities and the energetics of populations and communities within habitats too often get overlooked when taking an ecosystem-scale view of life.

A clear understanding of the fate of energy and materials in coastal regions is imperative, if one considers that the most rapid growth of the human population is occurring in developing nations, with concomitant increases in the degradation and destruction of coastal habitats. The need for this information will not abate as human encroachment on coastal resources is expected to continue well into the next century. An awareness of just how necessary, yet delicate, is the interplay of biological and physical forces among coastal ocean, land, and atmosphere has far-reaching consequences for making informed management decisions about sustainable development and conservation. The importance of land-sea-air exchanges, for instance, has recently become evident in recognizing that pollution problems in land drainage basins often translate into problems in the adjacent coastal zone.

The possible impacts of global climate change is another clear instance where understanding how the interconnections between land, sea, and atmosphere regulate coastal ecosystems is crucial. An ecosystem approach can be a powerful tool for conservation and management, if used wisely, with some understanding of its limitations (and extrapolations) and in concert with population regulation models that have already been developed, such as for coastal fisheries and marine mammals. Clearly, no one perspective can suffice.

I thank Drs. Russ Reichelt, Peter Moran, and Dave Williams for their forbearance and for giving me both the time and facilities to complete this work; Dr. Barry Clough for taking up the administrative reins; Tim Simmonds, Steve Clarke, and Karen Handley for their graphics and photographic expertise; Rhonda Lyons for her secretarial skills; the librarians at AIMS for chasing references; the CRC editors, Mike Kennish and Paul Petralia, for their help and encouragement; and my colleagues, Drs. Josephine Aller, Robert Aller, Barry Clough, Craig Johnson, Lawrence McCook, John Tietjen, Robert Twilley, and Clive Wilkinson, for reviewing parts of the manuscript. Any remaining errors or omissions are, of course, mine. Finally, I thank my wife Fiona for her patience, good humor, and understanding.

This work is Contribution No. 871 from the Australian Institute of Marine Science.

Daniel M. Alongi
Townsville, Queensland,
Australia

The Author

Daniel M. Alongi, B.S., M.A., Ph.D., is currently a principal research scientist at the Australian Institute of Marine Science in Townsville, Queensland, Australia. He obtained his B.S. degree *magna cum laude* in biology from The City College of New York in 1979. He received his M.A. in Marine Science from the School of Marine Science, Virginia Institute of Marine Science, College of William and Mary in Virginia in 1981. His Ph.D. degree was completed through the Skidaway Institute of Oceanography of the University of Georgia in 1984. He was awarded a Postdoctoral Research Fellowship by the Australian Institute of Marine Science in 1985 and has held the positions of Research Scientist and Senior Research Scientist at the Institute. He was on study leave when he held a visiting Associate Professorship in the Department of Microbial Ecology, Institute of Biological Sciences, Aarhus University in Denmark in 1992.

Research projects conducted by Dr. Alongi on coastal processes, including benthic trophic dynamics and biogeochemistry in intertidal, nearshore, continental shelf, coral reef, and deep-sea habitats, have been supported by the Institute and by grants from the National Science Foundation and from numerous Australian and other international funding agencies. Dr. Alongi's field research has been conducted in many parts of the world including the Atlantic coast of the U.S., Australia, Denmark, Papua New Guinea, Malaysia, and Vietnam.

Dr. Alongi is a member of the American Society of Limnology and Oceanography, Sigma Xi, and The Oceanography Society. He has served on several national and international committees and is currently an Associate Editor of *Deep-Sea Research* and on the editorial boards of the *Marine Ecology Progress Series* and *Coral Reefs*. He is the author or co-author of nearly 70 scientific publications. His current research is focused on sediment nutrient cycling in mangrove forests and other coastal habitats of Southeast Asia and the South Pacific, including the Great Barrier Reef.

TABLE OF CONTENTS

For My Mother and Father

Chapter 1

INTRODUCTION

"And the microscopic organisms ... those billions of animalicules, of which there are millions in a drop of water and eight hundred thousand of which are required to make one milligram — their role is no less important. They absorb marine salt, assimilate the solid elements in the water, and by making corals and madrepores, they build calcareous continents. Then the drop of water, deprived of its mineral element, becomes lighter, comes up to the surface again, absorbs the salts left by evaporation, becomes heavier, descends again, and brings to the animalicules new elements to absorb ... a double current, ascending and descending — a continuous movement and continuous life!"

Twenty Thousand Leagues Under the Sea (1870)

Jules Verne's description of life in the sea is antiquated but remarkably prescient in appreciating that the cycles of ocean life, energy, and matter are intertwined and greatly dependent on the activities of microbes. This should not be surprising if one considers that life began in primordial seas 3.6 billion years ago with the evolution of anaerobic prokaryotes ancestral to modern bacteria, Archaea, and Eukarya. It is surprising, however, that the participation of modern microbes in marine food webs and nutrient processes was not fully appreciated until only about 20 years ago, when improved methods and technology led to the discovery of their great abundance, growth, and productivity in seawater and sediments.[1]

The solar energy absorbed and fixed by various phototrophs is stored as biomass and eventually dissipated by a variety of small and large heterotrophs. Algae, cyanobacteria, and green and purple photosynthetic bacteria are among the most important primary producers in the sea, but where the seabed is shallow and the water clear enough to receive significant amounts of light, macroalgae (e.g., kelps) and vascular plants, such as seagrass, marsh grass, and mangrove, also contribute to the fixed carbon pool.

1

While some of this fixed carbon is directly consumed by herbivores, most carbon enters the detritus food web in the form of dissolved and nonliving particulate matter. The energetics of the pelagic food web is dominated by the microbial loop[2] — a complex network of autotrophic and heterotrophic bacteria, cyanobacteria, protozoa, and microzooplankton — within which a substantial share of fixed carbon and energy is incorporated into bacteria and subsequently dissipated as it is transferred from one consumer to another.

The laws of thermodynamics constrain the rates and efficiencies of energy transfer and transformation, but the requirements of living organisms impose further losses due to respiration (heat loss), egestion of unassimilated matter, and mortality.[3] The conversion of energy from one form to another (and from one trophic level to the next) cannot be 100% efficient (the second law of thermodynamics), and actual rates of energy transfer in aquatic and terrestrial food webs are in fact considerably less than the theoretical maximum limit — usually less than 20%, and more often, only 10 to 15%. The input and output pathways of energy and inorganic nutrients are different within ecosystems, with energy transfer being less efficient than for the cycling of inorganic nutrients. The main reason for the low efficiency of energy transfer is heat loss (respiration) — often greater than 50% of assimilated energy.[3] Nutrients, on the other hand, particularly those that are scarce and essential (e.g., nitrogen), are cycled much more efficiently, as most organisms have evolved physiological mechanisms for conservation of such elements. The distinction between living and nonliving components becomes somewhat blurred — as for trophic levels — in the energetics of ecosystems.

Despite the fact that most nonliving matter ends up on the sea floor, our understanding of the fate of this detritus within sediments is poor in comparison to our knowledge of planktonic processes. There are several reasons why this is so. The most obvious is that it is much more difficult to sample and isolate organisms and organic matter from inorganic sediment particles. There is also an enormous number of adsorption and absorption reactions between dissolved and particulate nutrients and clay, silt, and sand particles; many of the latter are often coated with organic compounds of varying reactivity. Also, much of the organic matter in sediments (for example, humic and fulvic acids) is refractory to immediate breakdown by microbes and other decomposers. Such organic compounds are often linked with more labile carbohydrates and proteins.[4] Isolating bacteria, ciliates, flagellates, fungi, nematodes, and other small organisms from sediments is difficult for the same reasons, as they are often intertwined with or stuck to organic coatings, including their own mucus. Recent techniques have somewhat circumvented these problems, but sedimentary biota are still more difficult to study than pelagic organisms.

The role of microbes and meiobenthos in marine sediments is now often equated to the role their pelagic counterparts play in the microbial loop. This idea has gained some credence as recent data indicate that sediment bacteria and protozoa are more abundant than believed less than a decade ago. Even to the present, the dual roles of sediment bacteria as (1) food for protozoans, invertebrates, and some vertebrates, and (2) as mineralizers of organic matter have not been fully reconciled as they have for pelagic bacteria. Both roles are closely interdependent but, until very recently, have been studied separately by ecologists and biogeochemists, respectively. The concept held by many benthic ecologists (and bordering as an act of faith) that sediment bacteria are heavily grazed by higher organisms is untenable and unrealistic, as it conflicts with our knowledge of their extensive participation in nutrient recycling and ignores the fact that complex physical and biogeochemical gradients exist in marine sediments, due in part to the activities of bacteria and other benthic organisms. Overemphasis on the grazing role developed from early work on trophic interactions and later work on cophrophagy, detritus aging, microbial gardening, resource limitation, and the applicability of optimal foraging theory to marine deposit-feeders.[5] Work on the trophic role of bacteria was essentially divorced from biogeochemical work on rates and pathways of organic matter decomposition.

More realistic views of the role of bacteria in sediments have since evolved,[5-7] taking into account the vertical distribution of consumers in relation to sediment redox chemistry and oxygen profiles, and the known decomposition pathways of organic matter. The role of sediment bacteria in benthic food chains and nutrient cycles can best be simplified as aerobic trophic links and anaerobic nutrient sinks. This infers that bacteria are linked trophically in surface aerobic sediments where most consumers reside, but not in deeper, anaerobic sediments, where life consists mostly of a variety of anaerobic bacteria and some protozoa. A notable exception is the oxidized lining of burrows and tubes of larger metazoans. Such structures may extend to considerable depths into the seabed.

Grazing of bacteria is discernible in aerobic environments depending on several factors, including the rate of detritus supply and the abundance of grazers. Significant depletion of bacterial biomass by grazers is observable in marine sediments if bacterial growth rates are less than or equal to rates of ingestion. If bacterial growth is equivalent to or outpaces consumption — as it appears to in organic-rich sediments — bacteria are likely to be regulated by other factors, such as temperature and nutrient supply.[5] The proportion of bacterial standing crop that is consumed depends not only on how fast they are eaten, but also on how quickly they can multiply. With the exception of suitable substrates (tube and burrow linings, leaf blades, sand grains),

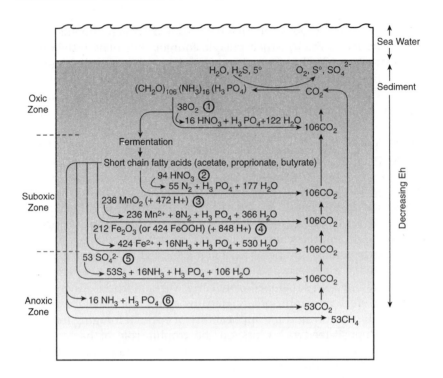

FIGURE 1.1

Model depicting the major oxidation reactions in the decomposition of organic matter with depth in marine sediments. Organic matter is represented by the formula $(CH_2O)_{106}$ $(NH_3)_{16} (H_3PO_4)$. The oxidation reactions are identified by number: (1) Aerobic respiration; (2) denitrification; (3) manganese reduction; (4) iron reduction; (5) sulfate reduction; and (6) methanogenesis. Note that nitrogen, phosphorus, and water are common metabolic products.

only a small fraction of bacteria is grazed, even in aerobic benthic habitats. Most bacteria in sediments must die and lyse naturally, with the next generation of bacteria consuming and mineralizing this material, either into new biomass or dissolved material. This situation is greatly different than for pelagic bacteria, which are more readily captured and consumed in greater proportion. Clearly, the distinction between feeding relationships and the participation of microbes in nutrient recycling is more complicated in sediments than in the water column, reflecting what is, in many respects, a more complex environment.

The decomposition and recycling of organic matter in sediments is orchestrated by a variety of bacterial types which use different electron acceptors sequentially as the geochemistry of sediment changes with depth. Three major zones of organic matter oxidation (Figure 1.1) are recognized:

- Oxic zone, in which oxygen is the major oxidant and limited by the depth to which O_2 can penetrate by diffusion, advection, and mixing
- Suboxic (or postoxic) zone, in which oxygen is available in bound form in nitrate, nitrite, and iron and manganese oxides
- Anoxic zone, in which sulfate reduction and methanogenesis are the primary oxidation reactions

This sequence coincides with decreasing free energy yield per mole of organic matter oxidized, but it is often not distinct, as sediments undergo spatial and temporal changes resulting in numerous microenvironments where some of these reactions can occur simultaneously within the same zone.

The sequence is ultimately constrained by the availability of specific organic compounds and oxidants, reaction kinetics and thermodynamics, and physiology of the bacteria. The common carbon endproduct is carbon dioxide. Other gases result from specific reaction processes: N_2 (from denitrification), hydrogen sulfide (from sulfate reduction), and methane (from methanogenesis). In the reaction sequence, organic matter is initially broken down by extracellular and membrane-bound hydrolytic enzymes produced by numerous microbes from large macromolecules or polymers (e.g., carbohydrates or proteins) to soluble compounds (e.g., alcohols) — dissolved organic matter (DOM). DOM is used by fermenting and acetogenic bacteria, resulting in the production of simpler compounds, such as acetate, lactate, and molecular hydrogen. These compounds are what is being oxidized by the various bacteria types that reduce nitrate, sulfate, and metal oxides to produce CO_2. Fermentative bacteria use these simple products as well. The fermentation reaction results in some energy release, but there is no net oxidation. The energy yielded by oxidation of these organic compounds to CO_2 by postoxic and anaerobic bacteria is incomplete, because some energy is shunted to the production of metabolic byproducts, such as ammonium, phosphate, dinitrogen, sulfide, reduced iron and manganese compounds, and methane (Figure 1.1). These metabolic products are available to fix carbon, either chemosynthetically (usually within the sediments) or photosynthetically (e.g., in the overlying water column).

Other organisms such as fungi, protozoans, and metazoans consume and oxidize organic matter with subsequent release of carbon dioxide via the tricarboxylic acid cycle, but bacteria mineralize the bulk of organic matter in the sea. Anaerobic pathways of decomposition in sediments — though thermodynamically less favorable than oxic reduction because of lower free energy yields — can equal, and often exceed, the amounts of organic matter decomposed by aerobic processes. This is evident in many coastal environments and has

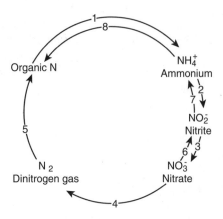

FIGURE 1.2
Model of the major pools and transformations of the nitrogen cycle. The processes are numbered as follows: 1= ammonification; 2 = nitrification (ammonium oxidation); 3 = nitrification (nitrite oxidation); 4 = denitrification and dissimilatory nitrate reduction; 5 = nitrogen fixation; 6 = assimilatory nitrate reduction; 7 = assimilatory nitrite reduction; 8 = immobilization and assimilation.

important consequences for the fate of organic matter and associated elements (e.g., sulfur) in the coastal ocean. The organic carbon cycle is linked to the cycling of elements such as nitrogen, phosphorus, and sulfur, as organic matter consists of many compounds (Figure 1.1). The cycling of nitrogen is particularly important, as it is often the major limiting element in coastal seas. The nitrogen cycle is more complex than those for the other elements, being mediated mostly by biological (primarily microbial) processes and having gaseous phases that are often difficult to measure. Also, by virtue of a wide range of oxidation states, nitrogen can serve as either a reductant or an oxidant in decomposition processes, as well as a nutrient for plants and other organisms.

The major transformation processes of the nitrogen cycle (Figure 1.2) are ammonification, nitrification, denitrification, nitrogen fixation, nitrate reduction, immobilization, and assimilation. Denitrification and nitrogen fixation are crucial processes, being the major pathways by which nitrogen is lost and gained, respectively, within ecosystems.

The impact of trophic interrelationships on nutrient cycling processes (and vice versa) within food webs was, until recently, considered separately by both terrestrial and aquatic ecologists. An integration of population and community ecology with energy and matter transformations and fluxes is critical for a proper understanding of food webs and ecosystem functioning.[8-10]

A good benthic example of how food webs and nutrient cycles are interlinked is illustrated in Figure 1.3. A variety of indirect and direct

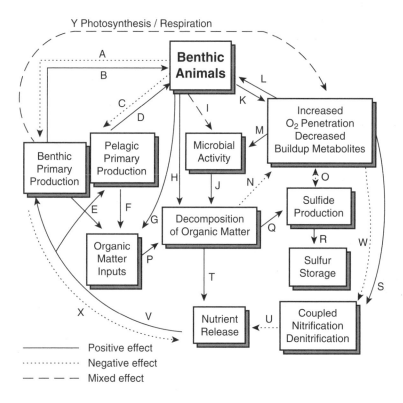

FIGURE 1.3

Model of the interrelationships between benthic organisms and biogeochemical processes in marine sediments. Solid lines represent positive interactions, dashed lines depict negative interactions, and dotted lines represent mixed outcomes. Letters are keyed to processes discussed in the text. (From Giblin, A. E., Foreman, K. H., and Banta, G. T., in *Linking Species and Ecosystems*, Jones, C. G. and Lawton, J. H., Eds., Chapman & Hall, New York, 1995, chap. 4. With permission.)

mechanisms control the extent to which changes in biogeochemistry lead to changes in the abundance and diversity of animals present, and vice versa. Benthic organisms can either stimulate or reduce primary production by either increasing nutrient availability (arrows H → T → V and → I → J → T → V) or by overgrazing (arrows A and C), respectively. Animals can directly alter rates of decomposition by assimilating organic matter (arrow H), whereas other organisms can increase the deposition of organic matter and its decomposition by capturing suspended matter in the overlying water column (arrows D → G → P). Stimulation of sediment metabolism and alteration of nutrient transformations can occur via vertical mixing of sediment particles and by ventilation of tubes and burrows (arrows K → S), grazing of microbes (arrow I), excretion, and physical breakdown of organic particles. Bioturbation in particular has subsequent effects, such as altering nutrient concentrations, distribution and availability of electron acceptors, redox status, oxidation of reduced compounds (arrows O → R), and element storage (arrow W). Some byproducts of metabolism, such as sulfides, are toxic to some organisms and function as a feedback control (arrows K → S → U → V → B) of various organisms and processes. So, there are numerous potential interactions and feedbacks between biogeochemical processes and organisms, both benthic and pelagic.

Recognition that physical forces exert strong control on both food webs and biogeochemical processes in sediments and in coastal waters is recent, reflecting the maturity of oceanography and a better understanding of variability in marine ecosystems. Much of the temporal and spatial variation of marine organism stocks can be explained by intrinsic factors (genetics, reproductive strategies, etc.), but these must be considered against the backdrop of environmental factors such as climate. Short- and long-term variations in environmental conditions exert their own control over living ocean resources, as evidenced by the biological consequences of the El Nino-southern oscillation (ENSO) phenomenon, upwelling, eddies and gyres, boundary layers, and fronts.

Food webs and nutrient processes in the coastal zone are therefore not divorced from processes occurring on land or in the atmosphere. For instance, deforestation has led to increased erosion of downstream catchment areas with attendant increases in suspended matter transported to adjacent coastal waters. Such increases can, in turn, result in shoaling of nearshore areas, smothering of benthic communities, and decreased light levels, resulting in a decline in primary productivity inshore. Another example is global warming, in which continued emissions of greenhouse gases into the atmosphere has resulted in a net shift in CO_2 flux from the pre-industrial scenario of the ocean as a CO_2 source to the ocean as a CO_2 sink. These problems have already had a drastic impact on many coastal environments, but how they actually affect coastal resources is poorly understood. It is imperative to understand such mechanisms and their role in shaping and altering food webs and nutrient cycles, in order to make the best management decisions to conserve the coastal ocean.

Many attributes of the coastal ocean of tropical latitudes are different from those of higher latitude (Table 1.1). These differences can be important because management decisions in the tropics are often based solely on information derived from temperate and boreal habitats. About 30% of the world's continental shelves lie in the tropics, so they can no longer be ignored or presumed to be identical to other climatic regions. In fact, a variety of evidence supports

TABLE 1.1

Some Major Environmental Characteristics Unique or Peculiar
to the Tropical Coastal Oceans

Habitats occurring mainly in tropics	Mangroves, coral reefs, hypersaline lagoons, and stromatolites
Temperature	Higher; narrow annual range; thermal maximum closer to ambient
Light intensity	Higher; narrow annual range
Climate	Monsoonal and dry; two seasons rather than four; greater incidence of storms (cyclones, typhoons, etc.)
Geological characteristics	Most of world's sediment discharge from continents; mud and coral more abundant inshore; shelves wide and shallow, with several dominated by carbonates; mixed terrigenous-carbonate facies; migrating fluid mudbanks; sabkhas
Hydrological and chemical	Lower dissolved nutrient and gas (O_2, CO_2) concentrations in most habitats; most of world's freshwater discharge to ocean; ENSO phenomenon; lower mean amplitude with slight increase near equator; permanently stratified waters; O_2 minimum layers; estuarization of shelf by river plumes; large buoyancy flux; highly variable salinity; strong tidal fronts; lutoclines and high salinity plugs in dry tropical estuaries

Note: See Table 6.13 for characteristics specific to coastal seas of the wet tropics.

Source: Data compiled from Alongi,[12] Saenger and Holmes,[13] and Hatcher et al.[14]

the supposition that the tropical marine biosphere is not as environ-
mentally stable as long believed. The tropical coastal ocean is not a
benign or uniform environment, but offers climatic and abiotic
conditions as inimical to life as the supposedly more inhospitable,
higher latitudes.[12]

Clearly, there is abundant evidence to reject the long-held belief
that the tropical ocean is simply a "Polynesian paradise". Monsoonal
rains, high temperatures, hypersaline and desiccated conditions, carbonate sedimentation and compaction, low and variable oxygen and
dissolved nutrient concentrations, chemical defenses of plants, smothering by massive riverine sedimentation, erosion of mudbanks and
tidal flats, and anoxia caused by impingement and stratification of

water masses are just some of the severe physical conditions that tropical marine organisms must face. This highlights the fact that many long-held beliefs of coastal ecological processes in temperate and polar latitudes are not universally true.

Temperate and polar environments may be more stressful, but subtropical and tropical habitats are subjected to more frequent and intense disturbances.[13] Tropical organisms may be vulnerable to physical and chemical stress, but are more resilient to disturbance events than their cooler-climate counterparts. Tropical organisms are closer to their physiological tolerances, considering the generally warm temperatures year-round and desiccation in many intertidal zones. The functional implication is that processes such as respiration, growth, decomposition, and nutrient cycling, are accelerated in the tropics compared to higher latitudes.[13-15] Marine evidence for a more rapid pace of life in the tropics is comparatively scarce compared with data for terrestrial environments in which a latitudinal gradient of shorter turnover times for soil nutrients from boreal to equatorial forests has been demonstrated and recently ascribed to more rapid microbial activity in the tropics.[15] More nutrients may indeed be tied up in tropical plant and microbial biomass than in temperate and polar environments. These generalizations are true for some tropical coastal habitats and food chains, but not for all.

Regardless of latitude, coastal ecosystems exist on a comparatively thin shell covering a cold, deep ocean mass. An inventory of the major reservoirs of carbon on Earth (Table 1.2) reminds us of just how miniscule and fragile living processes are in the scheme of things. About 99.94% of the total carbon on the upper crust of the Earth is inactive on an ecologically meaningful time scale, residing mostly as carbonate in sedimentary rocks. The largest active pool is dissolved inorganic carbon in seawater; other active pools are even smaller, with the total active pool less than 0.06% of the total inorganic and organic carbon on Earth. The largest pool of carbon vested in living organisms is terrestrial plant biomass, which is roughly 0.001% of the total carbon pool on Earth and only ~2.2% of the active carbon pool; the living pool of marine carbon is much smaller.[16]

Finally, it is worth noting that the study of ecosystem-level processes in the coastal ocean offers us one of several ways (albeit an important one) in which to elucidate the mechanisms underlying the functional role of the world's coastal oceans. The ecosystem approach will become an increasingly important management tool as the problems of environmental degradation multiply in the future, requiring solutions that have considered the consequences of physical, chemical, and geological forces on ecological processes. An integrated approach — as suggested in this volume — is an important step toward injecting

TABLE 1.2

Major Global Reservoirs of Inorganic
and Organic Carbon

Reservoir	Amount (10^{18} g C)
Sedimentary Rocks	
Inorganic	
Carbonates	60,000
Organic	
Kerogen, coal, and other organic carbon	15,000
Active Pools	
Inorganic	
Marine dissolved inorganic carbon	38
Soil carbonates	1.1
Atmospheric CO_2	0.66
Organic	
Soil humus	1.6
Terrestrial plant biomass	0.95
Marine dissolved organic carbon	0.60
Marine surface sediments	0.15

Source: From Hedges, J. I. and Keil, R. G., *Mar. Chem.*, 49, 81, 1995. With permission. Original references can be found in Table 1 of their paper.

some realism into our understanding of coastal ecosystems and offering exciting and robust answers to questions involving the sustainability and conservation of life in the coastal ocean.

Chapter 2

BEACHES AND
TIDAL FLATS

2.1   INTRODUCTION

Sandy beaches and intertidal flats are among the most physically dynamic environments in the sea. The energy expended in the movement
of water and sediment is pre-eminent compared to the energy dissipated through food webs. The geomorphology and nature of shorelines
vary greatly, regulated by wave and tidal energy, and sediment grain
size. Shorelines along many coasts reflect a continuum from high-
energy, coarse sandy beaches, to low-energy, sand- or mudflats where
physical forces are less dominant and where biological forces come into
play. The functional role of sandy beaches and, to a lesser extent, tidal
flats has lagged behind the study of other marine habitats.

Sandy beaches and tidal flats lack much of the spatial complexity
of rocky shores, mangrove forests, and coral reefs. Food chains and the
pathways of energy flow are thought to be correspondingly simple —
particularly on sandy beaches — and dependent upon the unidirectional flow of matter and energy from the sea. In truth, trophic pathways and the ecological energetics of such systems are subtle and
complex, as is the interplay among the physical, chemical, and geological factors that regulate them.

Physical settings are important, as functional differences among
beaches and tidal flats may be more a function of the degree of commonality of factors related to physical type than to other factors, such
as climate. Sandy beaches and their interstitial environment span a
range of types and are classified as:[17]

- Reflective
- Intermediate
- Dissipative

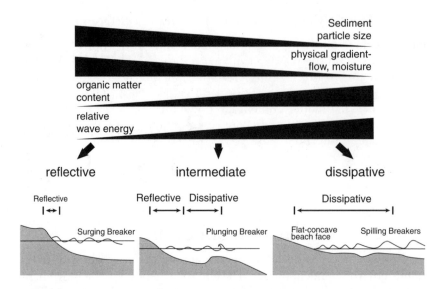

FIGURE 2.1

Gradations in the physical, chemical, and sedimentary structure of sandy beaches from reflective to dissipative types. (Adapted from Short[17] and McLachlan and Turner.[18])

In reality, these types represent a continuum (Figure 2.1). Reflective beaches are produced by waves of low height (<1 m). By wave action, such beaches rapidly filter and drain large volumes of water through the interstices, resulting in well-flushed and highly oxygenated, coarse sand deposits. Reflective beaches are found on open coasts with deep embayments and low wave shoaling and on low wave coasts in tropical and polar seas.

Dissipative beaches represent the other end of the continuum and are produced by a combination of high waves (>2.5 m) and fine sand deposits. These beaches have low surf-zone gradients and filter small volumes of water by tidal action. Slow flushing results in the buildup of organic matter and microbes, leading to sharp vertical gradients in oxygen, nutrients, and other geochemical factors, such as redox potential, sulfides, and pH.[18] Dissipative beaches are found on the west coasts of Australia and southern Africa, and seasonally on the west coast of the U.S. where high wave swells and fine sands are abundant. Intertidal sand- and mudflats fall roughly into this end of the continuum.

Intermediate beaches are more common worldwide, representing a mixture of characteristics of the two extremes such as medium sand and moderately high waves. They occur on coasts with moderate swell (many east coasts) and in trade wind and monsoonal regions. Such physical differences impact greatly on the dynamics of intertidal food webs.

2.2 FOOD CHAINS, ENERGY, AND CARBON FLOW

Early workers studying intertidal food chains and trophic relationships focused on the relative isolation of beaches and tidal flats from
food chains offshore.[19,20] Until recently, all intertidal habitats were
considered as open systems connected to adjacent coastal waters only
in terms of what matter and energy flowed from the sea to the shore.
There is a variety of ecological connections among land, the shore, and
the sea; not all intertidal habitats are simple open systems. Most tidal
flats vary over time as to whether or not they import or export organic
matter.[21] Tidal flats may oscillate between being "open" for a period
of time and "closed" the next, with respect to energy and material
exchange.

Brown and McLachlan[22] argue that adjacent surf zones must be
considered an integral part of sandy beach ecosystems. On this basis,
these beach surf-zone habitats are essentially semi-closed or closed
ecosystems, being fueled by surf-zone diatom blooms. Robertson[23]
clarified these terms on the basis of the extent of dependence by sandy
beaches on outside energy input (Table 2.1) in which semi-closed or
closed ecosystems are fueled mainly by *in situ* primary production
(usually diatoms) and where open systems are driven mostly by inputs
of material from offshore or from land.

2.2.1   Open Ecosystems

2.2.1.1 Sandy Beaches/Surf-Zone Systems

Reflective sandy beaches are open ecosystems in which food chains
receive most of their energy input from either the land or sea, or both.
Robertson[23] further considered these open systems to fall into two
categories:

- Sandy beaches receiving significant quantities of plant detritus from
  adjacent seagrass meadows, kelp beds, wetlands, or reefs
- Sandy beaches receiving little or no macrodetritus, dependent for
  energy subsidies on carrion, phytoplankton, or other small filterable
  particles or dissolved organic matter advected from offshore

On beaches receiving large inputs of macrophyte detritus, this
material is macerated and fragmented by a variety of scavengers within
the surf zone and at the drift-line onshore, driving macroscopic and
microbial food chains.[23,24] The rate of fragmentation and incorporation
of this material into food chains depends upon several factors, such as
the nutritional quality of the detritus, the location and rate of input,
and the physical milieu (e.g., variations in wave and tidal energy).

TABLE 2.1

Sandy Beach Types and Characteristics, Sources of Production, and
Major Macroconsumers

Category	Beach Type	Surf-Cell Circulation	Production Sources	Consumers
Semi-closed or closed	I/D	Yes	Diatom bloom, DOC, POC	Zooplankton, benthic filter feeders
Open				
Little or no macrophyte debris	D/I/R	No	Phytoplankton, POC, DOC, carrion	Filter feeders, omnivorous benthic scavengers
Macrophyte debris	D/I/R	No	Macrophytes, POC, DOC	Omnivorous benthic scavengers

Note: I = intermediate; D = dissipative; R = reflective; DOC = dissolved organic carbon; POC = particulate organic carbon.

Source: From Robertson, A. I., in *Marine Biology*, Hammond, L. and Synnot, R. N., Eds., Addison-Wesley, Longman, Australia, 1994. With permission.

Macrophyte detritus, detached and exported from mangroves, reefs, and salt marshes, strand on many beaches, but only the input of kelp and seagrass detritus to beaches on the west coasts of Australia and southern Africa have been well studied.[23,24] Microtidal, reflective beaches of low energy on the southwestern coast of Western Australia near Perth receive large subsidies of seagrass and seaweed debris that deposits as beach wrack along extensive lengths of coastline (Figure 2.2). Storms and heavy swell detach and transport this material from limestone reefs and seagrass meadows running parallel to the beach. The major components of these accumulations are the kelp, *Ecklonia radiata*, dead seagrasses, and several species of small red algae. The average standing crop of these accumulations in the surf zone is 20 kg dry weight per meter of coastline. Up to 1200 m³ of surf zone per kilometer of beach may be covered by detritus during certain times of the year.[25]

A food-chain model of the beach/surf-zone wrack system in Western Australia (Figure 2.3) shows that the main utilization pathway of the macrophyte detritus is via colonizing microbes and large populations of the surf-zone amphipod, *Allorchestes compressa*. Each year, an average of 72 kg C of macrophyte debris deposits on each meter of coast and is broken down on the beach and in the surf zone by wave action, sand abrasion, and microbial decay. This material is worked over thoroughly in the surf zone because of the very small tides that resuspend debris stranded on the beach. Robertson and Hansen[25] found that these wrack banks behave similarly to compost piles, with

FIGURE 2.2
Accumulations of detached algae and seagrasses forming stacks 2 m high on the sandy beach/surf zones of the southwest coast of Western Australia. (Photograph courtesy of A. I. Robertson.)

increased bacterial activity raising temperatures within the piles to 45°C. The standing mass of this debris turns over 12 to 14 times per year, equivalent to 20% of nearshore benthic primary production.

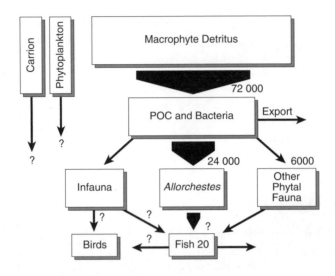

FIGURE 2.3
Carbon flow model of the reflective sandy beach/surf-zone ecosystem on the south-west coast of Western Australia which receives large amounts of macrophyte detritus. All flows are g C m^{-2} yr^{-1} and standing stocks are g Cm^{-2}. (From Robertson, A. I., in *Marine Biology*, Hammond, L. and Synnot, R. N., Eds., Longman, Melbourne, 1994, chap. 16. With permission.)

Dense populations of the amphipod, *Allorchestes compressa*, comprising more than 90% of the beach/surf-zone epifauna, feed on most of the small particles produced by fragmentation and decay,[26] processing 24 kg C or roughly one third of the detritus per meter of coast per year. An impoverished infauna, consisting primarily of donacid bivalves and hippid and ocypodid crabs, occupies the driftline, supported by feces from the amphipods and carrion (dead ascidians and sponges) from nearby reefs. Phytoplankton and zooplankton have a minor role in this system. The dense swarms of amphipods attract wading birds, crabs, and fishes, to the extent that these wrack deposits are important nursing grounds for several species of fish thought to be estuarine dependent.[27]

Amphipods and bacteria have a large role to play in other Australian sandy intertidal areas receiving plant detritus. On the tropical Queensland coast, macerated mangrove litter deposits as wrack on many adjacent, sheltered sandflats.[28,29] These litter piles shelter large populations of bacteria, protozoa, meiofauna, and the amphipod, *Parhyale hawaiensis*. Densities of these organisms and bacterial growth rates are significantly greater in the litter than in clean sand nearby. In laboratory experiments, Poovachiranon et al.[28] found that *P. hawaiensis* prefers mangrove litter when offered a choice of foods and ingests litter at high rates compared to other foods. Many sheltered sandflats in the tropics located adjacent to mangrove forests, seagrass beds, and reefs receive significant inputs of detritus derived from these habitats and likely are important sites for nutrient regeneration and energy flow. Unfortunately, most studies of these wrack sites have dealt with community structure and abundance of fauna rather than with rates of detritus decomposition and nutrient flow.

The pathways of carbon flow on sandy beaches receiving large quantities of macrophyte detritus depend greatly upon whether the material remains in the surf zone or on the beach for long periods of time. In contrast to the sandy beach ecosystem of Western Australia, kelp detritus transported to the beaches on the western coast of the Cape peninsula in South Africa remains on the beach rather than in the surf zone because the tidal range is greater than in Western Australia.[24] The rate of detrital input is also greater (116 kg C $m^{-1} yr^{-1}$) due to close proximity of the kelp beds. Because the kelp wrack is cast high upon the beaches, the macroconsumers are composed largely of semiterrestrial or terrestrial species.

A model of carbon flow within the kelp wrack beaches of western South Africa (Figure 2.4) shows that talitrid amphipods are major consumers of the kelp deposited onto the driftline, with kelp-fly larvae and herbivorous beetles consuming lesser amounts of detritus. These scavengers fall prey to birds, carnivorous beetles, and isopods. Carrion and phytoplankton debris contribute minor inputs of detritus to the beach and are consumed largely by carnivorous isopods and bivalve molluscs, respectively (Figure 2.4). The majority of kelp debris is processed initially by scavengers, with the remaining stranded detritus decomposed by bacteria, percolated through the sand interstices as DOM or particulate organic matter (POM), or washed back out to sea. Ultimately, the bulk of the detritus may be mineralized by the interstitial bacteria, protozoa, and meiofauna populations, if the low assimilation efficiencies of the major herbivores are correct; much of the material they consume is presumably returned to the sand column as excreta. Thus, only 0.5% of the original input goes to terrestrial food chains, with the remaining 99.5% either respired (mainly by bacteria) or exported to sea.

The fate of the material filtering into the sand is therefore an important pathway of carbon and energy flow. Using microcosms, Koop and Lucas[30] estimated rates of microbial decomposition of kelp detritus and defined the pathways of carbon flow by calculating a mass balance. Of an initial 100 g C of kelp detritus added to the sandy microcosms, 69.8% was respired, 28% converted into bacterial biomass (the carbon-conversion efficiency), and 2% lost to invertebrate grazers;

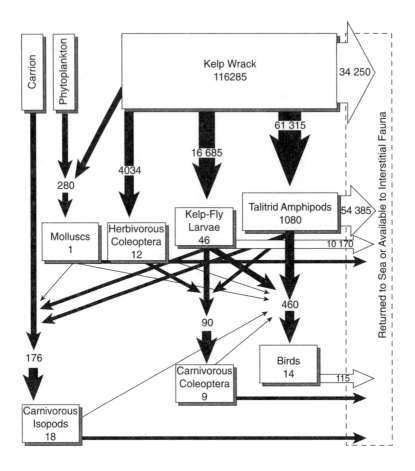

FIGURE 2.4
Carbon flow model of an exposed sandy beach on the west coast of Cape peninsula, South Africa, which receives large quantities of kelp detritus. All flows are as in Figure 2.3. (Adapted from Griffiths, C. L., in *Sandy Beaches as Ecosystems*, McLachlan, A. and Erasmus, T., Eds., Junk, The Hague, 1983, 547.)

only 0.2% was lost as leachate. 90% of the leachate was consumed by bacteria but dissolved material accounted for only a miniscule fraction of carbon flow. The bulk of kelp carbon deposited onto these sandy beaches is lost as respired CO_2.

On sandy beaches not receiving large quantities of macrophyte detritus or without significant *in situ* primary production, sources of organic matter include small particles of particulate organic carbon derived from offshore, phytoplankton advected from nearby upwelling areas, and mass strandings. The two latter sources are not thought to be major sources of energy for sandy beaches, being the exception rather than the norm. Mass strandings[31] may be significant, but episodic,

TABLE 2.2

Comparisons Between Indian and Scottish Beaches

Characteristics	Indian	Scottish
Sand temperature (°C)	32–45	7–15
Mean tidal range (m)	1.0	3.2
Interstitial salinity	34–35	23-34
Sand carbon (μg g^{-1})	200–400	200–300
Surf water carbon (μg g^{-1})	900–3700	500–1000
Sand chlorophyll a (μg g^{-1})	0.02–0.07	0.4–1.7
Carbon input (g C m^{-2} yr^{-1})	1070	500
Respiration (g C m^{-2} yr^{-1})	164	42
Microbial production (g C m^{-2} yr^{-1})	72	15
Meiofauna biomass (mg dry weight m^{-2})	24–60	273–523
Macrofauna biomass (g dry weight m^{-2})	0.8	2.8

Source: Data from Steele[33] and Munro et al.[34]

and difficult to quantify over time, and few beaches occupy shorelines adjacent to upwelling conditions.[32]

Much of the source for carbon on beaches devoid of significant benthic diatom production is derived from the inflow and uptake of plankton, POM, and DOM from the tidal water either via filtration or by percolation into the sand column (Table 2.1). The most comprehensive studies of such beaches were conducted from 1968 to 1973 as part of the International Biological Programme (IBP) Scottish/Indian project.[33,34] The purpose of the project was to compare and contrast energy flow through temperate Scottish and tropical Indian beaches. Such comparisons are problematical considering that the Indian beaches are low energy, reflective to intermediate types, whereas the Scottish beaches are low energy but dissipative.[22] Microbial methods were limited during this time, with most techniques underestimating bacterial standing stocks and activity.

Nevertheless, there are some clear differences between the two beach locations that cannot be explained simply by differences in beach type or poor methodology (Table 2.2). First, microbial production is nearly five times greater in the Indian beach than in the Scottish beach, although the rate of carbon input to the Indian beach is only double the input to the Scottish beach. Second, meiofaunal biomass is ten times lower in the tropical sands than in the temperate sands, and macrofaunal biomass is nearly four times greater on the Scottish beach than in India. Microbial production was calculated using dissolved organic carbon (DOC) uptake data from experimental sand columns. Nevertheless, it is reasonable to conclude that there is proportionally more carbon flow through microbial pathways of decomposition on a per area basis in the tropical sands than in the temperate beach. This is a reasonable conclusion, considering the higher temperatures, higher rates of water percolation, and higher organic carbon levels in the surf water at the Indian location. Such conclusions are not universal for tropical vs. temperate beach energetics, but the small amount of available data[12] suggests that bacterial activity is typically greater in tropical sediments than in temperate sediments of identical grain size (see also Table 2.4).

2.2.1.2 Tidal Flats and Mudbanks

Sandy beaches are not the only open intertidal systems, as tidal flats and mudbanks of many estuaries are so considered. Intertidal mudbanks persist in many large river deltas, but they have been largely ignored by ecologists. Algal mats frequently fail to develop on these unstable banks due to frequent episodes of sediment deposition and erosion, but many banks are sites of active decomposition of organic matter. In the estuarine delta of the Fly River in Papua New Guinea, for instance, Alongi[35] examined microbial decomposition and nutrient diagenesis in several such intertidal mudbanks. X-radiographic analyses at several mudbanks indicated a depositional history of significant quantities of mangrove detritus derived from adjacent forests. Flux studies indicated little or no *in situ* microalgal production. A few benthic organisms were found, mainly on the sediment surface. Sediment bacteria were abundant, as were rates of bacterial carbon production and growth; fluxes of dissolved nutrients across the sediment-water interface were undetectable. Alongi[35] proposed that the rapidly growing bacterial flora in these muds are fueled mainly by the deposition of mangrove-derived detritus. The bacteria mineralize and sequester most of the labile organic matter to the extent that these mudbanks act as nutrient sinks rather than as net exporters of nutrients to adjacent coastal waters.

Other tropical rivers, such as the Amazon, contain massive intertidal mudbanks that migrate along shore, acting as storage sites for large quantities of sediment and organic matter. These mudbanks may operate similarly to those in the Fly River, but energy and material flow within the intertidal muds of nearly all tropical rivers remain unexplored. This is unfortunate, as mudbanks may have a significant role to play in mass balances of elements and materials in the tropical ocean, especially if one considers the massive amounts of dissolved and particulate material transported to tropical river deltas and deposited onto river banks and adjacent shores (see Section 6.7).

In contrast, on tidal flats, benthic macroorganisms are more abundant and meet their energy and nutritional needs by feeding on benthic microalgae and labile organic matter settling out of the overlying water; suspension-feeders are greatly dependent on seston. The dominant flux of energy and nutrient flow is through microbial food chains, both aerobic and anaerobic. Owing to the more quiescent nature of intertidal flats, biological interactions (competition, predation, amensalism, etc.) are well developed among tidal flat residents; such interactions can affect energy flow. The trophic importance of bacteria, protozoa, microalgae, and meiobenthos in intertidal food webs has long been recognized,[20] but actual rates of production, consumption, and respiration within these groups and their contribution to the energetics of the entire food web have been accurately estimated only recently.

The traditional concept of food chains on tidal flats being energetically dominated largely by herbivorous macrobenthos with microbes serving solely as food has been replaced by more realistic concepts in which bacteria and protozoa play a larger role in utilizing and mineralizing sediment organic matter. This paradigm shift reflects recognition of the role of microbes in these processes. Such recognition has come as a result of greater attention being paid to biogeochemical ideas and techniques and from advances in methodology that indicated that bacterial biomass, growth, and productivity in sediments are greater than previously believed. Further, advances in biogeochemistry now indicate that the oxidation reactions and pathways of microbial decomposition in sediments are complex.

Nearly all models of energy and carbon flow in sandflat and mudflat ecosystems drastically underestimate the role and complexities of microbial food chains. For instance, Warwick et al.[36] described carbon flow on an intertidal mudflat in the Lynher River estuary in England by means of a steady-state descriptive model. Thirteen major macrofaunal populations were treated separately, but here they are combined into two groups representing benthic filter-feeders and deposit-feeders (Figure 2.5). The principal source of primary production is contributed by diatoms on the mud surface, estimated at 143 g C m^{-2} yr^{-1}. When the mudflat is covered at high tide, filter-feeding benthos (primarily the bivalves *Mya* and *Cardium* and the polychaetes *Manayunkia* and *Fabricia*) consume nearly 25% of the estimated phytoplankton production of 82 g C m^{-2} yr^{-1}. Some combination of the remaining phytoplankton carbon is presumably exported by the tide and deposited on the sediment surface. Bacteria in the tidal water account for nearly half of the carbon input to filter-feeders. The deposit-feeding assemblages feed largely on the benthic diatoms. Sediment particulate organic carbon (POC) is estimated to contribute a nearly equal proportion of carbon (111 g C m^{-2} yr^{-1}), although only 10% is considered readily assimilable. Bacteria are presumed to contribute only a minor amount of carbon to higher trophic levels (2.8 g C m^{-2} yr^{-1}) and are considered unimportant in carbon turnover. Respiration is similarly underestimated. Meiofauna account for 65% of metazoan production, of which less than 13% is consumed by predatory infauna, particularly *Nephtys* and *Protohydra*, with the bulk available for consumption by birds,

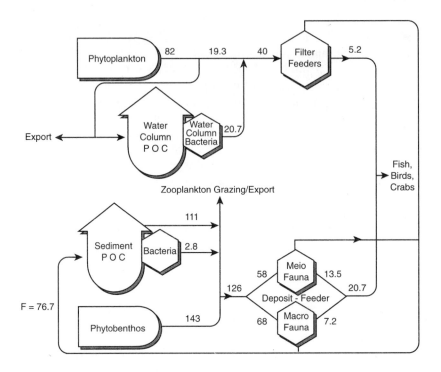

FIGURE 2.5

Carbon flow model of a mudflat of the Lynher estuary in Cornwall, England. Fluxes are g C m^{-2} yr^{-1} and standing stocks are g C m^{-2}. F = recycled organic carbon via feces. (Adapted from Warwick, R. M., Join, I. R., and Radford, P. J., in *Ecological Processes in Coastal Environments*, Jeffries, R. L. and Davey, A. J., Eds., Blackwell Scientific, Oxford, 1979, 429.)

crabs, and fish. Nearly 34% of total primary production is returned to the sediments as feces.

In contrast, Ellison[37] calculated that roughly 84% of benthic primary production (107 g C m^{-2} yr^{-1}) is eventually lost as bacterial and meiofaunal respiration on a mudflat of the nearby Tamar River. Benthic respiration was measured, but meiofauna production (45 g C m^{-2} yr^{-1}) — calculated to exceed bacterial production (11 g C m^{-2} yr^{-1}) — was estimated by the use of P:B ratios. Bacterial production was greatly underestimated by the use of very old data on bacterial turnover in sediments. The role of macrobenthos was not considered.

In tidal mudflats in the Wadden Sea, Kuipers et al.[38] recognized that high rates of oxygen consumption indicate that the bulk of carbon and energy flow is similarly incorporated into microbial food chains (Figure 2.6). They estimated that nearly 75% of the annual organic input (350 g C m^{-2} yr^{-1}) onto the tidal flat is shunted through the "small food web" — bacteria, meiofauna, and some small macrofauna — with

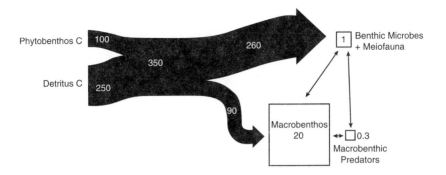

FIGURE 2.6
Carbon flow through benthic food chains in the tidal flat system of the western Wadden Sea. Fluxes and standing stocks are g C m⁻² d⁻¹ and g C m⁻², respectively. (Adapted from Kuipers, B. R., de Wilde, P. A. W., and Creutzberg, F., *Mar. Ecol. Prog. Ser.*, 5, 215, 1981.)

the remainder flowing to the macrobenthos. They hypothesized that the bulk of detrital mineralization is performed by subsurface, anaerobic bacteria that are unlikely to be consumed by higher trophic levels, thus serving as a "last link" in the benthic food chain.

During the past decade, our understanding of the role of bacteria in intertidal food chains has grown, leading to more realistic depictions of carbon flow in intertidal flat sediments.[39-41] In the northern Wadden Sea, Asmus and Asmus[39] calculated energy flow through a seagrass meadow and through sandflats dominated seaward by the lugworm *Arenicola marina* and shoreward by *Nereis diversicolor-Corophium volutator* assemblages. On the *Arenicola* sandflat, nearly 46% of total net primary production (phytoplankton: 28 g C m⁻² yr⁻¹; microphytobenthos: 99 g C m⁻² yr⁻¹) is eventually respired, with bacteria and other microbenthos accounting for nearly one third of the respiration. Similarly, in the *Nereis-Corophium* sandflat sediments, nearly half of the net microphytobenthic primary production (there is no phytoplankton production) is eventually respired, with the bulk of the respiratory losses attributed to microbial respiration; empirical measurements of microbial biomass and productivity were not made.

Partitioning of production and respiration among benthic-size groups shows that bacteria and microalgae are usually the most productive groups in tropical and temperate intertidal sediments.[12,40-42] Schwinghamer et al.,[40] using allometric equations relating production-to-biomass (P:B) ratios to mean individual body mass, found that in Canadian mudflat sediments (Table 2.3), annual production and contribution to community aerobic respiration decreases with increasing size, from bacteria to macrofauna. The bacterial and microalgal estimates are crude, given that respiration was assumed to equal production and that only 1 to 10% of bacterial biomass was assumed to be

TABLE 2.3

Average Secondary Production, Biomass, and
Production-to-Biomass (P:B) Ratio for Different
Benthic Trophic Groups in Intertidal Mudflats
at Bay of Fundy, Canada

Trophic Group	Production (g C m^{-2} yr^{-1})	Biomass (g C m^{-2})	P:B Ratio
Bacteria	12.3–122.8[a]	0.04–0.4	292
Microalgae	47.8	2	24
Meiofauna	12.3	1	12
Macrofauna	2.4	1	2

[a] Assumes 1 to 10% of bacteria are metabolically
   active.

Note: Original values of kcal were converted to g C
       assuming 1 g C = 12 kcal.

Source: Data adapted from Schwinghamer, P. et al.,
Mar. Ecol. Prog. Ser., 31, 131, 1986.

metabolically active. These assumptions are conservative, as they do
not account for respiration and production of anaerobic bacteria, such
as sulfate reducers; however, even with these limitations, it is clear that
the proportional distribution of benthic activity favors bacteria and
microalgae.

The advent of radiotracer methods to measure the synthesis of
nucleic acids in bacteria has provided more accurate measurements of
their productivity in sediments. Studies using these methods confirm
that bacterial production constitutes a large fraction of benthic secondary production in intertidal sediments. Several workers measured rates
of bacterial productivity exceeding rates of microalgal production (Table
2.4).

2.2.1.3 Sediment Bacteria: Aerobic Links, Anaerobic Sinks

The traditional ecological concept of grazing control on sediment bacteria by protozoans, meiofauna, and macrobenthos has recently been
called into question.[5,43] Alongi[5] and Kemp[43] have attempted to reconcile the traditional grazing control view with the biogeochemist's view
of bacteria as mineralizers of organic detritus. Both views suggest that
the regulation of bacterial biomass and growth in sediments is complex, controlled by factors such as the ratio of predator-to-prey, the
quantity and quality of organic matter input, and environmental
changes (e.g., temperature). Grazing is an important trophic link at the
sediment surface and within tubes and burrows, where benthic metazoans are most abundant; however, in deeper, anoxic layers, bacteria

TABLE 2.4

Comparisons of Gross Primary Production (GPP), Sediment
Respiration, and Bacterial Production in Different Tropical and
Temperate Intertidal Flats

Location	GPP	Respiration	Bacterial Production	Ref.
Tropical mudbanks, Papua New Guinea	71	67	206	35
Tropical sandflat, Queensland, Australia	45	94	110	29
Temperate sandflat, Maine	28	76	13	42
Temperate mudflat, Maine	29	76	56	42
Temperate sandflats, Wadden Sea, The Netherlands	100[a]	107	70	41

[a] Assumed from earlier data from nearby sandflats (from several references in van Duyl and Kop[41]).

Note: Units are g C m^{-2} yr^{-1}. Bacterial production estimates are averages of several sites or sampling periods, standardized using the same conversion factors.

Source: Data compiled from Alongi,[29,35] van Duyl and Kop,[41] and Cammen.[42]

may be consumed by protozoans and micrometazoans not linked further to larger consumers. The bulk of bacterial biomass remains ungrazed, mineralizing organic matter. This is supported by recent evidence of high rates of bacterial growth and productivity in surface and subsurface sediments.

The idea that benthic microbial food chains are a trophic sink has sparked research into the applicability of the microbial loop concept to benthic food chains, particularly in intertidal muds and sands. Grazing experiments using fluorescently labeled bacteria support the view that the relationships among bacteria, flagellates, ciliates, microalgae, and meiofauna are interwoven and complex.[44-46] Heterotrophic nano-flagellates and ciliates, only when very abundant, can graze a significant portion of bacterial biomass. The portion they graze is also dependent upon the rate of bacterial productivity. For instance, Kemp[43] found that in salt marsh and saline pond sediments, ciliate bacterivory accounts for the daily removal of less than 4% of bacterial biomass per day, which he attributed to low ciliate abundance relative to bacterial biomass and rapid rates of bacterial productivity. Hondeveld et al.[44] found that grazing rates of heterotrophic nanoflagellates from intertidal flats of the Wadden Sea ranged from 0 to 104 bacteria per flagellate hr^{-1} — not enough to keep pace with bacterial production. In more sophisticated experiments using organisms from a temperate mudflat, Epstein and Shiaris[45] found that rates of bacterivory vary widely among

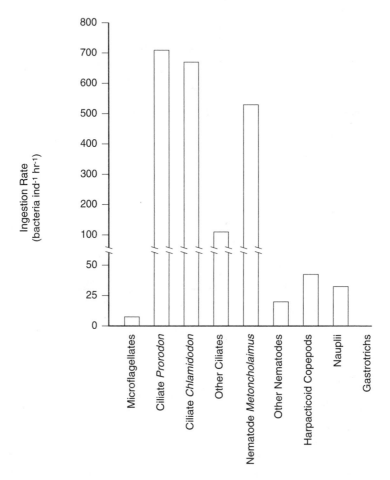

FIGURE 2.7
Grazing rates on sediment bacteria by various benthic protozoans and meiofauna from a temperate mudflat in Massachusetts. (Data from Epstein, S. S. and Shiaris, M. P., *Appl. Environ. Microbiol.*, 58, 2426, 1992.)

protozoans and meiofauna (Figure 2.7). Some ciliates and nematodes show high rates of bacterivory, but their low densities in the mudflat result in only a small portion of bacterial standing stock being consumed. It was calculated that microflagellates, ciliates, and nematodes daily consume only 0.2%, 0.1%, and 0.03% of bacterial biomass.

Empirical evidence supports the view that, while bacteria are intensively grazed in burrow linings, on tubes, and within the first few millimeters of sediment, on the whole, bacteria are not regulated by grazing. From a food chain point of view, this is not surprising considering that protozoans, meiofauna, and macrobenthos feed on a variety

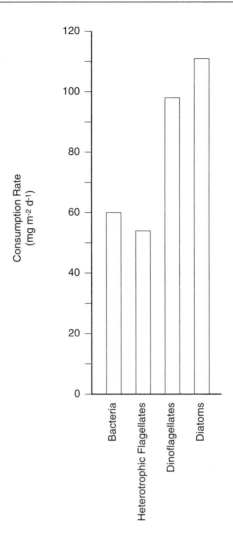

FIGURE 2.8
Consumption rates of bacterial, protozoan and microalgal prey by ciliates in a temperate sandflat on the White Sea in northern Russia. (Data from Epstein, S. S., Burkovsky, I. V., and Shiaris, M. P., *J. Exp. Mar. Biol. Ecol.*, 165, 103, 1992.)

of foods other than bacteria in order to sustain a balanced diet.[5] For instance, Epstein et al.[46] found that on a sandflat ciliates feed on flagellates, dinoflagellates, bacteria, diatoms, and other ciliates but feed faster on the dinoflagellates and diatoms than on bacteria (Figure 2.8). Similarly, Montagna et al.[47] observed that meiofauna taxa vary their grazing rates of microalgae on mudflats in response to changes in microalgal production. Microbial and microfaunal food chains are a

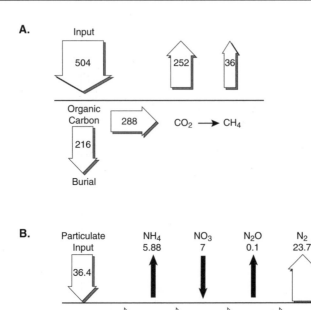

FIGURE 2.9
Pathways of carbon (A) and nitrogen (B) flux to a depth of 0.5 m in mudflat sediments
in the Scheldt estuary, The Netherlands. Fluxes and standing stocks are g C or N m^{-2}
yr^{-1} and g C or N m^{-2}, respectively. (Adapted from Middelburg, J. J. et al., *Hydrobiologia*,
311, 57, 1995.)

consortium of trophic types exhibiting a variety of facilitative, mutu-
alistic, and inhibitive interactions other than predation [48] that affect the
transfer of energy and materials in intertidal sediments.

A mass balance approach confirms that bacteria, using a variety of
electron acceptors, mineralize the bulk of organic material deposited
onto intertidal sediments. Many studies have quantified individual
carbon pathways (e.g., sulfate reduction, methanogenesis) in intertidal
sediments, but few have synthesized the various measures of carbon
oxidation into a coherent picture of bacterial carbon cycling at a given
location. Middelburg and his co-workers[49] provide a comprehensive
budget of bacterial carbon cycling in mudflat sediments near Doel at
the mouth of the Westerschelde estuary on the Dutch coast (Figure
2.9A). Roughly 504 g C m^{-2} is delivered annually to these sediments, of
which nearly 43% is buried, with the remainder mineralized and released

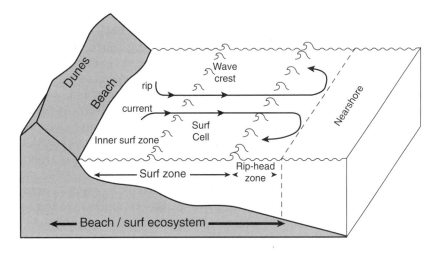

FIGURE 2.10

Idealized view of the structure and maintenance of rip currents and surf-cell circulation within an intermediate-type sandy beach/surf-zone ecosystem. (Adapted from Talbot, M. M., Bate, G. C., and Campbell, E. E., *Oceanogr. Mar. Biol. Ann. Rev.*, 28, 155, 1990.)

as carbon dioxide and, to a lesser extent, methane. Further temporal and spatial analyses[50] in the estuary indicate that mineralization peaks in summer, correlating positively with increasing temperatures, and decomposition rates decline in mudflat deposits towards the estuary mouth, probably due to aging and lower nutritional quality of riverine matter. The mudflats in the estuary represent a combined sink of ~29,700 ton C yr^{-1}, accounting for roughly 30% of the total estimated retention of carbon in this estuary. Bacterial decomposition and respiration of carbon in mudflat sediments can, therefore, represent a significant loss of materials and energy from some estuarine systems.

2.2.2 Closed Ecosystems

All tidal flats receive some inputs of material from the land or sea or from both directions, but some systems are self sustaining. Sandy beaches occurring in several areas of the world, such as on the east coast of South Africa, the southeast coast of Brazil, the northwest coast of New Zealand, and the northwest coast of the U.S., may be considered closed ecosystems. These systems have wide, dissipative surf zones where extensive diatom blooms occur and are retained within the beach/surf-zone area by surf-zone circulation patterns.

The formation and maintenance of surf cells and their circulation patterns off of intermediate and dissipative beaches are related to the formation of rip currents (Figure 2.10). When waves approach the beach perpendicular to the shore, water flowing back off the beach tends to flow into areas of low wave height, giving rise to rip currents. These currents ebb in strength with distance from the beach as they are met by incoming waves of higher energy, change direction as water motion slows, to form a cell circulation pattern. Where tidal range is very large, tidal motion may play an important role in determining cell circulation patterns. These patterns and associated diatom blooms are specific to sandy beaches. Several reasons have been advanced to explain this phenomenon, including beach hydrography and morphology, wind, nutrient supply, and rainfall. Talbot et al.[51] propose that surf-zone diatoms accumulate into dense patches along exposed sandy coasts in relation to climate, requiring pulses of high wave energy coupled with conducive beach morphology (grain size, uninterrupted beach length) and freshwater runoff.

Energy and carbon flow within high-energy beach/surf-zone ecosystems are dominated by high rates of diatom production, most of which is retained within the system.[52,53] On the southeast coast of Africa, the exposed beaches of Algoa Bay are characterized by continual heavy wave action. The shore has a well-developed surf zone that functions together with the beach as a self-sustaining ecosystem.[53,54] The beach is 50 to 100 m wide backed by extensive dunes; seaward, the surf zone is 100 to 300 m wide followed by a rip-head zone of roughly equal width. This rip-head zone is the outer limit of the surf circulation cells. The surf-zone phytoplankton is dominated by the surf diatom, *Anaulus australis*, with a mixed community of diatoms and flagellates in the rip-head zone.

Primary production averages 1205 mg C m^{-2} d^{-1}, and fecal input into the suspended POC pool is 112 mg C m^{-2} d^{-1} for a total available carbon input of 1317 mg C m^{-2} d^{-1}. Other sources of food are allochthonous detritus derived from macrophytes detached from nearby estuaries, rocky shores and reefs, and carrion (mainly cnidarians). Inputs from these sources are unknown but are thought to be minor. High diatom productivity drives microbial food chains in surf waters and sediments and macroconsumers such as benthos, zooplankton, fishes, and seabirds (Figure 2.11). Assuming a total available carbon pool of 1317 mg C m^{-2} d^{-1}, nearly 40% is shunted through pelagic microbial food chains, 20% is consumed by the interstitial fauna, and 5% is consumed by macrofauna; the remainder is available for export as detritus to adjacent coastal waters.

The macrofaunal food web is most dynamic in the surf zone, where longshore currents concentrate diatom patches exploited by zooplankton. Zooplankton feeding on phytoplankton accounts for roughly 10% of the total available carbon pool (Figure 2.11). The surf-zone zooplankton assemblages are composed of copepods, ostracods,

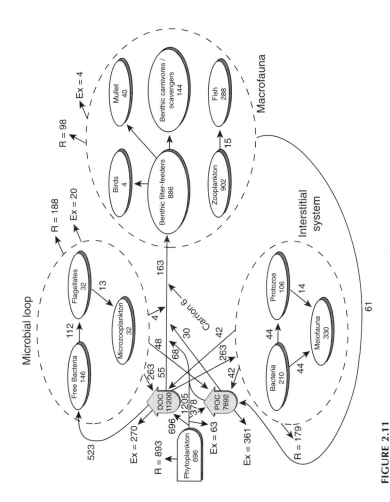

FIGURE 2.11

The main pathways of carbon flow in the Sundays River sandy beach/surf-zone ecosystem, South Africa. Fluxes are mg C m^{-2} d^{-1}, and stocking stock units are mg C m^{-2}. R = respiration; Ex = export. (Adapted from McLachlan and Romer[53] and Heymans and McLachlan.[54])

cladocerans, chaetognaths, siphonophores, and medusae, but mysid shrimps and small penaeid prawns make up 90% of the fauna. Swarming by the mysids attract substantial predation by many pelagic species of juvenile fish. Filter-feeding bivalves are the next conspicuous macroconsumer group feeding on diatoms. *Donax serra* is the most abundant benthic filter-feeder, comprising nearly 70% of macrobenthic biomass. This species and other shore dwellers attract marine predators (crabs and fishes) and terrestrial predators, such as birds and mammals. The crabs are the main scavengers on carrion washed up on the beach. The mullet *Liza richardsoni* is the main fish predator on surf diatoms. Mullet and many of the small fishes that feed on benthos and zooplankton are, in turn, a source of food for larger, predatory fish.

The interstitial fauna is aerobic, driven mainly by organic matter flushed into the sand by wave action, but carbon and energy fluxes among members of the intertidal food web are poorly understood (see previous section). Trophic interrelationships are known, with bacteria fueled by DOM (including diatom mucus) and by particulate detritus. Protozoans feed on bacteria and DOM. Meiofauna feed on a range of microbial foods and POM. McLachlan and Romer[53] and Heymans and McLachlan[54] concluded that the interstitial fauna constitute a carbon sink in the system with little, if any, carbon transfer to other trophic groups. Using network analysis, Heymans and McLachlan[54] constructed a carbon budget and identified key trophic links in the Sundays beach/ surf-zone system. Their results indicate that the interstitial communities are energetically second in importance to the pelagic microbial loop. The productivity of higher trophic levels, benthic and pelagic, are greatly dependent on recycled carbon with detritivory amounting to an average carbon flux of 868 mg C m^{-2} d^{-1} — equal to 92% of recycled carbon entering higher trophic levels.

The microbial loop in surf-zone waters, comprised of a consortium of bacterial, protozoan, and microzooplankton assemblages, is the main pathway for carbon flow in the South African system (Table 2.5). Many of the flow rates are based on assumptions in the literature that may or may not be applicable to these particular assemblages. McLachlan and Romer [53] assumed a bacterial growth yield of 25% to estimate a total bacterial carbon requirement of 120,668 g C m^{-1} yr^{-1} which is 98% of the carbon required to sustain the entire microbial loop. Extracellular release of dissolved compounds from phytoplankton is a large source (~45%) of available carbon; mucus generated by the diatom *Anaulus* alone may provide up to 50% of bacterial carbon requirements.

Naked flagellates, dinoflagellates, tintinnid ciliates, copepod nauplii, and copepodites dominate the heterotrophic nanoplankton and microplankton (Table 2.5). Phagotrophic flagellates of 2 to 20 µm in size are the principal predators on the bacterioplankton, and their

TABLE 2.5

Carbon Flow Through the Microbial Loop in Surf-Zone and Rip-Head Waters in a South African Sandy Beach/ Surf-Zone Ecosystem

Trophic Group	Biomass (μg C l^{-1})	Production (μg C l^{-1} d^{-1})	Total Carbon Requirements (g C m^{-1} d^{-1})
Bacteria	96[a]	109	120,668[b]
Flagellates	14[a]	19	20,440[c]
Microzooplankton	18	5[d]	3148[e]

[a] Assumes a bacterial carbon-to-volume ratio of 2.2×10^7 μg C μm^{-3}.

[b] Assumes bacterial growth efficiency of 25%.

[c] Assumes flagellate growth efficiency of 40% and a daily carbon ration 3.5× body carbon.

[d] Assumes 33% growth efficiency for ciliates and 0.2 d^{-1} growth rate of nauplii and copepodites.

[e] Assumes that tintinnids ingest 43% and other micro-metazoans 50% of their own body carbon per day.

Source: Data from McLachlan, A. and Romer, G. S., in *Trophic Relationships in the Marine Environment,* Barnes, M. and Gibson, R. N., Eds., Aberdeen University Press, Aberdeen, 1990, 356.

consumption of bacterial biomass may equate to ~68% of daily bacterial production. In contrast, the microzooplankton are estimated to remove pico- and nanoplankton biomass equivalent to only 6% of total carbon fixed daily, playing a minor energetic role in losses in the system.

Considerable losses of energy and materials occur within the microbial loop, particularly if the low growth efficiency of bacteria is accurate. McLachlan and Romer[53] propose that the relative importance of the microbial loop increases from its virtual absence within reflective beach systems to its dominance in intermediate and dissipative beach/surf-zone ecosystems. They suggest that the energetic role of interstitial fauna and flora is similar among beach types, but that the importance of macrofauna increases from reflective to intermediate to dissipative systems, commensurate with a general increase in food web complexity.

2.3   NITROGEN CYCLING

Information on nitrogen cycling on sandy beaches and tidal flats is poor compared with data on energy and carbon flow. Specific processes, such as nitrogen fixation and denitrification, have been measured individually in several intertidal locations. These studies have been extensively reviewed in Blackburn and Sorensen.[55] Here, recent advances on nitrogen cycling in microbial mats on mudflats are examined briefly, as well as recent efforts to construct nitrogen budgets for sandy beaches and tidal flats. Various aspects of element cycling on stromatolites and microbial mats in other habitats have been described recently in Cohen and Rosenberg[56] and in Stal and Caumette.[57]

2.3.1   Microbial Mats on Mudflats

Microbial mats are often a productive component of intertidal mudflats, persisting because of tight coupling and exchange of essential elements among highly diverse autotrophic and heterotrophic microorganisms (mainly cyanobacteria and eubacteria) that are considered an important source of new nitrogen. The extent to which this is true depends upon whether or not a balance exists between the rate of input (nitrogen fixation) and the rate of output (denitrification) and the degree to which mineralized nitrogen is retained within the mat matrix.

Rates of nitrogen fixation are often rapid, but losses due to denitrification have not been measured frequently. Simultaneous nitrogen fixation and denitrification has been observed in microbial mats on mudflats in Tomales Bay in California.[58,59] On average, rates of nitrogen fixation exceed denitrification rates in the daytime, but nighttime rates of denitrification equal or exceed rates of nitrogen fixation. Diel fluctuations may be related to daily variations in oxygenic photosynthesis and subsequent changes in oxygen concentration. Depending on time of year, denitrification may offset new nitrogen inputs via nitrogen fixation, with mats being a sink for fixed nitrogen in summer but a source for fixed nitrogen in fall and spring. This temporal pattern is partially related to seasonal inputs of limiting nitrate and carbon derived from rainfall.[58] On an annual basis, these mats lose 1.5 g N m^{-2} but, by difference, retain 9.8 g N m^{-2}. Extrapolating to the entire mudflat and assuming Redfield stoichiometry, these mats can support the fixation of nearly 64 g C m^{-2} yr^{-1}. Microbial mats may, therefore, enhance the input of carbon and nitrogen to tidal flats, although their potential role in supporting neighboring food chains should be further examined.

2.3.2   Nitrogen Budgets

Aspects of nitrogen cycling have been examined on only a few beaches, but some nitrogen pathways can be traced for systems receiving large quantities of stranded kelp detritus.[30,60] A complete budget has recently been constructed for a closed beach/surf-zone system.[61]

Some evidence suggests that sandy beaches, even those receiving macrophyte input, may not accumulate nitrogen in the long-term. Using a mass balance approach, Koop and Lucas[30] traced the flow of nitrogen from kelp debris in sandy beach microcosms. Unlike carbon flow, where most is lost as respiration, most nitrogen (94.2%) is converted into bacterial biomass, with a small remainder lost to the atmosphere by denitrification (2.3%) and immediate export to the sea (1.5%). These experiments were run for only 8 days, so nitrogen immobilized by bacteria in the natural environment may eventually leach back to the sea. Koop and Lucas[30] calculated that 95% of kelp nitrogen deposited onto the beach would ultimately be returned to adjacent coastal waters. On a sandy beach receiving high (14,338 g N m^{-1} yr^{-1}) inputs of red algae, McLachlan and McGwynne[60] followed net changes in biomass and groundwater input over the course of a year and found that the beach did not accumulate nitrogen. The main reason appears to be rapid turnover of the stranded wrack, as measured from litter bag experiments.

Nitrogen pathways appear to be more complex within sandy beach/surf-zone systems and on well-protected beaches and tidal flats, where quiescent conditions permit enough organic matter to accumulate and aerobic and anaerobic food chains to develop more fully. Nitrogen is tightly recycled and thus conserved, being a limiting element. In contrast, carbon is not efficiently retained, as a large proportion is lost to the atmosphere by respiration.

A complete nitrogen budget constructed for a sandy beach/surf-zone ecosystem in the Eastern Cape region of South Africa indicates that most nitrogen is recycled (Figure 2.12). Groundwater, estuarine inputs, and rain are the main sources of nitrogen. Phytoplankton, mainly diatoms, are the major consumers of nitrogen. Coupled with nitrogen recycled in the surf zone, total available nitrogen exceeds phytoplankton primary production requirements by ~4 kg N m^{-1} yr^{-1}, reflecting either some export from the system or a degree of expected error. Losses as a result of burial or denitrification appear to be minor, considering the extent of sediment reworking and the dominance aerobic sand. Recycled nitrogen supplies nearly 87% of phytoplankton requirements, contributed in roughly equal proportions by the pelagic and benthic microbial assemblages and by macroorganisms. No exchange of nitrogen is presumed to occur between the interstitial food web and the other food chains, reflecting the fact that benthic microbial and meiofaunal food chains are virtual trophic sinks, as discussed earlier. Nearly one quarter of the nitrogen required by the microbial loop is supplied by the macrofauna. The entire fauna and flora recycle 99% of the nitrogen required for phytoplankton production supporting

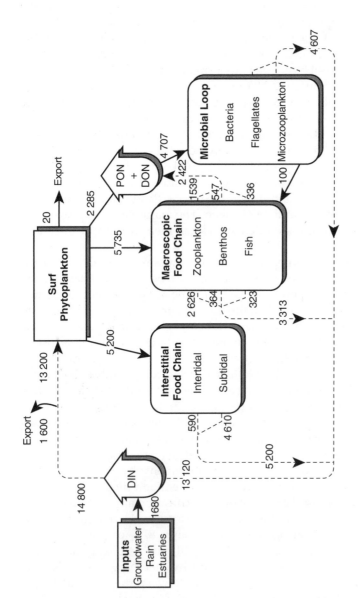

FIGURE 2.12

Nitrogen flow model of the Sundays River sandy beach/surf-zone ecosystem, South Africa. Solid lines indicate grazing pathways; broken lines indicate recycling pathways. Fluxes are g N m^{-1} yr^{-1}. (From Cockcroft, A. C. and McLachlan, A., *Mar. Ecol. Prog. Ser.*, 100, 287, 1993. With permission.)

the hypothesis of McLachlan and Romer[53] that sandy beach/surf-zone ecosystems are self sustaining.

Mudflat sediments, in contrast, are an important sink for nitrogen. For example, Middelburg et al.[49] constructed a sediment nitrogen budget for the intertidal muds near Doel in the Scheldt estuary (Figure 2.9B). Each year, about 36 g N m^{-2} yr^{-1} of particulate nitrogen are delivered to these muds, in which roughly 42% is buried and the remainder mineralized by benthos to ammonium. Some ammonium is released from the sediment, but the bulk is immediately nitrified. Approximately 23.8 g N m^{-2} yr^{-1} is ultimately denitrified, with 30% of the NO$_3^-$ supplied from the tidal water to the sediment by diffusion. Nearly all of the denitrified nitrogen escapes as N$_2$ gas. Proportionally less nitrogen is lost (14%) from the estuary than carbon (30%). Conservation of nitrogen is a persistent characteristic of nearly all marine habitats.

2.4   LINKAGES TO PHYSICAL PROCESSES

Because of their physical settings, sandy beaches and intertidal flats and their associated food webs are greatly affected by physical processes. Models and budgets of energy and nutrient flow do not reflect the large-scale or fine-scale complexities or subtleties of change, inherent to tidal environments. The rhythmic movements of interstitial water and fauna induced by tides or, at the other extreme, movements of parcels of water, energy, and materials between intertidal and adjacent subtidal ecosystems are not adequately conveyed in static depictions but are undoubtedly of great importance. Recent appreciation and examination of some large- and small-scale processes, such as intertidal-offshore exchanges and tidal water-sediment interactions, emphasize that sandy beaches and tidal flat ecosystems do not function in isolation from land and sea.[21]

2.4.1   Water-Sediment Interactions: Effects of Tides, Waves, and Storms

The movements of tidal water greatly affect changes in interstitial water levels and chemistry. When intertidal sediments are exposed, surface sediments are drained of water. The depth of drainage is determined largely by grain size and elevation; beaches and sandflats drain more quickly and thoroughly than do mudflats. When covered by tides, oxygen again becomes limiting in fine sediments. These tidal movements cause oscillations in the redox state which affect the rates and pathways of energy and nutrient flow. In an intertidal mudflat of the Elbe estuary, Kerner and Wallman[62] and Kerner[63] found, for instance, that exposure to air and submersion during a tidal cycle produces alternating aerobic and anaerobic layers in the surface muds, shifting microbial decomposition of organic matter from mainly oxygen and nitrate respiration to fermentation and metal reduction. This shift also occurs seasonally, with a greater proportion of energy flow shunted through aerobic pathways in winter when low rates of microbenthic production, low temperatures, storms, and ice re-oxidize the upper sediment layers.

Tides, wind-induced waves, and storms also play roles in altering beaches and tidal flats. The extreme example of beach and tidal flat erosion by storms is well documented, but storms may induce less drastic changes by:[64]

- Altering the distribution and concentration of particulate material in surface sediment and tidal water
- Increasing advection of porewater
- Disrupting the stability of surface sediments affecting the abundance and productivity of benthic organisms, including primary producers

On an intertidal sandflat in Delaware, Bock and Miller[64] followed the effects of a storm on particulate organic matter and found that storms increased suspended POM and bedload transport. Further, the storms diluted the protein and chlorophyll content of the suspended matter. Overall, the storms increased the quantity but lowered the quality of particulate food for benthic suspension-feeders.

Resuspension of surface sediments and associated microphytobenthos occurs frequently on tidal flats, caused by tides and wind-induced waves that are not necessarily related to storms. Resuspension can be critical in affecting food availability and eroding surface organisms, thus lowering productivity, and is also important in determining whether or not intertidal flats are sources or sinks of suspended matter for adjacent habitats. de Jonge and van Beusekom[65] noted the ecological consequences of resuspension in a study of wind-driven resuspension in the Ems estuary. They observed that benthic and pelagic diatoms are continually mixed in relation to wind speed, and that the exchange of suspended matter and some surface-dwelling organisms between the tidal flats and overlying water is very rapid, contributing to a mosaic of unstable and stabilized sediment patches. Spatial and temporal heterogeneity driven by such physical processes undoubtedly generates variability in food chain dynamics and energy flow on tidal flats and beaches, but such processes have not been adequately examined.

2.4.2   Exchanges with Land and Sea

Intertidal systems have been categorized here as either open or closed, but the distinction is blurred because beaches and tidal flats clearly interface between land and sea and are controlled by a variety of physical processes. Closed systems (e.g., sandy beach/surf zones in South Africa) undoubtedly "open" when storms break down surf cells.

DeAngelis[9] suggests that ecosystems are open with respect to hydrodynamic forces and energy flow, but closed with respect to nutrient cycles. This is not true in many shoal habitats. Some closed sandy beach/surf-zone systems are greatly reliant on groundwater for a significant fraction of their nutrient input.[66] Some tidal flats simultaneously serve as a local reservoir for heat energy[67] and as a source of dissolved salts[68] and nutrients[69] for offshore waters. Sandy beaches and tidal flats are, therefore, an important link between land and sea, playing a critical role in the exchange of energy and materials between terrestrial and marine ecosystems.

Chapter 3

MANGROVES AND SALT MARSHES

3.1 INTRODUCTION

For a part of each day, tidal waters bathe meadows of marsh grasses and forests of mangrove trees inhabiting vast expanses of the world's coastlines. Both of these salt-tolerant plants develop best under a quiescent physical milieu, where low wave energy and shelter facilitate accretion of soft sediments that enable mangrove trees and marsh grasses to establish roots and grow. Mangrove forests — more structurally complex than marsh grass communities — are architecturally simple compared to tropical rain forests, often lacking an understory of shrubs or ferns. Salt marshes and mangrove forests are similar in having low diversity of species compared to other plant communities and ecosystems.

Mangrove forests and salt marshes are valuable resources for multiple economic and ecological reasons, as they provide:

- Recreation
- Education
- Habitats for many estuarine and marine assemblages
- Control of floods, storm surges, and coastal erosion
- Filters for nutrients, pollutants, and some pathogens

In the tropics, many mangrove forests are not only a source of edible finfish and shellfish, but they also provide shelter, wood for fuel, and a variety of natural products (see Chapter 8). Marshes and mangroves fulfill nearly identical roles, but as explored in this chapter, the differences in their structural and functional attributes outweigh their similarities.

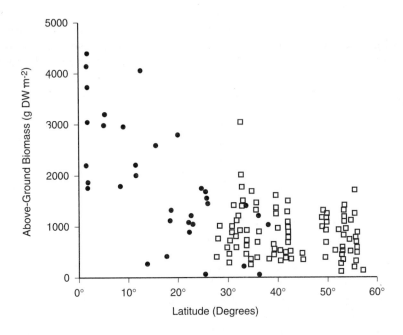

FIGURE 3.1
Global trends in above-ground biomass estimates of salt marshes (open squares) and mangrove forests (closed circles) with latitude. (Adapted from Turner,[70] de Leeuw and Buth,[71] Twilley,[72] and Clough.[73])

3.2   GLOBAL TRENDS IN PLANT BIOMASS AND PRIMARY PRODUCTION

Salt marshes develop most fully in temperate latitudes, whereas mangrove forests are most luxuriant in the tropics. From a global perspective, the data indicate that mangrove forests attain peak biomass near the equator with a decline towards temperate latitudes (Figure 3.1). Salt marshes and mangrove forests co-occur along a latitudinal gradient of roughly 10° (27 to 38°) with marshes replacing mangroves as the dominant coastal vegetation by approximately 28° latitude, with some notable mixed stands along the gulf coast of the U.S. and along the southeastern coast of Australia.

The pattern of above-ground biomass of salt marshes with latitude is not clear (Figure 3.1). This probably reflects the more complete database that exists for marshes than for mangroves, encompassing the real variation either within a given latitude, among continents, or among species. In his original analysis of the standing biomass of *Spartina alterniflora* marshes along the east coast of North America, Turner[70] recorded a negative correlation between latitude and above-ground biomass and suggested that the latitudinal gradient in *S. alterniflora* was related to increased solar radiation from north to south. When the European marshes, composed largely of *Spartina anglica*, are included,[71] the correlation is not significant (Figure 3.1). Latitudinal gradients exist for individual species, but latitudinal variations at a generic or multi-species level are large, reflecting variations among continents and even among biogeographical regions.

The standing crops of mangrove forests are generally greater than those of salt marshes (Figure 3.1). This is not surprising if one considers that many of the mangrove forests bordering the equator are immense (Figure 3.2), with above-ground biomass often exceeding 2000 g DW — equivalent to the standing crop of many of the world's tropical rainforests.[74]

Lacking a woody stem, salt marsh plants are smaller and less structurally complex than mangrove trees. Most of the above-ground biomass of mangroves is tied up in stems and prop roots, with a much smaller amount of total standing crop vested in leaves and branches (Figure 3.3). For *Rhizophora* species, the allocation of biomass of prop roots proportionally increases with an increase in tree size, presumably for further support, and gives some indication of the accumulation of above-ground biomass as the trees and forests age. Mangrove tree stems and prop roots are long-lived, stable structures, and only a small proportion of standing biomass consists of dead trees, although the proportion varies with species and age of the forest.[73] Salt marsh grasses are different in this respect, being non-woody plants with a much greater proportion of dead-to-live standing biomass.[75] Temporal changes in above-ground, living biomass are seasonal for salt marshes with little interannual variation.[76] The few data available for mangrove forests indicate year-to-year, and even decadal, changes in mangrove tree densities and living biomass, although litterfall patterns are seasonal.[77] This is expected because of the longer lifespans of trees than of grasses.

The proportion of biomass vested below-ground differs between mangroves and marshes, reflecting the structural differences between trees and grasses. In *Spartina* marshes, two to four times more biomass is tied up as below-ground roots and rhizomes, rather than as above-ground shoots and standing dead material.[75] Much less data are available for mangroves, but most estimates[72,78] indicate that below-ground biomass accounts for roughly 50% of total forest biomass.

Salt marshes and mangrove forests are commonly believed to be highly productive ecosystems. Much evidence supports this supposition, but not all salt marshes and mangrove forests are highly productive,

FIGURE 3.2
A mature mangrove forest in Galley Reach, Papua New Guinea, where *Rhizophora apiculata* trees reach 30 m in height. (Photograph courtesy of B. Clough.)

especially at their respective latitudinal limits or climatic extremes.[73] In the arid tropics, mangrove forests are small in area and the trees are stunted due to physiological stress.[73] Neither is it correct to infer that

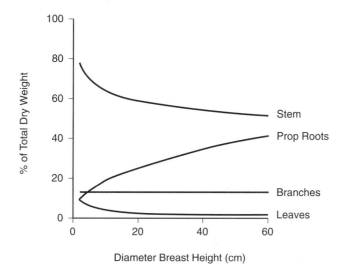

FIGURE 3.3
Partitioning of above-ground biomass in relation to tree size for *Rhizophora apiculata* and
R. stylosa in Northeastern Australia. (Reprinted from Clough, B. F., in *Tropical Mangrove
Ecosystems,* Robertson, A. I. and Alongi, D. M., Eds., American Geophysical Union,
Washington, D.C., 1992, chap. 8. With permission.)

a luxuriant forest is productive; biomass rarely mirrors rates of primary
productivity. In Malaysian mangroves, for instance, Gong et al.[79] found
that maximum biomass occurred in a 20-year-old *Rhizophora* forest, but
that productivity was higher in a 15-year-old forest.

Numerous estimates of above-ground, net primary production
indicate that, on average, mangroves are more productive than salt
marsh grasses (Table 3.1). All of these estimates must be considered
with caution owing to the different methods used. Net photosynthetic production by mangrove canopies has almost certainly been
underestimated. Using three independent indirect and direct methods, Clough et al.[82] and Gong et al.[83] calculated that previous measurements using the light interception method — in common use
throughout Southeast Asia — underestimate true rates of photosynthesis by an order of magnitude. Earlier measurements are conservative, as the rates of net production obtained using these methods
are not sufficient to account for the observed accumulation of above-
ground biomass. Rates of net production of marsh grasses are also
likely underestimates given that some losses, such as leachates from
roots and rhizomes, are not known.[76]

Wood production, ranging from 3 to 25 metric tons ha^{-1} yr^{-1}, accounts for roughly 60% of mangrove net primary production, with the

TABLE 3.1

Estimates of Above- and Below-Ground Net Primary Production (NPP) of Some Selected Species/Community Types of Salt Marsh Grasses and Mangroves

Community Type	Location	Above-Ground NPP (g dry weight m^{-2} yr^{-1})	Below-Ground NPP (g dry weight m^{-2} yr^{-1})
Marsh Grasses			
Distichlis spicata	Pacific coast	750–1500	—
	Atlantic coast	—	1070–3400
Juncus roemerianus	Georgia	2200	—
	Gulf coast	4250	1360–7600
Spartina alterniflora	Atlantic coast	500–2000	550–4200
	Gulf coast	3250	279–6000
Spartina patens	Gulf coast	7500	—
	Atlantic coast	—	310–3270
Mangroves			
Rhizophora apiculata	Southeast Asia	1900–3900	—
Mixed *Rhizophora* spp.	New Guinea	1750–3790	—
	Indonesia	990–2990	—
Bruguiera sexangula	China	3500	—

Source: Adapted from Kennish[80] and Saenger.[81]

remaining 40% accounted for by litter (leaves, twigs, and flowering parts).[72,73] Seasonality of mangrove production is not known but has been assumed given some evidence of seasonality and latitudinal variation in litterfall patterns.[72,77] Litterfall rates average 8 to 10 metric tons dry weight ha^{-1} yr^{-1}, with a peak usually occurring immediately before or during the monsoon season. The ease with which litterfall can be measured has resulted in a comparative wealth of information on this component of net primary production, but it is erroneous to equate litterfall with net production. There is no evidence that litterfall and net primary production are correlated.[73] Neither is there any reliable information on below-ground production for mangroves.

Net below-ground production of salt marsh grass is equivalent to, and frequently greater than, above-ground production (Table 3.1). In *Spartina* marshes in Georgia, below-ground production is roughly 1.6 times greater than above-ground production.[75] Similar results were obtained by Hackney and de la Cruz[84] in U.S. marshes bordering the Gulf of Mexico. Seasonally, below-ground biomass and production builds up from late autumn to early winter when above-ground biomass is smallest, and peaks prior to maximum

TABLE 3.2

Contribution of Different Autotrophs to Annual Net Primary Production
in Some Salt Marshes and an Australian Mangrove Forest

Component	South Carolina	New York	Massachusetts	Australia
Marsh/mangrove	1707	692	1056	2969
Macroalgae	200	75	—	—
Microalgae	200	50	120	104
Phytoplankton	100	50	24	150
Epiphytes and neuston	100	12	25	260
Total	2307	879	1225	3483

Note: Units are g C m^{-2} yr^{-1}.

Source: Data from Vernberg,[85] Valiela and Teal,[86] Robertson et al.,[87] and
Alongi.[88]

shoot emergence. When above-ground growth is rapid in spring,
below-ground biomass declines. This temporal asymmetry suggests
that marsh grasses seasonally redistribute and store biomass. Again,
nearly all below-ground production is underestimated, as the methods do not include DOM released by roots and rhizomes and the
sloughing of some material.

Macroalgae, benthic microalgae, phytoplankton, epiphytes, and
neuston contribute to total net primary production in salt marshes
and mangroves. These autotrophs, frequently overlooked, may *in
toto* account for more than half of total net primary production in
some systems.[73,85] In the *Spartina* marshes of North America, however, *Spartina* contributes 74 to 86% of the net production, with
other autotrophs accounting for the remaining production (Table
3.2). In mangrove forests with a well-developed canopy, trees contribute most of the production; other autotrophic contributions may
be minor due to severe light limitation (Table 3.2). This is not true
for all forests, as edaphic algal and epiphyte production is greater
in fringing mangroves due to sufficient light penetrating the less
luxuriant canopy.[87] In contrast, phytoplankton production is low to
moderate (10 to 700 mg C m^{-3} d^{-1}) in most marsh and mangrove
creeks and estuaries, varying with geomorphology, water motion,
turbidity and nutrient delivery.[87] Plankton production may also be
reduced by the release of tannins by plant roots and decomposing
wood and leaf detritus. High rates of phytoplankton production (1
to 5 g C m^{-3} d^{-1}) occur mainly in lagoons or small embayments
fringed by mangroves and marshes, particularly those receiving
significant quantities of nutrients from coastal human populations.
It is now recognized that algae play a significant role in primary
productivity and are a major source of food for many organisms in
salt marshes and mangroves.

3.3   FACTORS LIMITING PLANT PRODUCTION AND GROWTH

The relationship between these halophytes and edaphic characteristics is fundamental to understanding what limits plant growth and the subsequent transfer of organic matter to detritus food chains. Both marsh grasses and mangroves have very different photosynthetic characteristics, resulting in very different physiological tolerances of common physicochemical factors, such as temperature, nutrients, solar radiation, oxygen, and water (rainfall). Soil conditions (pH, Eh, grain size) determine the extent of availability of macro- and micronutrients, oxygen, and freshwater to the plants. These factors operate synergistically and antagonistically, depending upon local conditions.

3.3.1   Temperature and Light

Mangroves and marsh grasses grow over a wide range of temperatures and light conditions. Their latitudinal distribution doubtless reflects some overlap in many physicochemical requirements, but there are clear individual differences in physiological tolerances to temperature and solar radiation. Most salt marsh plants[89] utilize the C_4 pathway of carbon fixation producing oxaloacetic acid (OAA), a four-carbon compound during the dark photosynthesis reactions. In contrast, mangroves[90] utilize C_3 photosynthetic biochemistry, in which phosphoglyceric acid (PGA), a three-carbon compound, is produced during the dark reactions. C_4 plants have a specialized leaf anatomy which maximizes internal CO_2 concentrations, thereby allowing the plants to maintain a higher rate of photosynthesis at higher temperatures and at stronger light intensities than C_3 plants. For instance, the temperature optimum for the C_4 grass, *Spartina alterniflora*[89] is 30 to 35°C whereas the temperature optimum for the C_3 mangrove, *Rhizophora stylosa*,[90] is 25 to 30°C. Moreover, mangrove trees become light-saturated at quantum flux densities of 30 to 50% sunlight. This means that the photosynthetic rates for C_3 plants level off at lower temperatures or at lower light intensities than for C_4 plants. These differences in carbon fixing abilities result in different carbon isotope signatures for mangrove and marsh plants. Common marsh grasses[91] exhibit $\delta^{13}C$ values ranging from –12.7 to –26.0%; mangrove foliage[90] ranges in composition from –24.6 to –32.2%.

Why mangroves are C_3 plants is puzzling considering the higher temperatures and more intense light levels in the tropics. The C_3 vs. C_4 dichotomy is not well understood, as there are similar striking anomalies among other vegetation.[74] All woody plants are C_3 plants, so the use of the C_3 pathway is a likely reflection of the terrestrial ancestry of mangroves. Mangroves may, however, be evolving towards C_4 biochemistry. There is evidence[92] that some species possess the enzymes necessary for C_4 photosynthesis. Mangroves are also conservative in their use of water compared to other C_3 plants, possessing thick cuticles, wax coatings, and sunken or hindered stomata to aid in minimizing water loss.

The C_3 tolerance to low light intensity may have a selective advantage for mangroves, as light attenuates rapidly under most forest canopies. Members of the genus *Rhizophora* are remarkably adaptive in being able to change the angle of their leaves with position in the canopy. Leaves at the top of the canopy are inclined vertically to avoid absorption of excessive radiation, with leaves becoming progressively more horizontal with depth into the canopy to maximize interception of the decreasing light. Salt marsh grasses in contrast may be seasonally limited by light, particularly in boreal latitudes, and dense culms of grass may limit light for edaphic microalgae in some systems.

3.3.2   Salinity

Most species of marsh grasses and mangroves are classified as either facultative or obligate halophytes but vary widely in ability to tolerate salt. The cordgrass *Spartina alterniflora* is an obligate halophyte that grows best at a salinity of 10 but survives in salinities up to 70.[93] The tolerances of the mangroves *Avicennia marina, Rhizophora stylosa, R. mangle, R. apiculata, Lumnitzera racemosa,* and *Xylocarpus granatum* are similar, all growing best at soil salinities of 10 to 20 and varying in their ability to tolerate high salinity.[94] *Avicennia marina,* for instance, grows as a scrub on hypersaline pans in the arid tropics, whereas *Nypa fruticans* grows in freshwater swamps and on river banks, having no obligatory requirement for NaCl beyond trace amounts.

Both halophyte groups are affected identically by high salinity stress and have developed nearly identical adaptive mechanisms.[93,94] High salt concentrations either damage cell membranes, leading to increased permeability and subsequent loss of nutrients/ions to the surrounding sediment, or result in reduced water intake. Loss of water balance limits growth because plants cannot fix carbon without losing water. Salt marsh grasses and mangroves have the selective capacity to exclude or concentrate preferred inorganic ions. Tips of *S. alterniflora* blades and the underside of *A. marina* leaves possess salt glands. Other species, such as *Spartina patens* and *R. apiculata,* actively slow the uptake of toxic ions via their roots. Exclusion of salt may act as a feedback mechanism, whereby excluded salts concentrate in the sediment to the extent that the rate of water uptake becomes severely limiting. Passiourna et al.[95] suggest that this may explain why mangroves, and perhaps most halophytes, have low transpiration rates.

3.3.3   Anoxia and Water Movement

Salt marsh grasses and mangroves grow in sediments that are waterlogged by tides and are anaerobic a few millimeters below the surface. The interactions between plant and soil are complex, with several interrelated factors affecting growth and production. Plants modify the soil immediately surrounding their roots and rhizomes in order to:

- Avoid toxic ions
- Maximize availability of oxygen and nutrients

Nutrient concentrations, redox potential, pH, oxygen levels, and other factors, such as interstitial salinity, porosity and permeability, are in turn affected by sediment grain size. On a larger-scale, these factors are determined partly by:

- Frequency of tidal inundation
- Climate
- Geomorphology

For instance, in the arid tropics of northern Western Australia, a large tidal regime (≥5 to 8 m) and historically minute rainfall result in mangrove forests of small height inhabiting sandy sediments. Conversely, in the wet tropics of Southeast Asia, luxuriant mangroves occupy sediments that are largely silt-clay. Edaphic characteristics, such as permeability, dissolved oxygen and nutrient concentrations, pH, and redox potential, are largely dependent upon sediment grain size.

Tidal and interstitial water motion and rainfall greatly affect plant growth. It is a common observation that zonation and productivity of marsh grasses and mangroves is enhanced by increased tidal wetting.[70,73,78] Low marshes and low intertidal mangroves are usually more productive than plant communities growing higher on land. The causal mechanisms are not fully understood, but it is likely that more frequent flooding and drainage replenishes sediment nutrients and dissolved oxygen and flushes out toxic compounds, such as salts, sulfides, and other byproducts of microbial anaerobic activities. Less frequent wetting may result in the buildup of toxic sulfides, drying, and

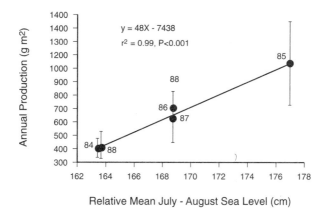

FIGURE 3.4

The positive relationship between annual above-ground *Spartina* production and changes in mean sea level (July to August measurements only) in South Carolina over the period of 1984 to 1988. (From Morris, J. T., Kjerfve, B., and Dean, J. M., *Limnol. Oceanogr.*, 35, 926, 1990. With permission.)

subsequent increases in interstitial salinity regulated by evapotranspiration and frequency of precipitation.

Marshes and mangroves higher on land accumulate less organic matter transported via tides, and usually reside in sediments coarse in texture. Stand size and production may be due not only to geophysical factors related to intertidal position, but also to long-term changes in hydrography. In South Carolina, a positive correlation was found between yearly above-ground production of *Spartina* and changes in mean sea level (Figure 3.4). Morris et al.[96] attributed this phenomenon to amelioration of high porewater salinities in summer by the increase in tidal inundation of estuarine water. Primary productivity did not correlate with temperature but did correlate positively with rainfall, although to a much lesser degree than change in sea level. In Dutch marshes, de Leeuw et al.[97] found that, over a 13-year period, above-ground biomass correlated negatively with rainfall deficit; higher soil salinities during dry years inhibited plant growth.

The movement of interstitial water and its effect on plant growth have not been measured frequently in salt marshes and not at all within mangroves. On Sapelo Island, Wiegert and his colleagues[98] experimentally increased subsurface drainage within a stand of *S. alterniflora* and found an increase in mean shoot height and above-ground production. They attributed the enhanced growth to increased oxygen availability and a reduction in soil sulfides.

The degree of anoxia influences pH, the availability of oxygen and nutrients, and the buildup of toxic compounds in marsh and mangrove sediments.[99,100] It is well known that marsh grasses and mangroves translocate oxygen to their roots, oxidizing the soil proximate to their roots and rhizomes. Clear evidence for this activity comes from measurements of sediment redox potential. Howes et al.[99] measured lower redox levels in bare muds than in tall marsh sediments in the Great Sippewissett Marsh. Similar findings have been made for some, but not all, mangrove sediments. Over a 5-year period, Alongi[101] found no significant difference in redox potential between *Rhizophora* muds and adjacent bare creek-bank muds of Hinchinbrook Island in Australia. He attributed this result to the very high spatial heterogeneity of mangrove roots and the oxidizing effects of creek drainage.

The effects of soil oxidation by mangroves has been clearly observed on a smaller scale. Andersen and Kristensen,[102] using microelectrodes in the rhizosphere of *Avicennia marina*, noted that the oxic zone of the sediment around the roots was ~0.5 mm, suggesting that very little oxygen is released from *Avicennia* roots. McKee et al.[103] may have reconciled earlier conflicting evidence in noting a negative correlation of redox potential and porewater sulfide concentration, with pneumatophore density underscoring the patchiness in redox status in mangrove sediments. Subsequent experiments with mangrove seedlings by McKee[104] imply a reciprocal effect in that adults oxidize surrounding soils, but their ability to colonize and grow in a particular location depends upon spatial and temporal variations in soil redox and sulfide levels.

It has been suggested for some time that sulfides negatively affect plant growth by inhibiting nutrient uptake, but some causal mechanisms have been demonstrated only recently.[94,105] The idea of uptake inhibition of nutrients came from early research on rice plants which found slower uptake of ammonium under anaerobic conditions than in aerobic soils. In laboratory experiments, it has been shown that sulfide concentrations (as low as 250 μM) significantly inhibit uptake kinetics of NH_4^+ by *S. alterniflora*.[93] Moreover, the presence of sulfide inhibited uptake to a greater extent than the absence of oxygen alone, suggesting some biochemical interference. Indeed, the absence of oxygen in cultures of *S. alterniflora* causes a switch from aerobic to fermentative metabolism in root cells.[105] Addition of dissolved sulfide further shuts down the fermentative metabolism in the roots, leading to an energy deficit; uptake of nitrogen is inhibited because it is an energy-dependent process. No such evidence exists for mangroves.

3.3.4 Bioturbation

By virtue of their feeding, walking, and burrowing activities, crabs greatly affect:

- Sediment chemistry[106]
- Abundance of other benthic organisms[87]
- Growth and survival of marsh grasses[107] and mangrove forests[108]

Crabs are the most abundant and conspicuous benthic macro-invertebrates in most salt marshes and mangrove forests. Crab burrows increase sediment drainage, shift soil redox to more oxic conditions by increasing sediment surface area, increase decomposition of below-ground organic matter, and enhance primary productivity of most marsh grasses and mangrove trees. In a New England salt marsh, experimental reductions in densities of the fiddler crab *Uca* led to a 35% increase in root biomass and to a 47% decrease in above-ground plant biomass over a single growing season.[107]

More recent studies[106,108] show that crabs can affect plant growth and survival by influencing interstitial water chemistry and fluxes across the sediment-water interface. In removal experiments in a mixed *Rhizophora* forest in northern Australia, Smith et al.[108] found that in plots without crabs (mostly the grapsids, *Sesarma messa* and *S. semperi longicristatum*), porewater sulfide and ammonium concentrations increased compared to control sediments. Moreover, forest growth (as measured by stipule fall) was significantly reduced in the absence of crabs. They attributed these results to the aeration effects caused by burrowing.

Crabs are known to affect surface microbial biomass and taxonomic composition and to enhance total microbial activity,[87] but no data are extant on how crabs affect the rates and pathways of subsurface bacterial mineralization — that is, whether sulfate reduction is dampened, and oxic or postoxic pathways are enhanced, by bioturbation.

3.3.5 Nutrient Availability

The physicochemical properties of sediments are critical in determining the extent of macro- and micronutrient availability to marsh grasses and mangroves. Most nitrogen and phosphorus is bound to humic and fulvic acids, clays, minerals, and metal complexes.[109] A further complication is that the composition of the organic complexes is very poorly understood, so it is not known what proportion of the dissolved and particulate pools of organic matter are readily available. For instance, dissolved organic nitrogen (DON) is often the largest dissolved nitrogen pool in mangrove[101] and marsh[110] muds, but the definition of DON is operational, considered as the difference in concentration between the total and oxidizable pools. The actual components of DON are undoubtedly complex. The same scenario is true for the dissolved organic carbon (DOC) pool; only a small fraction of DOC has been clearly identified.

Most of the inorganic phosphorus in marsh and mangrove sediments is adsorbed or strongly bound to calcium and iron, severely limiting its availability to plants. Ammonium is nearly always the dominant form of dissolved inorganic nitrogen (DIN) in anoxic porewater in salt marshes[110] and mangroves,[101] followed by lesser amounts of nitrite and nitrate. Ammonium is produced by bacterial breakdown of organic nitrogen and by animal excretion and is stable in anoxic waters. Nitrite and nitrate are intermediates in the oxidation of ammonium to the eventual release as di-nitrogen via denitrification, so these pools turn over rapidly and are small. In salt marsh sediments, at least half of the ammonium pool is bound;[109] a slightly larger proportion of ammonium is free in the porewater in mangrove sediments.[101] Ammonium is readily exchangeable with other cations, so the high concentrations of sodium in saline soils tends to swamp the cation exchange sites, displacing NH_4^+. The ion is readily mobile and susceptible to being leached by heavy rain. It was earlier thought that drainage through creek banks was a significant pathway of ammonium loss from wetlands, but recent measurements[110] in salt marshes suggest that this route is minor.

Changes in the dissolved nitrogen pools may relate to plant uptake, temporal changes in microbial decomposition, and to variations with tidal height. Salt marsh grasses utilize ammonium,[93] but it is not clear if most mangroves do.[100] Boto et al.[112] found that cultured *Avicennia marina* seedlings prefer nitrate as their sole nitrogen source, but Naidoo[113] grew *A. marina* seedlings using ammonium. In subsequent experiments with *Bruguiera gymnorrhiza*, Naidoo[114] found that this species grows on both forms of nitrogen, suggesting that the preferred form is species specific.

Under field conditions, the dissolved nitrogen pools in marsh soils change mainly in response to seasonal patterns of microbial decomposition and, to a lesser extent, to seasonality of marsh grass growth and production.[115] Few such data exist for temporal changes in sediment nitrogen pools in mangrove forests. On Hinchinbrook Island, Boto and Wellington[116] ascribed autumn-spring differences in the size of the sediment ammonium pool to seasonality in mangrove growth. Subsequent work [101] in this same forest indicates that temporal variations are inconsistent, as are seasonal patterns in plant growth and microbial decomposition. The clearest field evidence of the influence of grass and tree growth on nutrient levels has been in comparisons between vegetated and bare sediments;[101,115] dissolved nutrient concentrations are usually higher in the bare sediments.

Both salt marshes[115,117] and mangroves[101,116] often exhibit zonational patterns in sediment nutrient concentrations. In most locations, concentrations of most particulate nutrients decline, but most dissolved nutrient species are highly variable, with increasing tidal height. Zonation changes in nutrient concentrations have been attributed to tidal height differences in:

- Frequency of tidal inundation
- Interstitial salinity
- Redox potential
- pH
- Plant growth

The heterogeneous nature of marsh and mangrove sediments, combined with the complex interaction between physicochemical characteristics and availability of nutrients, obscures our understanding of whether or not these halophytes are nutrient limited.

Several field experiments have shown that additions of nitrogen or phosphorus can increase plant biomass and growth, but other field evidence indicates large differences with tidal height, age of the plant communities, and even between different locations dominated by the same species.[117,118] The bulk of the salt marsh evidence indicates that short and tall marsh grasses are limited by nitrogen, not by phosphorus.[115,117] The response of mangroves to nutrient additions similarly indicates nitrogen-limiting conditions, but some evidence suggests that some mangroves are also phosphorus limited. In field experiments, Boto and Wellington[118] found that additions of nitrogen elicited positive growth responses in mixed *Rhizophora* forests, regardless of tidal height, but the addition of phosphorus resulted in increased tree growth only within the high intertidal. A regional-scale survey of mangrove forest net primary production and soil nutrient levels in northern Australia and Papua New Guinea showed that low productivity forests in some areas of Australia coincided with the lowest levels of soil phosphorus whereas the high productivity forests, particularly in Papua New Guinea, coincided with the highest concentrations of sediment extractable phosphorus. Australian sediments tend to be phosphorus deficient, so phosphorus limitation is likely to be a regional phenomenon in the tropics coinciding with the most highly weathered soils. Indeed, the origin of a particular sediment deposit may determine whether a salt marsh or mangrove forest is nitrogen or phosphorus limited, or not limited by either element. Comparisons of nutrient concentrations between salt marsh and mangrove sediments are difficult because of large differences even within habitats or regions. Nevertheless, concentrations of most dissolved nutrients are lower in mangrove sediments (low μM range) than in salt marsh sediments (high μM to low mM range); particulate carbon and nitrogen are equivalent, but phosphorus concentrations are generally lower in mangrove sediments.

Why concentrations of some nutrients in marine sediments are, on average, lower in the tropics than in higher latitude systems is not clear. Several factors proposed to explain a similar dichotomy between tropical rain forests and temperate deciduous forests can be offered to explain similar differences between marshes and mangroves:[12]

- Nutrient concentrations are lower, but turnover is more rapid, in tropical sediments.
- A greater proportion of nutrients is tied up in microbial and forest biomass in the tropics.
- Tropical sediments are more highly weathered and leached of elements than are sediments of higher latitudes.

The extent to which these apparent differences may be expressed in rates and pathways of energy and nutrient flow between mangroves and salt marshes is the focus of the remainder of this chapter.

3.4   FOOD WEBS AND DECOMPOSITION PROCESSES

Pelagic and benthic food chains in salt marshes and mangrove forests are fueled by living and dead plant matter. Until the mid-1970s, the view was held that marsh grass and mangrove detritus were the only important sources of energy for organisms inhabiting these systems.[85,87] This was not an unreasonable view, considering the vast expanses of marsh grasses and mangrove trees in most locations. A more sophisticated view has now emerged, suggesting that many organisms obtain their sustenance not only from detritus, but also from a variety of living micro- and macroalgal foods in order to obtain a balanced diet This view is more realistic, considering the refractory nature of vascular plant and wood detritus compared to more nutritious algae. Nevertheless, salt marsh and mangrove detritus — by the sheer size of its standing mass — is the main source of energy and carbon for decomposers that process this material and govern the recycling and exchange of nutrients and energy flow between these systems and adjacent coastal waters. The relative roles of detritus and algae in food chains and nutrient cycles are considered below.

3.4.1 Grazing Food Webs

3.4.1.1 Herbivory on Plant Tissue

Marsh grasses and mangroves lose little of their total standing crop by direct grazing.[119,120] Rates of herbivory are highly variable among plant and animal species and locations. In salt marshes, phytophagous insects and a variety of other wildlife feed directly on young shoots, stem bases, seeds, and roots and rhizomes. In their excellent review of *Spartina* grazers, Pfeiffer and Wiegert[119] calculated that consumption of living *Spartina* tissue by aerial herbivores (largely the multivoltine plant hopper *Prokelisia marginata* and the grasshopper *Orchelimum fidicinium*) accounts for roughly 9% of net *Spartina* production, equivalent to herbivory in bulrush (*Scirpus*) marshes and in sugercane plantations. The proportion of equivalent primary production directly consumed by grazers is small in other American and in European marshes.[85]

In mangroves, insects and arboreal crabs are the major grazers on mangrove leaf tissue.[120] In Australian *Rhizophora* forests, only ~2% of canopy production enters the grazing food chain. There are, however, instances in which that proportion may be greater. Robertson[120] noted that mangrove forests in close proximity to adjacent terrestrial forests may lose up to 35% of leaf area to insects; forests receiving nutrient inputs from allochthonous sources may therefore be subjected to greater rates of leaf herbivory. The extent of direct grazing is also species specific. Differences in leaf chemistry and palatability among marsh and mangrove species are well known. For instance, leaves of *Avicennia marina* are more heavily grazed than *Rhizophora* species, and insect defoliation of seedlings of the *Xylocarpus* species is common in the Australasian region.[120]

Direct grazing, although proportionally small compared to net production, may have a greater, indirect impact on primary production.[119] Sublethal damage or mortality of seeds, roots, and leaves may delay reproduction or slow plant growth. Seed predators may remove new potential individuals, having a direct impact on production. Other effects may be more subtle, not directly linked to energy and material losses; sapsucking insects and other herbivores may:

- Transmit pathogens or harmful salivary secretions
- Insert eggs that may cause premature senescence
- Consume quantities of plant sap sufficient to depress potential primary production

These potential impacts have not been adequately studied in salt marshes or mangrove forests.

3.4.1.2 Direct Consumption of Algal and Vascular Plant Matter

In contrast to mangrove forests where most algae are epiphytic on prop roots, salt marsh algae inhabit mostly creek waters and the sediment surfaces of creek banks. The bulk of carbon fixed in marsh and mangrove creek waters is produced primarily by small autotrophs, such as cyanobacteria, dinoflagellates, and other autotrophic protozoans, but comparatively few data are available.[85,87] These autotrophs may attract a wide range of benthic and pelagic herbivores and omnivores. In salt marsh (and presumably, mangrove) waterways, pelagic autotrophs are consumed by a variety of organisms ranging in size from protists to fish. In the classic notion of marine food chains, copepods are the dominant grazers; however, in most coastal waters, members of the microbial loop are now understood to dominate both grazing and detrital food chains. Most suspension-feeding copepods, gelatinous mucus-net feeders, rotifers, cladocerans, and many fish larvae feed more effectively on nanoplankton than on micro- or net plankton. Grazing and excretion rates are known for many planktonic organisms, but the amount of carbon flowing through these grazing pathways in mangrove and marsh waters is very poorly understood. It is thought that little transfer of carbon occurs between the microbial loop and larger pelagic consumers.

Grazing by benthos and nekton on edaphic microalgae and aufwuchs communities and on attached macrophytes growing on mangrove prop roots and grass shoots, as well as benthic filter-feeding on water-column microalgae, are significant routes of energy flow in many marsh and mangrove systems.[85,87] In salt marshes, beds of the mussel *Geukensia demissa* and reefs of the oyster *Crassostrea virginica* commonly develop, functioning as biological and material filters, removing and processing enormous amounts of dissolved and particulate materials, including bacteria and algae, from creek waters.[85]

Grazing experiments have shown that a variety of benthic epifauna and infauna can similarly process large quantities of algal matter.[121,123] Grazing on sediment surface and epiphytic algae can result in localized patches of depleted standing stocks, reducing availability to other consumers. Infaunal bivalves, such as the deposit-feeder *Macoma baltica*, inhibit development of microalgal communities in some marshes, although algal concentrations depend on bivalve densities.[121] Feeding by epifaunal molluscs, particularly gastropods, can maximize small-scale patchiness of edaphic microalgae in marshes and mangroves.[85] Smaller organisms, such as harpacticoid copepods and nematodes, feed on a variety of microalgal foods, with selectivity depending upon cell size. Epigrowth-feeding nematodes mainly ingest chlorophytes and diatoms, including larger forms that they break open to suck out cell juices; however, even algal-feeding nematodes consume smaller microbes by rasping bacterial and chlorophyte cells off detrital particles.

Differences between grazing and detrital food webs are often blurred, as most estuarine organisms are omnivores, feeding on a variety of foods in order to balance their nutrition. Further, diets of individual species often change from larval to juvenile to adult stages, or while migrating, or seasonally.

Several studies using stable isotope ratios of carbon, nitrogen, and sulfur have evaluated the relative importance of plant matter and algae in marsh[122-124] and mangrove[125-127] food webs. This approach involves the measurement of the relative abundance of different light and heavy isotopes of each element that naturally occur when the organic matter is created. These differences, or ratios, are signatures incorporated into the consumers that ingest the organic material, recording long-term nutrition. Isotope ratios differ among plants, animals, and microbes.[122,123] This method has limitations,[122,123] as wide variations in isotope ratios for a given organism are common; also the ratio may not be conserved because of further fractionation along a food chain. Care needs to be taken in interpreting a given value as it represents the average of all foods consumed. The multiple use of stable isotope ratios of carbon, nitrogen, and sulfur aids interpretation and avoids the ambiguities of using a single isotope. Sullivan and Moncreiff[124] used multiple stable isotopes in an irregularly flooded salt marsh in Mississippi to identify trophic relationships of the fish and invertebrate communities of the marsh with benthic and planktonic algae and living *Spartina* and *Juncus*. Nearly 88% of the consumers exhibited δ ^{13}C values falling within the range for edaphic algae and zooplankton but distinct from the carbon isotope composition of the vascular plants. Plots of the δ ^{13}C against the δ ^{34}S values confirm that the fauna clustered very closely to edaphic algae and zooplankton. They concluded that benthic and planktonic algae are the main food items for most consumers in the marsh. The contribution of *Spartina* and *Juncus* appears to be minor.

Other marsh studies[121,123] using multiple isotopes indicate that the main food consumed can vary with location within the marsh/estuarine system. In the Great Sippewissett Marsh, for example, the isotope signature for a population of the ribbed mussel, *Geukensia demissa*, located deep within the marsh, is close to the *Spartina* signature; those mussels living nearer to an adjacent bay have values closer to phytoplankton.

Most multiple isotope studies indicate that detritus and algae are equally important food sources for macroconsumers. The stable carbon isotope study by Rodelli et al.[125] in Malaysian mangroves indicates

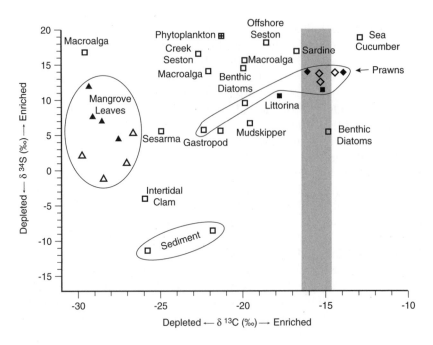

FIGURE 3.5

Relationship between δ S and δ C values for various primary producers, detritus, and consumers taken from mangrove forests and from adjacent coastal waters off the west coast of peninsular Malaysia. Circled areas enclose values for mangrove leaves (left), sediment (bottom), and prawns (right). Stippled area is grand mean δ ^{13}C ratio for natural diatoms in he literature. (Adapted from Newell, R. I. E. et al., *Mar. Biol.*, 123, 595, 1995.)

roughly equal consumption of mangrove and algal carbon by many suspension-feeding bivalves, gastropods, polychaetes, crabs, and prawns. A subsequent study by Newell et al.[126] at the same location using multiple isotopes clarified the relative importance of microalgae, phytoplankton, and mangroves in the diet of the penaeid prawns *Penaeus merguiensis* and *Parapenaeopsis sculptilis* and some other invertebrates as a function of age and location. Newell et al.[126] found that juvenile *P. merguiensis* living in mangrove creeks have stable isotope signatures indicative of a mixed diet of benthic microalgae and mangrove detritus, but adults migrating offshore have values suggesting a diet composed more of phytoplankton and benthic microalgal material and progressively less of mangrove detritus. The sympatric *P. sculptilis* had isotopic ratios very similar to those of *P. merguiensis* juveniles and adults also caught offshore (righthand circled area in Figure 3.5). A plot of the δ ^{13}C and δ ^{34}S values (Figure 3.5) shows progressing dietary change by the penaeids and confirms that the crab *Sesarma versicolor* feeds mainly on mangrove leaves. Other dominant invertebrates, such as the gastropod *Telescopium mauritsi* and the bivalve *Polymesoda erosa* have strong mangrove carbon signatures, but the gastropod *Littorina melanostoma* has a stronger δ ^{13}C signal than that measured by Rodelli et al.,[125] indicating a greater dietary dependence on benthic microalgae than on mangrove detritus. This work is further supported by data from a stable isotope study in riverine mangroves of the central Philippines[128] which shows that juveniles of the penaeids *Metapenaeus ensis, Penaeus indicus, P. merguiensis,* and *P. monodon* feed preferentially on phytoplankton and epiphytic algae rather than on mangrove leaf detritus. Stable isotope studies clarify the results of many earlier gut-content studies that showed utilization of unidentifiable remnants of plant detritus.

Benthic grazers are conspicuous on mud, prop roots, and shoot surfaces and can be equally important in consuming autotrophs in creek waters.[127] These and other benthic and pelagic herbivores and omnivores are not energetically efficient, however, producing dissolved and particulate waste products that ultimately link grazing food webs to decomposers in sediments and in the water column.

3.4.2 Decomposer Food Webs

Most living mangrove and marsh plants are not grazed heavily, so when these plants die, most of their mass is eventually incorporated into a complex web of decomposer food chains. Ultimately, all of the material not leached as dissolved organic matter, buried, or transported out of the system is invaded and transformed by microbes. Some proportion of this microbe-rich detritus is further utilized by a variety of consumers, depending upon such factors as:

- Detritus source
- Detritus age
- Rate of supply

This is overly simplistic, as there may be considerable direct consumption of detritus with little or no microbial modification. Differences and similarities between decomposer food webs in mangroves and salt marshes are influenced not only by differences in rates of primary productivity and climatic conditions, but also by inherent biochemical differences in detritus composition, species composition, and abundance of consumers. Clear differences in detritus food chains between habitats that appear structurally similar occur because local factors (e.g., hydrography, soil type) come into play. Such subtle, but important, differences are just beginning to be understood.

3.4.2.1 *Direct Consumption of Litter*

Recent work[85,87] reveals that a significant proportion of detritus in mangrove forests and salt marshes is consumed directly or initially processed by benthic scavengers, with no or relatively little microbial decomposition. This finding is somewhat contrary to the traditional belief that vascular plant debris requires microbial enrichment before consumption by detritivores, as suggested in early models of mangrove[129] and salt marsh[130] food chains.

In salt marshes, consumption of standing dead plants and newly fallen fragments of litter appears to be done mainly by shredder snails.[131] Crabs assist in the initial processing of litter by burying fallen fragments in the soil during burrowing, rather than by direct consumption.[132] Recent studies by Newell and his colleagues[131,133,134] demonstrate that the standing-dead crop of *Spartina alterniflora* in Georgia salt marshes — which accounts for more than half of total plant mass — is shredded by salt marsh periwinkles (*Littorina irrorata*). Where abundant, the periwinkles act as conveyors of shoot fragments to the marsh soil surface, rasping colonizing fungus and plant tissue, and producing feces containing undigested bits of this material. Shredders are inactive in winter, allowing fungus and bacteria to further decay the tissue. This process likely occurs in other salt marshes given the ubiquity of periwinkles and congeners, but data from other sites are lacking. Experiments are urgently needed to estimate rates of direct consumption of marsh litter, especially standing detritus, as opposed to consumption of microbial/detrital complexes, for which a considerable amount of data exists.

In contrast to salt marshes where very little fallen litter is found on the marsh surface and very little transport of large particulate shoot material is observed, the floors of many mangrove forests are littered with fallen trunks, twigs, flowering parts, and leaves.[87] The amount of litter lying on a forest floor depends upon:

- Tidal amplitude
- Stand age and species composition of the forest
- Abundance and species composition of scavengers

Recent work in Australia, Southeast Asia, Africa, the Caribbean, and South America shows that a large (30 to 80%) proportion of leaves, propagules, and other litter on the floors of mangrove forests is consumed or hidden underground by crabs.[87,135] Retention and consumption of litter by crabs conserve carbon and nitrogen within forests by reducing the amount of litter available for export. The proportion of litter consumed or buried has been measured in

TABLE 3.3

Litter Processing by Crabs and Their Contribution to Nitrogen Conservation in Various Mangrove Forests of North Queensland, Australia[a]

Parameter	Rhizophora	Bruguiera/ Ceriops	Ceriops	Avicennia
Litterfall	556	1022	822	519
Litter standing stock	2	6	6	84
Litter consumed by crabs	154	803	580	173
Microbial decay	5	5	5	168
Export	397	252	194	107
Litter nitrogen consumed by crabs	0.9 (3%)	4.8 (64%)	2.7 (53%)	1.5 (11%)

[a] All forests are high intertidal, except for *Rhizophora* (mid-intertidal).

Note: Values in parentheses are percentage of nitrogen required for forest primary production, as processed by crabs. Units are g dry weight m^{-2} for standing mass and g dry weight $m^{-2} yr^{-1}$ for fluxes.

Source: Adapted from Robertson, A. I., Alongi, D. M., and Boto, K. G., in *Tropical Mangrove Ecosystems,* Robertson, A. I. and Alongi, D. M., Eds., American Geophysical Union, Washington, D.C., 1992, chap. 10.

some Australian forests (Table 3.3). In low- and mid-intertidal *Rhizophora* forests, the grapsid crab *Sesarma messa* annually processes at least 28% of the yearly leaf litter fall, but this figure does not include consumption or burial of flowers, propagules, or other litter. In high intertidal *Bruguiera* or *Ceriops* forests, crabs remove more than 70% of the total annual litter fall. Most of this material is pasted to burrow walls, presumably to foster microbial colonization and to facilitate leaching of tannins; only a fraction is consumed immediately. In contrast, high-intertidal *Avicennia* forests are dominated by microphagous ocypodids rather than sesarmid crabs, so these forests are subject to less litter consumption. The standing stock of litter lying on the floors of *Avicennia* forests is therefore high. The retention of litter in these forests by crabs serves to conserve nitrogen for forest primary production (Table 3.3).

Sesarmid crabs are remarkable in their ability to consume mangrove litter without significant alteration to its chemical composition.[135] Nearly all other mangrove and salt marsh detritivores prefer ingesting foods that are either leached of polyphenolic acids or aged and modified by colonizing bacteria and fungi. In a series of experiments with *Sesarma messa* and *S. smithii,* Micheli[135] found that *S. messa* is a nonselective feeder, ingesting all species of mangrove leaves offered but consuming more decayed than senescent leaves. *S. smithii* prefers *Rhizophora stylosa* leaves. Neither species selects leaves based on any apparent chemical differences, such as tannin content and carbon-to-nitrogen (C:N) ratio. Field observations found that the crabs consume the litter in their burrows within 2 weeks, before any significant change in the C:N ratio of the litter. Both species also scrape material — presumably bacteria and microalgae — from the mud surface which Michelli[135] postulates must be an important nitrogen source considering the poor nitrogen content of the leaf litter. Moreover, she found that the growth, survival, and reproduction of *Sesarma messa* is cued to the quantity and species of mangrove leaf litter, highlighting a close link between forest production and secondary production of sesarmid crabs.[136]

3.4.2.2 Decomposition of Leaves, Roots, Shoots, and Wood

The decomposition of detritus occurs in three phases:

- Leaching of soluble compounds
- Microbial degradation
- Fragmentation or consumption by larger heterotrophs

Rates of detritus decomposition by microbes depend upon a range of factors, including:

- Temperature
- pH
- Oxygen availability
- Degree of wetting
- Microbial abundance and composition
- Detritus particle size
- Detritus source

These factors are interrelated, with some operating synergistically and others acting antagonistically. Phytoplankton and seaweed detritus decays much faster than vascular and woody detritus because of the lack of structural tissue. Such detritus has lower C:N:P ratios,[137] resulting in greater bacterial growth efficiencies and subsequently more rapid rates of decomposition. Rates of decay of salt marsh and mangrove detritus are roughly equivalent but slow compared to most other types of marine plant detritus. Decomposition rates must be viewed with caution, as there are often significant differences in decay rates among studies using different experimental methods.

Although their stable isotope signatures and amino acid composi-
tions differ,[138] mangrove and salt marsh detritus undergo a similar
progression of decay phases, beginning with significant losses of soluble
compounds (e.g., sugars, simple tannins) over the first 10 to 14 days.
Tannins are a major fraction of the dissolved organic matter (DOM)
pool, but the loss of tannins does not inhibit microbial colonization and
decay, except at high concentrations. After significant amounts (30 to
50%) are leached, this DOM can further stimulate mineralization of the
remnant lignocellulosic components.[139] Benner et al.[140] found that lignins
are lost at the same rate as polysaccharides, and that leaching is the
major pathway of lignin loss in the decay of *Rhizophora mangle* leaves.

Further decomposition of the material is mediated by a consortium
of bacterial, fungal, and protozoan communities.[141] This decomposition occurs largely by hydrolysis by extracellular enzymes released by
the microorganisms, followed by uptake and subsequent incorporation of solubilized compounds into microbial biomass. Some of the
carbon not incorporated into microbial growth is lost by respiration
and further leaching of DOC from the microbial cells. Bacterial growth
may be accelerated by grazing and disturbance activities of nematodes
and other small invertebrates, if they are abundant enough.

The chemical composition of the particulate detritus changes as a
result of preferential decomposition and microbial colonization, with
a gradual increase in the percentage of nitrogen relative to carbon (i.e.,
reduced C:N ratio).[138-141] Microbes preferentially immobilize nitrogen,
but they account for only a small proportion of the total nitrogen
content of the aging material. Bacteria usually account for 1 to 2% of
the nitrogen but, as shown in *Spartina* studies,[141] fungal biomass can
account for 12 to 22% of the total detritus nitrogen. Microbial extracellular proteins and exudates may account for some of the remaining
nitrogen, but most may be immobilized as proteinaceous material
bound to phenolic compounds. A contrary view has been put forward
more recently by White and Howes,[142] in which they estimate that by
the midpoint of decomposition of *S. alterniflora* detritus, 50 to 65% of
the total detrital nitrogen pool is "externally derived", probably resulting from microbial immobilization. The mechanisms of nitrogen immobilization are not known, so the composition of the total nitrogen
pool of both salt marsh and mangrove detritus remains poorly understood.

Even after significant decay and microbial enrichment, mangrove
and salt marsh detritus are poor foods compared to phytoplankton,
seagrass, and seaweed detritus.[143] Aged *Spartina* litter is clearly more
assimilable than fresh detritus, but growth of most detritivores fed this
material is still poor unless nitrogen is supplemented.[143] Growth of
benthic protozoans and invertebrates fed fresh and aged mangrove

detritus is even poorer, with many populations unable to sustain reproduction.[87] Poorer yields of bacterial and invertebrate growth on detritus may be not only a function of lower nutritional quality, but may also be do to the inhibitory effect of tannins. The low nutritional quality of mangrove and marsh detritus partly explains why most resident organisms eat other, more nutritious, foods.

A large proportion of above-ground standing biomass in salt marshes and in mangrove forests consists of standing-dead shoots and wood, respectively, but decomposition of these components has been ignored until recently. In *Spartina* marshes, dead leaves and shoots undergo substantial decomposition by fungi while still attached and standing.[131,133,144] Decomposition can be rapid once the shoots are moistened by rain, dew, tides, or even high humidity. Organic losses can exceed 60% of the original material over one season. Fungal productivity exceeds bacterial productivity in the warmer seasons, but fungal activity declines toward autumn; in late winter/early spring, bacterial productivity equals fungal productivity. Nearly all detrital nitrogen is immobilized into fungal biomass. A preliminary budget (Figure 3.6) shows that of the original dead-leaf carbon, nearly equal proportions are (1) lost as leachate and respired CO_2, and (2) fragmented into small particles and retained as attached shreds. The remainder is incorporated into fungal and bacterial biomass. Dying tagged leaves are low in nitrogen compared to living leaves; the lost nitrogen is probably conserved by either translocating to the rhizomes prior to the death of the shoots or by immobilizing into fungal mycelia.

A comprehensive study by Currin et al.[134] shows that standing-dead *Spartina* matter is an important food source for many organisms (Figure 3.7). The isotope signatures of most organisms overlap, however, further supporting the view that most salt marsh animals feed on a variety of living and detrital foods.

Wood decomposition in mangrove forests is slower[97] but similarly dependent upon frequency of tidal wetting and climate. In logged Malaysian forests, stumps of *R. apiculata* breakup after only 2 years, probably due to the small sizes of the trees, high rainfall, and high humidity.[83] In tropical Australia, trunks of *Rhizophora* spp. decompose rapidly compared to wood in terrestrial forests, but decay more slowly than the Malaysian wood, with 20% of the original carbon remaining after nearly 16 years.[87] Leaching of soluble tannins and microbial decomposition occurs, but teredinid molluscs are the major decomposers, rapidly consuming wood with the aid of symbiotic cellulolytic, nitrogen-fixing bacteria. In mature *Rhizophora* forests, where the mass of fallen timber is large (344 g C m^{-2}), decay of wood-derived carbon (44 g C m^{-2} yr^{-1}) is nearly as important as leaf consumption by crabs

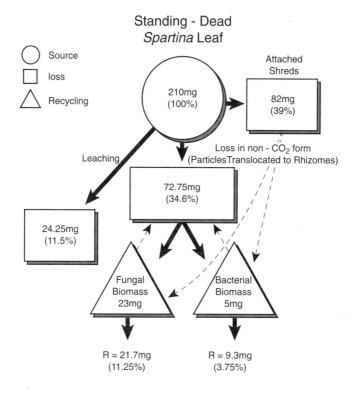

FIGURE 3.6

Carbon flow from decomposition of standing-dead leaves of *Spartina alterniflora* in a Georgia marsh in autumn. Units are in mg C. Values in parentheses are percentages of the original carbon. R = respiration. (Data from Newell, S. Y. and Barlocher, F., *J. Exp. Mar. Biol. Ecol.*, 171, 39, 1993.)

(66 g C m^{-2} yr^{-1}). Carbon flow through wood decay is minor (4 g C m^{-2} yr^{-1}) in younger, developing forests, where the mass of fallen timber is low (28 g C m^{-2}). The decomposition of wood lying on the floor of mature forests in the wet tropics may therefore be a major contributor to detrital carbon flow in mangroves.[87]

Below-ground decomposition of marsh grass roots and rhizomes can be rapid, equivalent to and sometimes greater than rates of decomposition of above-ground biomass. In a Dutch estuary, Hemminga et al.[145] found that *S. anglica* root material decomposed at equal rates at various soil depths and at the sediment surface, with an average turnover time of 2.0 to 3.9 years. Relative nitrogen levels increased over the course of the experiment. In American marshes, *S. alterniflora* roots and rhizome material decays at roughly similar rates.[85,142] On Sapelo Island, Georgia, below-ground biomass lost 55% of its original mass over 18 months. Biochemical analyses indicate that polysaccharides

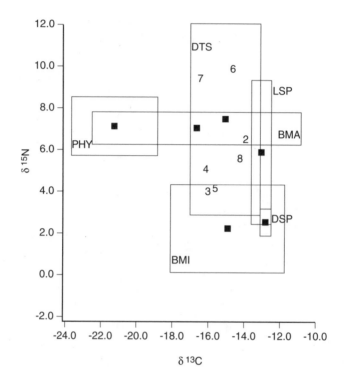

FIGURE 3.7
The relationship between δ ^{15}N and δ ^{13}C values for primary producers and consumers from various salt marshes. Primary producers: PHY, phytoplankton; BMA, benthic macroalgae; BMI, benthic microalgae; LSP, live and senescent *S. alterniflora*; DSP, stand-ing-dead *S. alterniflora*; DTS, sedimentary detrital *S. alterniflora*. The large boxes around each primary producer depict the standard error of mean values. Consumers: 2 = *Ilyanassa obsoleta*; 3 = *Uca pugnax*; 4 = *U. pugilator*; 5 = *Littorina irrorata*; 6 = *Fundulus heteroclitus*; 7 = *F. majalis*; and 8 = meiofauna. (From Currin, C. A., Newell, S. Y., and Paerl, H. W., *Mar. Ecol. Prog. Ser.*, 121, 99, 1995. With permission.)

are preferentially lost compared to lignin, although significant lignin degradation occurs under anoxic conditions.[146] A major fraction (>50%) of the initial nitrogen is lost during the first four months at a rate of ~10 μg N per g tissue per day, followed by accumulation of exogenous nitrogen at a faster rate of ~19 μg N per g tissue per day. The immo-bilization of nitrogen is tightly linked to rates of microbial activity. In the end, more nitrogen accumulates than was present in the initial mass of material.

Decomposition of mangrove below-ground biomass is also rapid, but only two studies have attempted to estimate root and rhizome decay, and only for *Avicennia marina*.[87] The fibrous (~1-mm diameter) roots of this mangrove lose 30 to 52% of original mass after 154 days, with more rapid decay occurring under aerobic conditions; main roots

(1- to 2-cm diameter) lose 60% of their initial weight after 270 days. Both studies took place at the southern-most limit of mangrove distribution (New Zealand and southern Australia), so decomposition rates are likely to be more rapid closer to the equator.

3.4.2.3 Sediment Carbon Cycling

Decomposition of most of the above- and below-ground dead mass of marsh grasses and mangroves reflects transformation, assimilation, and conversion of much of this material into bacterial biomass and respired carbon dioxide.[148,149] Indeed, some of the highest rates of bacterial carbon cycling have been measured in salt marsh and mangrove sediments, fueled by the large standing amounts of detrital material.

Organic detritus produced or deposited on the sediment surface supports the growth of aerobic decomposers, rapidly using up the available free oxygen. Aerobic decomposition is usually more rapid than the rate of oxygen diffusion into surface sediments and into animal tubes and burrows, so oxygen quickly becomes limiting. Most decomposition is carried out by anaerobic bacteria below the upper few millimeters of sediment. With the exception of some bacterial anoxygenic photosynthesis and carbon fixation, most of the fixed carbon in subsurface sediments is derived from several sources:[150]

- Particulate organic carbon (POC) deposited from tidal import
- Excreta from living and dead organisms
- POC and dissolved organic carbon (DOC) derived from the grasses, trees, and algae growing at the sediment surface

Not surprisingly, the bulk of carbon input comes from the plants via several pathways:

- Litter accumulation and burial
- Root decay
- DOC released by living roots
- Excreta and mucus produced by organisms

Anaerobic decomposition in marshes and mangroves is performed by a wide variety of bacterial types utilizing a variety of electron acceptors to oxidize carbon. The rates of diagenesis by each of these pathways is just beginning to be understood, particularly in mangrove sediments. It was earlier thought that aerobic decomposition is more efficient than anaerobic decay, accounting for most sediment diagenesis. This may not be true in organic-rich deposits typical of salt marshes and mangroves. In their review of heterotrophic microbial

TABLE 3.4

Contribution of Bacterial Decomposition Pathways to Total Carbon
Oxidation in Two Salt Marsh and One Mangrove Ecosystem (Western
Australia)

	Sapelo Island	Sippewisssett	Western Australia
Aerobic respiration	390	390	60
Denitrification	10	3	6
Manganese and iron reduction	Negligible	Negligible	Negligible
Sulfate reduction	850	1800	200
Methanogenesis	40	1-8	5
Total	**1290**	**2201**	**271**

Note: Units are g C m^{-2} yr^{-1}.

Source: Data compiled from Howarth[149] and Alongi.[151]

activity in salt marsh soils, Howarth and Hobbie[148] noted that most
organic production (largely from below-ground roots and rhizome
production) is decomposed anaerobically, and they postulated that
anaerobic microbial activity is fueled by the large pool of dead below-
ground biomass and by rapid release of labile organic compounds
excreted by live roots. A number of studies conducted subsequently in
salt marshes and mangroves have found that sulfate-reducing bacteria
and related fermentative bacteria oxidize most of the sedimentary
organic carbon (Table 3.4).

Rates of total carbon oxidation (as estimated by CO_2 release) between marsh and mangrove sediments are roughly equivalent, but
rates of sulfate reduction are, on average, greater in salt marsh (144 to
1800 g C m^{-2} yr^{-1})[149] than in mangrove (10 to 500 g C m^{-2} yr^{-1}) sediments.[151-154] In both ecosystems, most of the radiolabel becomes incorporated into pyrite (FeS_2) and elemental sulfur (S^0). The extent to which
the reduced sulfur becomes incorporated into these pools rather than
into acid-volatile pools of soluble sulfides (H_2S, HS^-) and iron
monosulfides (FeS) depends on several interrelated factors, including:[149]

- Bioturbation
- Sediment grain size
- Redox oscillations
- Iron chemistry
- Temperature
- Tidal inundation frequency
- Sulfate and organic carbon availability

Sulfate reducers are fueled by a limited number of organic compounds, principally low-molecular-weight alcohols and fatty acids resulting from associated fermentative breakdown of POM and from DOM released from roots and rhizomes. When roots metabolize anaerobically, they produce compounds such as ethanol and malate which can be used directly by sulfate reducers.[148]

Temporal and spatial patterns in rates of sulfate reduction relate closely to the distribution and/or seasonality of plant production.[154-156] In salt marshes, rates of sulfate reduction are faster in the warmer months, not only because of higher temperatures, but also because of more rapid plant activity in which aerial plant parts and rhizomes release DOM and gases into the soil, altering soil redox. Sulfate reduction rates are variable in marshes of greater tidal height, being cued to dessication-saturation cycles caused by variations in tidal flooding and rainfall.[155]

In mangroves, sulfate reducers are also fueled by DOC released from roots or POC derived from decaying below-ground roots and rhizomes or from litter buried by crabs. As in salt marsh soils, there is rapid formation of pyrite and elemental sulfur.[152,153] Seasonality in rates of sulfate reduction has not been adequately examined. Differences in sulfate reduction rates and total carbon oxidation in mangroves have been observed with tidal height (as in marshes), forest type, and stand age. In both Thailand and Pakistan, Kristensen and his colleagues[152,153] measured higher rates of sulfate reduction and total carbon oxidation in low-intertidal forests than in fringing or high-intertidal forests. Decomposition rates are, on average, faster in sediments of *Rhizophora* forests than in *Avicennia* forests.[151-154] The reasons are unclear, but sediments colonized by *Rhizophora* are usually of finer texture, more anoxic, and more organic rich than those inhabited by *Avicennia*. Other factors may account for such differences, but these have not been examined.

Of the various transformation processes of the nitrogen cycle, only denitrification is closely linked to organic carbon mineralization, but the available data from marshes and mangroves indicate that proportionally little carbon is consumed by denitrifiers, accounting for only a small fraction of total carbon oxidation (Table 3.4). Denitrification, however, is one of the major pathways of nitrogen loss in both mangroves and salt marshes (see Section 3.5).

True rates of methane formation have not been measured within salt marsh or mangrove sediments, but methanogenesis is also thought to be a minor fraction of total carbon oxidation. Measured rates of methane released from marshes[148-150] and mangroves[157] range from 0.2 to 40 g C m^{-2} yr^{-1}, but rates of methane release across the sediment-air interface do not account for methane that may be produced and subsequently oxidized at the sediment surface or used and transformed by other anaerobes. The extent to which this occurs in marsh and mangrove sediments remains unknown.

Recent studies[149,158] suggest dynamic iron and manganese oxidation-reduction cycles in marsh and mangrove sediments, although metal reduction appears to account for a minor fraction of total carbon oxidation (Table 3.4). Ferric ion (Fe^{+3}) can be reduced by hydrogen sulfide without biological mediation, and present evidence suggests that most iron reduction is chemical. Manganese reduction may also be mediated by reactions with readily available sulfides, but recent data indicates that metal reduction is predominantly microbially mediated. Spratt and his colleagues[158] measured the rate of microbial manganese oxide production in salt marsh and mangrove sediments by measuring oxidation of a dye, leuco crystal violet, by manganese oxides to produce crystal violet. Comparison with kill controls shows significantly higher oxidation rates that indicate microbial participation. Rates of microbial manganese oxidation ranged from 3 to 119 pmol per mg dry weight per hr in mangroves and from 0.5 to 2.3 nmol per mg dry weight per hr in a salt marsh. Converted on a carbon basis, these values support the idea that manganese oxidation-reduction reactions account for only a small fraction of total carbon oxidation. In any case, more work and new methods are necessary to define clearly the significance of metal reduction in carbon mineralization processes in salt marsh and mangrove sediments.

Most of the carbon mineralized in marsh and mangroves is lost as CO_2, largely as a consequence of microbial respiration. This is seen most clearly in a carbon budget for short-form *S. alterniflora* sediments in the Great Sippewissett Marsh (Table 3.5). In these sediments, with a carbon input rate of 810 to 936 g m^{-2} yr^{-1}, 86 to 89% is lost as CO_2. Other losses from the system are comparatively small, but burial of carbon is significant. Carbon burial may be significant in other salt marshes and in mangroves, but burial is ameliorated by intense crab bioturbation and the presence of extensive roots and rhizomes. A few studies[160,161] indicate sedimentation rates in mangroves and salt marshes are on the order of 1 to 5 mm yr^{-1}, with higher accretion rates in deltaic or basin mangroves and low elevation marshes than in fringing forests and high elevation marshes.

Data obtained using other methods (e.g., nucleic acid synthesis) support the biogeochemical estimates of very rapid carbon mineralization and incorporation into bacterial biomass in marsh and mangrove sediments. Bacterial growth rates and production, estimated by uptake of ^{3}H-labeled thymidine, are usually rapid in surface and subsurface marsh[7] and mangrove[12] sediments. Bacterial biomass and rates of production in Australian mangrove sediments are among the highest recorded for aquatic sediments. Flux chamber experiments in these same mangroves indicate that DOC and amino acids are rapidly and completely sequestered by

TABLE 3.5

Below-Ground Carbon Budget for Short-Form
Spartina alterniflora in the Great Sippewissett
Marsh, Massachusetts

Carbon Form Measured	Carbon Flux	
CO_2 evolution	720–804	(86–89%)
Carbon burial	88.8	(9–11%)
CH_4 emission	1.2–3.6	(<0.5%)
Volatile C-S compound emission	0–3.6	(<0.5%)
Calculated DOC export	0–36	(0–4%)
Total carbon input	**810–936**	

Note: Units are g C m^{-2} yr^{-1}. Ranges in parentheses
are percentages of carbon input. DOC =
dissolved organic carbon.

Source: Data from Howes, B. L., Dacey, J. W. H., and
Teal, J. M., *Ecology*, 66, 595, 1985.

bacteria at the sediment-water interface.[5] Bacterial production outpaces
consumption, supporting the argument that bacteria in mangrove sediments are a sink for carbon. This conclusion is further supported by the
fact that most mangrove sediments have low densities of protozoans,
meiobenthos, and macroinfauna (excluding crabs, which feed largely on
litter) compared to salt marshes and tidal flats.[87] In any case, most of these
organisms feed in aerobic microzones in mangrove and marsh sediments
and cannot be expected to consume more than a small fraction of total
sedimentary bacterial biomass. A large proportion of the sediment bacteria, therefore, remains ungrazed, dying of natural mortality and continually being turned over by other bacteria. Such recycling may serve to
retain labile carbon within the system.[5,87]

3.4.2.4 Pelagic Detrital Processes

Tidal waters in salt marsh and mangrove creeks and waterways exhibit variable rates of primary productivity depending on turbidity
and the extent of freshwater and nutrient input.[162] This authochthonous
material, mixed with particulate and dissolved material, floated off or
out of vegetated surfaces by tidal exchange or by precipitation, is
processed by communities of heterotrophic bacteria, nanoplankton,
microzooplankton, net plankton, and some nekton. The bulk of material and nutrient flow circuits through the microbial loop.[162,163] The
species composition of these communities is controlled by the degree
of freshwater flushing and seasonal variation in salinity and temperature, but the structure and function of such assemblages have not been
investigated often in mangrove waters.[87]

Bacteria are the most abundant and most productive heterotrophs in marsh and mangrove waters.[162,163] In mangrove-dominated estuaries, standing stocks and productivity vary widely, largely in relation to tidal cycles. In the Gambia River estuary in West Africa, bacterial densities vary from 1 to 2×10^6 cells ml^{-1}, with only 20% attached to detrital particles.[162] Seasonal changes are minor, but concentrations of attached bacteria peak during rising and flood tides when suspended solid concentrations are high. Maximum densities of free bacteria often occur in areas of high chlorophyll concentration, suggesting bacterial uptake of phytoplankton DOM. Radiolabel experiments indicate, however, that attached bacteria are the most metabolically active plankton members.[162,163] In the Fly River estuary in Papua New Guinea,[162] bacterial numbers are low ($\sim 10^5$ cells ml^{-1}), but bacterial production ranges from 20 to 498 mg C m^{-3} d^{-1}, similar to rates within temperate estuaries.[163] In both the Gambia and Fly River estuaries, total pelagic mineralization rates exceed water-column primary production. Presumably, protozoa and microzooplankton participate in these mineralization processes in mangrove waters, but their activities (including their role as predators) have yet to be quantified.

Pelagic mineralization processes are better understood in salt marsh estuaries.[163] Bacteria in the water column are fueled by DOC derived from phytoplankton and, in some cases, by DOM leached from marsh grasses.[165-167] Bacterial numbers and productivity appear to be regulated by:

- Nutrient supply
- Availability of substrates
- Tides
- Predation by plankton and benthic filter-feeders
- Changes in climate (e.g., temperature)

Bacterial densities are higher ($\sim 10^9$ cells ml^{-1}) in salt marsh waters than in the few mangrove waterways that have been sampled, but rates of bacterial productivity are within a narrower range[163] of 97 to 116 mg C m^{-3} d^{-1}. Bacteria also account for the bulk of mineralization in salt marsh waters, but the ratio of bacterial to primary productivity is lower, averaging <20% in most estuaries.[163] As in mangroves, bacterial densities are highest in tidal creeks and channels, declining toward river mouths and offshore waters.[166] In a marsh creek in Chesapeake Bay, Shiah and Ducklow[166] found that temperature and substrate supply regulate bacterial growth rates, with temperature being the dominant factor (Figure 3.8). Substrates are limiting in summer due to high rates of bacterial growth. In other seasons, temperature regulates bacterial

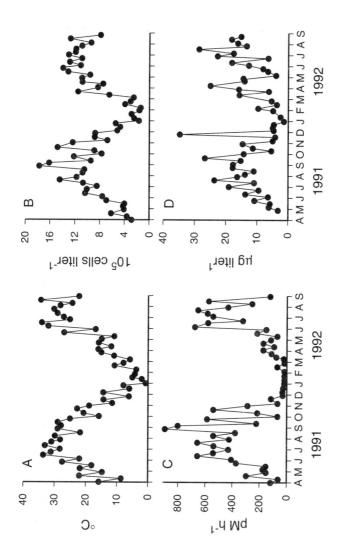

FIGURE 3.8

Seasonal changes in water temperature (A) regulating annual cycles of bacterial standing stocks (B), productivity (C, thymidine and incorporation rate), and chlorophyll *a* (D) in a salt marsh tidal creek. (Adapted from Shiah, F.-W. and Ducklow, H. W., *Limnol. Oceanogr.*, 40, 55, 1993.)

growth (Figure 3.8); substrates and their rates of supply are high during these periods compared to bacterial demand.

Superimposed on environmental changes is predation by other members of the microbial loop. An enormous literature has detailed the complex web of predatory controls within pico- and nanoplanktonic assemblages, but relatively few studies have been conducted in salt marsh waters. In the marsh-dominated Duplin River estuary, it was found that small (<20 μm) protozoa consume bacteria equivalent to 40 to 45% of their rates of production.[167] The standing crop of protists is frequently out of phase with bacterial standing stocks. These oscillations suggest that protozoan predators exert considerable control over bacterial densities in salt marsh waters.[167]

3.5   NITROGEN FLOW

A wide range of oxidation states permits nitrogen to serve either as a reductant or an oxidant in diagenetic reactions. The nitrogen cycle is therefore complex, involving many transformation processes. All of these reactions in mangroves and salt marshes have been exhaustively reviewed by Valiela[168] and Alongi et al.,[169] so only recent advances are discussed here, particularly with respect to plant-soil-microbe relations and ecosystem budgets.

3.5.1   Nitrogen Flow Through Plants and Sediments

Nitrogen compounds transform and flow rapidly within sediments and between sediments, plants, air, and tidal water.[168,169] The rate of nitrogen flow depends upon several factors, including:

- Rate of nitrogen supply
- Types and abundance of microbes and their rates of growth
- Rates of plant uptake
- Temperature
- Soil redox
- Sediment type

A budget of nitrogen cycling in sediments of the *Rhizophora* forests on Hinchinbrook Island in Australia (Figure 3.9) reveals very dynamic pools of ammonium, nitrite, nitrate, and DON and the influence of plant uptake on nitrogen cycling. This is seen most clearly for the ammonium pool in which the known rate of input

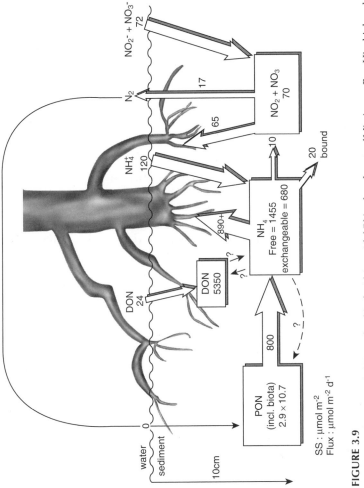

FIGURE 3.9

Nitrogen cycling in muds (10-cm depth) of mid-intertidal *Rhizophora* forests of Missionary Bay, Hinchinbrook Island, Australia. Units are mmol N m⁻² for standing stocks and mmol N m⁻² d⁻¹ for fluxes. (Updated from Alongi, D. M., Boto, K. G., and Robertson, A. I., in *Tropical Mangrove Ecosystems*, Robertson, A. I. and Alongi, D. M., Eds., American Geophysics Union, Washington, D.C., 1992, chap. 9.)

from ammonification and uptake from tidal water does not balance with the known rate of output to the nitrite and nitrate pools. Exchange of ammonium with the DON pool and uptake/immobilization into the particulate organic nitrogen (PON) pool is unknown. Nedwell et al.[154] suggest that organic nitrogen derived from roots supports high rates of ammonium production and nitrogen fixation in mangrove sediments. The input from the DON pool may be significant, as measured rates of amino acid flux are rapid within the Australian sediments, but there was no flux across the sediment-water interface (Figure 3.9).[169] Volatilization of ammonium is probably small at the pH range (6.5 to 8.2) measured in these sediments.

This budget does not consider the contribution of excretion by crabs and other biota, which may be considerable; other losses, such as sorption to inorganic particles are likely to be minor. In order for the budget to balance, it is reasonable to suggest that most of the ammonium is taken up by mangrove roots. The PON pool is large, but does turn over rapidly, as the entire dissolved nitrogen pool accounts for only 10 to 15% of the nitrogen required to sustain the measured rates of bacterial production. Calculations by Alongi et al.[169] suggest a turnover time of ~10 days for the PON pool. Further support for these calculations comes from the fact that bacterial biomass (as nitrogen) accounts for nearly 8% of the PON pool. This figure may be an underestimate as conservative conversion factors were used, but this value is much greater than the bacterial contribution to the PON pool in salt marsh sediments.[148,149]

The nitrogen budget shows that most of the available $NO_2 + NO_3$ pool is taken up from the tidal water rather than from nitrification. The flux of nitrogen via denitrification and nitrogen fixation in these sediments is minor. Nitrogen fixation was not measured within the rhizosphere but may be considerable, considering fixation rates measured in other systems.[154] Equally slow rates of denitrification have been measured in other mangroves.[170,171]

Similarly rapid turnover of nitrogen has been measured in mangrove sediments in Thailand[172] and in Jamaican[154] forests. Nedwell et al.[154] measured rapid turnover of ammonium in *Avicennia* and *Rhizophora* forests in Jamaica, estimating a net ammonium availability to the plant roots of 10 mmol m^{-2} d^{-1}, which gives a productivity of 2000 g C m^{-2} yr^{-1}. This supports earlier work by Blackburn et al.[172] in *R. apiculata* forests in Thailand.

The cycling of nitrogen in salt marshes shows similarly rapid transformations between nitrogen pools and a close association between bacterial mineralization and uptake of nitrogen by marsh grasses.[173,174] Most of the inorganic nitrogen transformed by bacteria

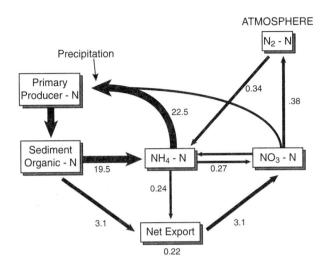

FIGURE 3.10
The annual nitrogen budget for an English salt marsh. Units are mol N m⁻² yr⁻¹.
(Adapted from Azni b Abd. Aziz, S. and Nedwell, D. B., *Estuarine Coast. Shelf Sci.,* 22, 689, 1986.)

from the organic form is taken up by marsh grasses, with the balance necessary for sustained plant growth obtained by nitrogen fixation occurring within the rhizosphere. In English marshes dominated by *Puccinellia maritima* and *Halimione portulacoides*,[173] the amount of organic nitrogen mineralized to ammonium in the sediment nearly balances with the nitrogen required for plant growth, most of which occurs below-ground (Figure 3.10). This budget does not include possible inputs of nitrogen from precipitation or groundwater, but it does show that ammonium is the main source of nitrogen for the marsh grasses, with some nitrogen contributed by biological fixation; little ammonium is lost by nitrification and denitrification or by tidal exchange.

Using ¹⁵N to trace nitrogen turnover in *Spartina alterniflora* roots and rhizomes, White and Howes[174] also found a close coupling between plant growth and microbial remineralization. The nitrogen budget for short *Spartina* (Figure 3.11) indicates that translocation of nitrogen from above- to below-ground biomass during senescence equals nearly 38% of above-ground production. Leaching equates to about 10% of above-ground production. The largest flux of nitrogen is via microbial remineralization, equating to ~80% of below-ground plant production, with the remainder being buried, denitrified, and exported. Input from nitrogen fixation in the rhizosphere can be substantial, roughly equating to half of below-ground production. These results

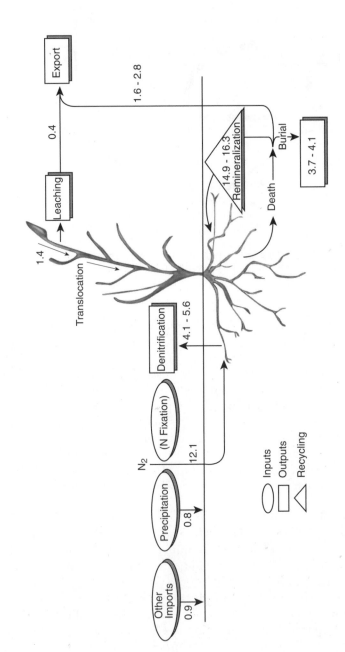

FIGURE 3.11

The annual nitrogen budget for short-form *S. alterniflora* in a New England salt marsh. Units are mol N m⁻² yr⁻¹. (Data from White, D. S. and Howes, B. L., *Limnol. Oceanogr.*, 39, 1878, 1994.)

agree with most other studies which indicate that, over 1 to 3 years, decaying *Spartina* loses the bulk of its original nitrogen by microbial decomposition and, to a much lesser extent, by leaching and burial. Inputs account for only 60% of the annual plant demand; losses via denitrification, burial, and export remove 40 to 60% of plant nitrogen demand. These losses are compensated for by retaining nitrogen through remineralization and translocation, both of which equate to supplying roughly 70 to 80% of the annual plant nitrogen requirements.

These results suggest that translocation from above- to below-ground plant parts, as well as nitrogen incorporation into organic matter and through microbial mineralization pathways, all act to conserve nitrogen within the system, minimizing loss to the atmosphere and tidal waters. In this sense, mangrove forests and salt marshes behave similarly, retaining and conserving nitrogen by a variety of recycling mechanisms.

3.5.2 Nitrogen Budgets For Whole Ecosystems

Various transformation and exchange processes of nitrogen have been measured in numerous salt marshes and mangroves, but only two locations offer complete budgets (Table 3.6) that permit some comparisons between ecosystems: Great Sippewissett Marsh in Massachusetts and the mangrove forests of Missionary Bay in Queensland, Australia. The Great Sippewissett Marsh system, located on the western shore of Cape Cod, MA, is small (483,800 m²) compared to the Missionary Bay forests (64 km²). The marsh has a single channel through which seawater from Buzzards Bay flows, flooding the marsh twice daily to a maximum tidal range of 1.6 m.[86] The total area of the marsh includes muddy and sandy creek bottoms (166,600 m²), algal mats (13,000 m²), low marsh dominated by short (122,500 m²) and tall (91,100 m²) *S. alterniflora*, and high marsh colonized by *S. patens* and *Distichlis spicata*.

Missionary Bay, in contrast, is a large embayment on the northern end of Hinchinbrook Island within the Great Barrier Reef lagoon in north Queensland, Australia.[169] The island is a heavily forested and mountainous national park adjacent to the mainland and lies on a sharp climatic gradient receiving an average yearly rainfall of 2500 mm. The western end of Missionary Bay receives some additional freshwater input from mountain runoff, but most of the system is tidally dominated and marine. Tides are semidiurnal, with an average range of 2 m. The total area of the system includes saltpans (7.5 km²), 11 waterways (14 km²), and 42.5 km² of mangrove forests, composed mainly of mixed *Rhizophora* spp. in the mid-intertidal (75%) and low-intertidal (17%) zones.[169]

TABLE 3.6

Nitrogen Budgets of an American Salt Marsh and
an Australian Mangrove Ecosystem by the Various
Processes and Their Contribution to Inputs,
Outputs, and Net Exchange

	Great Sippewissett Salt Marsh	Missionary Bay Mangrove Forest
Inputs		
Precipitation	380	30
Groundwater flow	6120	30
N$_2$ fixation	3280	36,830
Tidal exchange	26,200	168,600
Other	10	0
Total	**35,990**	**205,490**
Outputs		
Tidal exchange	31,600	192,430
Denitrification	6940	2900
Sedimentation	1300	?
Other	30	?
Total	**39,870**	**195,330**
Net exchange	**–3880**	**10,160**

Note: Units are kg N yr^{-1} for each entire system. Losses
are shown as negative numbers on the net exchange
line.

Source: Data compiled from Valiela and Teal[86] and
Alongi et al.[169]

Nitrogen enters the Great Sippewissett Marsh through groundwater, nitrogen fixation, and, to a lesser extent, by precipitation (Table
3.6). The tides exchange much more nitrogen, but with a net loss. The
second major loss of nitrogen is via denitrification, followed by sedimentation. Other inputs/outputs, such as ammonia volatilization,
shellfish removal, sediment loss, and fecal deposits (mainly by birds),
are minor. In contrast, nitrogen enters the Missionary Bay ecosystem
by nitrogen fixation with little contribution from rainfall and groundwater. Similar to the Great Sippewissett Marsh, tidal exchange dominates, but there is also net loss (Table 3.6). Other fluxes such as ammonia volatization have not been measured, but these are presumably
minor. Unlike the salt marsh, denitrification losses are small in this
mangrove system compared to tidal exchange.

The net exchange of nitrogen (last line, Table 3.6) is a small loss to
the Great Sippewissett Marsh and a small gain to the Missionary Bay

TABLE 3.7

Annual Exchanges of the Major Forms of Nitrogen for the Great
Sippewissett Marsh and the Mangrove Forests of Missionary Bay

Form of N	Great Sippewissett Salt Marsh			Missionary Bay Mangrove Forest		
	Input	Output	Net Exchange	Input	Output	Net Exchange
NO_3-N	3420	1220	2200	5300	6300	−1000
NH_4-N	3150	3550	−400	11,140	5360	5780
DON	19,200	18,500	700	152,220	104,450	47,770
Particulate N	6750	8200	−1460	<<<1	76,320	−76,320
N_2	3280	6940	−3660	36,830	2900	33,930
Total	35,800	38,410	−2620	205,490	195,330	10,160

Note: Units are kg N yr^{-1}. DON = dissolved organic nitrogen.

Source: Data compiled from Valiela and Teal[86] and Alongi et al.[16]

mangroves. Both ecosystems are roughly balanced between autotrophy and heterotrophy with respect to nitrogen, considering extrapola-
tion errors and possible contributions from unmeasured fluxes.

An examination of the annual exchange of the major forms of
nitrogen (Table 3.7) suggests that both systems function by very different means. In the Great Sippewissett Marsh, inputs are dominated by
DON, particulate nitrogen, and N_2 (via nitrogen fixation), with roughly
equal contributions from nitrate, ammonium, and N_2, with losses mostly
of DON and particulate nitrogen. The largest exchange is net loss of N_2
by offsetting rates of nitrogen fixation and denitrification. In the mangrove system, the largest inputs are also in the form of DON, but there
is no significant input of particulate nitrogen, reflecting import of
nitrogen in dissolved form (mostly as DON) and export in particulate
form (e.g., leaves, twigs, bark). Much nitrogen is lost via denitrification.

Both systems have evolved efficient mechanisms to conserve nitrogen. In the Great Sippewissett Marsh, there is a very tight coupling
among plants, soil, and rhizome bacteria. Other mechanisms include
net storage of nitrogen as sedimented material and as standing-dead,
shoot-fungal complexes. In Missionary Bay, where water and sediment nitrogen concentrations are much lower, the plant-microbe link
is also close. The large, above- and below-ground mass of living trees
and dead wood lying on the forest floor, litter processing by crabs, and
low rates of denitrification all serve to conserve nitrogen. Both ecosystems similarly conserve nitrogen by flushing material that is refractory
and in an advanced stage of diagenesis. Both may share other conservation mechanisms currently attributed to the other, such as litter

processing by crabs (for marshes) and net sedimentation (for mangroves).

Overall, major similarities include:

- Dominance of physical processes controlling exchange
- Net balance of nitrogen, considering probable errors
- Tidal exchanges and nitrogen fixation being major inputs
- Tidal exchange as the major output

There are, however, clear differences:

- Biotic mechanisms contribute 9% of the inputs and 18% of the outputs in the salt marsh, whereas biotic mechanisms contribute 18% of the inputs but only 1% of the outputs in Missionary Bay.
- Rates of N_2 fixation far exceed rates of denitrification in the mangroves, whereas they are nearly equal in the marshes.
- Inputs from groundwater and precipitation are greater in the marsh.
- There is little particulate input into Missionary Bay, whereas there is significant particulate input into the Great Sippewissett Marsh.

It is improbable that comparisons between other salt marsh and mangrove ecosystems would reveal the same similarities and differences, given the inherent variations within each type of ecosystem. Other mangrove systems may be more similar to salt marshes with regard to import-export pathways. Although most mangrove systems appear to export nitrogen as litter,[175] the Missionary Bay system is atypical, having little freshwater input. Other mangroves may be very different. The fringe mangroves in the Terminos lagoon in Mexico, for example, act as a sink of inorganic nitrogen and as a source of dissolved and particulate nitrogen.[175] The same is true for salt marshes, such as the Bly Creek system in South Carolina, which exports nitrogen mainly as DON.[176]

These budgets ignore subtle, yet complex, variations within each system. High marshes and mangroves may not exchange nutrients at the same rate (or direction) as low marshes and mangroves. In the Bly Creek system, the low marshes colonized by tall-form *Spartina* have higher uptake and release rates of dissolved nitrogen compared with short-form *Spartina* in high marshes. These differences are due to differences in plant growth rates, sediment types, and tidal inundation frequency.[176] In Missionary Bay, low-intertidal mangroves import dissolved nitrogen by sediment uptake at faster rates than mangroves in the mid-intertidal, where most of the nitrogen is taken up by epiphytes on prop roots.[169]

In any case, nutrient budgets for other salt marshes and mangrove forests in different climatic settings are urgently needed to determine whether or not any inherent functional differences found between these temperate and tropical ecosystems are universal and whether or not most salt marshes and mangrove forests are in net autotrophic-heterotrophic balance.

3.6   OUTWELLING

No other hypothesis has stimulated as much coastal research as the concept of "outwelling", originally proposed by Eugene Odum and defined as the export of nutrients or organic detritus from fertile estuaries to support productivity of offshore waters. Recent reviews[72,177,178] indicate that while most mangroves and marshes export some net material, many others do not. Some marshes and mangroves show net import of organic matter. This is true for many European salt marshes lacking significant grass production in low-intertidal zones and for many semi-enclosed or lagoonal mangroves with restricted water circulation. This situation is also true for salt marshes in some South African estuaries, where plant production occurs mostly in high marshes which do not export substantial amounts of organic carbon.[179]

Hopkinson[178] concluded that nearly all salt marshes are net autotrophic, exporting significant quantities of organic detritus to offshore waters. Modifying and updating Hopkinson's comparisons[178] (Table 3.8), we find that most marshes and mangroves are net autotrophic and export organic matter and do so in proportion to their total net autotrophic production. For instance, the open, expansive Barataria Bay system exports an average of 27 g C m^{-2} yr^{-1} — equivalent to only 1% of net primary production. Other marshes, in contrast, export more organic carbon, both in absolute and relative terms, to net primary production (Table 3.8).

Sufficient and high-quality data exist for only two mangrove systems: Rookery Bay in Florida and Missionary Bay in Australia. The variation between these two systems is as great as the variability among salt marshes. Both mangrove systems export organic carbon within the same proportion relative to net primary production — in the range of 6 to 22% of net primary production. The only apparent difference between the salt marshes and mangrove forests is the low heterotrophic respiration relative to net primary production in mangroves. These values are underestimates as they do not include respiration from heterotrophs decomposing wood on the forest floor and on prop roots. These systems, therefore, may be less net autotrophic than they now appear.

TABLE 3.8

Some Physical Characteristics and Rates of Carbon Flow in Several Salt Marsh and Mangrove (Missionary Bay and Rookery Bay) Ecosystems[a]

	Sippewissett	Flax Pond	Sapelo Island	Barataria	Missionary Bay	Rookery Bay
Total area (m²)	4.8×10^5	5.7×10^5	1.3×10^7	5.1×10^9	6.4×10^7	4.0×10^5
Tidal range (m)	1.1	1.8	1.4–3.2	0–0.3	2	0.55
Water volume (m³)	4.5×10^4	2.8×10^4	8.4×10^6	4.2×10^9	1.5×10^7	3.8×10^7
Rainfall (m)	1.2	1.0	1.5	1.5	2.5	1.3
GPP	1695–2140	1275	3941	4261	—	3318
R_{auto}	906	696–1190	2149	2287	—	2219
R_{auto}/GPP (%)	42–53	40–68	55	54	—	67
NPP	941–1386	535–1029	1791	1974	1500	1099
R_{hetero}	772	235–729	710	1773	130	197
R_{hetero}/NPP (%)	56–82	44–68	40	89	9	18
NPP – R_{hetero}	169–614	280	1081	201	1370	902
$Exp_{measured}$	80	100	379	9–30	332	64
$Exp_{measured}$/NPP (%)	6–9	10–19	21	1	18	6
$Exp_{calculated}$	80–525	100	1052	27	325	64
$Exp_{calculated}$/NPP (%)	6–38	10–19	59	1	22	6

[a] Units are g C m⁻² yr⁻¹.

Note: GPP = gross primary production; R_{auto} = plant respiration; NPP = net primary production; R_{hetero} = heterotrophic respiration; $Exp_{measured}$ = directly measured export; $Exp_{calculated}$ = export calculated from mass balance.

Source: Adapted and updated from Hopkinson[178] with mangrove data from Twilley et al.[177] and Alongi et al.[169]

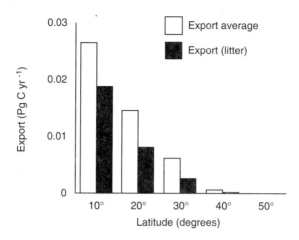

FIGURE 3.12

Global estimates of organic carbon export from mangroves with latitude. (From Twilley, R. R., Chen, R. H., and Hargis, T., Water, *Air, Soil Pollut.*, 64, 265, 1992. With permission.)

Not all outwelling ecosystems export both DOC and POC. Taylor and Allanson[179] recently summarized the literature and found that the proportion of exported total organic carbon that is composed of DOC and POC for salt marshes varies greatly. All salt marshes show some net export of DOC, ranging in rates from 13 to 328 g C m^{-2} yr^{-1}. POC exchange varies, however, from net import — such as in the Flax Pond (61 g C m^{-2} yr^{-1}), Ems-Dollard (140 g C m^{-2} yr^{-1}), and Bly Creek (31 g C m^{-2} yr^{-1}) systems — to net export. Some marshes export carbon mostly in dissolved form, whereas others export mostly particulate material. This range of variation is true also for mangroves. For instance, the Missionary Bay system exports POC but imports a small amount of DOC. Mangroves in south Florida, in contrast, export both, but mostly DOC.

Globally, export of organic carbon from mangrove forests ranges from 2 to 420 g C m^{-2} yr^{-1} and averages 210 g C m^{-2} yr^{-1}. Figure 3.12 shows that carbon export from mangroves varies with latitude, not only from more luxuriant and productive forests closer to the equator, but with greater rainfall. With more extensive information, Hopkinson[178] noted a similarly wide range of carbon export (27 to 1052 g C m^{-2} yr^{-1}) from salt marshes and was able to test for significant relationships of the amount of organic carbon exported from the various American marsh estuaries with ecosystem metabolism, burial, and gross and net primary production. Excluding larger coastal ecosystems less dependent upon salt marshes, Hopkinson[178] found a positive relationship of carbon export with gross and net primary production and a negative

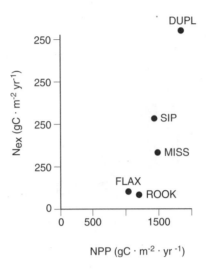

FIGURE 3.13
Relation between rates of carbon export and net primary production of some salt marsh and mangrove ecosystems. Systems: DUPL, Duplin estuary, Georgia; SIP, Great Sippewissett Marsh, Massachusetts; FLAX, Flax Pond, New York; ROOK, Rookery Bay, Florida; MISS, Missionary Bay, Australia. (Adapted from Hopkinson, C. S., in *Coastal-Offshore Ecosystem Interactions*, Jansson, B.-O., Ed., Springer-Verlag, Berlin, 1988, 122.)

relationship with burial. This means that as production of organic matter increases or storage decreases, the export of organic carbon increases. The relationship between carbon export and net primary production is strengthened when the two mangrove systems from Table 3.8 are added (Figure 3.13), suggesting that the same factors influence organic matter export from salt marsh and mangrove ecosystems.

The mass balance estimates do not reflect temporal and spatial variations in export. The quantity and quality of exported organic carbon may vary over consecutive tidal cycles, as observed in Brazilian mangroves,[180] or the bulk of exchange may occur during storm events, as found by Roman and Daiber[181] in Delaware marshes. The role of climate plays a strong role in the tropics where litterfall partly relates to rainfall and frequency of cyclones.

The amount of organic matter available for export is influenced not only by rate of production, but also by the geomorphological, hydrological, and hydrodynamic characteristics of each system, to the extent that each system is fairly unique. Specific characteristics include:

- Tidal range
- Ratio of wetland to watershed area

- Water circulation
- Total wetland area
- Frequency of storms and rainfall
- Volume of water exchange

By these mechanisms, salt marshes[85] and mangroves[101] exchange other nutrients such as phosphorus and silicon. Tidal exchange of these nutrient elements is considerably less understood compared to carbon and nitrogen, but it appears that the direction of transport is equally variable among marshes and mangroves. Some systems exhibit net export and others require some net import. In the Missionary Bay mangroves,[101] phosphorus and silicon are rapidly cycled within the forests, and between the forests and offshore waters. In fact, the largest dissolved fluxes measured in this system involve silicon. On a whole-system basis, ~14,340 kg of silicon are imported annually into the forests, compared with an annual import of 3270 kg of nitrogen and 800 kg of phosphorus. Presumably, this silicon is required for algal growth (e.g., diatoms), but algal production accounts for only ~20% of the imported silicon, suggesting that most silicon is required for tree production. The silicon requirement of mangroves and most marsh grasses is not known, but these data illustrate that substantial amounts of other frequently overlooked elements may be imported or exported from salt marshes and mangroves.

The role of micro- and meso-scale hydrodynamics has recently been recognized as a key to material exchange processes in salt marsh and mangrove systems. Calculations of tidal exchange using only creek dimensions could result in severe error. For instance, tidal flow at the marsh/mangrove edge is slower due to friction with the irregular marsh/mangrove surfaces than water motion in the middle of a creek. Ignoring this factor could result in a large overestimation of export. Water motion is complicated, slowed by friction among marsh shoots and prop roots to the extent that there is significant lateral trapping and mixing of flooding and ebbing water within mangroves[182] and salt marshes.[183] This can greatly alter transport of dissolved and particulate material, including suspended sediments. Figure 3.14 summarizes how hydrodynamic constraints have ecological consequences. It is therefore clear that hydrodynamics can play a major role in influencing the transport of material between wetlands and coastal waters.

The fate of organic matter exported from salt marshes and mangrove forests will be discussed in Chapters 6 and 7 within the context of their contribution to adjacent coastal and continental shelf ecosystems. Salt marshes and mangroves are clearly not isolated systems. Furthermore, the movement and interdependence of

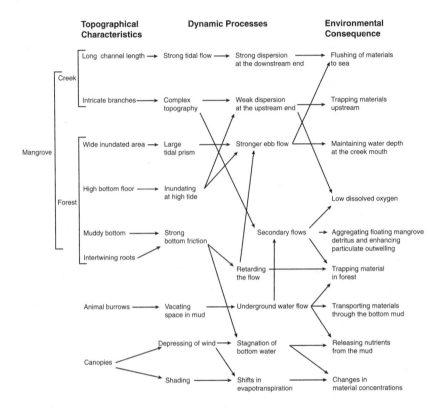

FIGURE 3.14

Summary of connections among physical, chemical and biological processes in mangroves. (From Wolanski, E., Mazda, Y., and Ridd, P., in *Tropical Mangrove Ecosystems,* Robertson, A. I. and Alongi, D. M., Eds., American Geophysical Union, Washington, D.C., 1992, chap. 3. With permission.)

migratory organisms, particularly fish and crustaceans, between wetland and adjacent offshore habitats is not reflected in these mass balance exercises. Salt marshes and mangroves are therefore in myriad ways trophically and energetically linked to adjacent coastal ecosystems.

Chapter 4

SEAWEED AND SEAGRASS ECOSYSTEMS

4.1 INTRODUCTION

Seaweeds and seagrasses thrive on many sheltered to fully exposed coasts, occupying rocky shores, coral reefs, and many other coastal habitats. Seagrasses, being rooted angiosperms, commonly flourish as meadows in soft sediments at and below the intertidal zone. In temperate regions, *Zostera* (eelgrass) is often the most common genus, whereas the genera *Thalassia* (turtlegrass), *Halodule*, and *Syringodium* are frequent co-dominants in the subtropics and tropics.

Seaweeds are common inhabitants of rocky intertidal zones, attaching themselves by means of a holdfast, but some species are commonly found free-floating in many coastal waters. Some seaweeds such as kelp thrive as forests in cool, clear waters in higher latitudes and in subtropical areas are associated with upwelling, such as within the Peru, California, and Benguela currents.

Seaweeds are marine algae belonging to three classes: the Chlorophyceae, Rhodophyceae, and Phaeophyceae. Members of the last class dominate many shores; kelps (Laminariales) live subtidally, and rockweeds (Fucales) live mostly on rocky intertidal shores of cool temperate regions. Kelp forests are often dominated by three genera: *Laminaria*, *Macrocystis*, and *Ecklonia*. *Laminaria* is a common genus, attaining most luxuriant growth and biomass in the North Atlantic and off the coasts of Northeast Asia. *Ecklonia* is found mainly off Australia and South Africa, whereas *Macrocystis* grows mostly off the western coast of North America, off the southern coast of South America, and off New Zealand, Tasmania, and South Africa.

In the tropics, rocky shores are desolate at first glance, lacking species of *Fucus* and *Ascophyllum* that are so common to temperate rocky shores. This is apparently due to of high temperatures, desiccation, and high light intensities.[184] Blue-green algae, filamentous green algae, and encrusting lichens survive and grow dense and turf-like close to the substrata on tropical rocky shores; productivity is presumably quite high, as substantial communities of grazers actively feed on the tufted algae. It is usually only below the spray zone and in subtidal waters that some macroalgae, such as *Sargassum* and *Turbinaria*, can live on tropical rocky shorelines.

The contribution of seagrass meadows and seaweed-based ecosystems to the global ocean is disproportionate to their small area. Seagrasses and seaweeds are highly productive, and with other marine macrophytes, account for roughly 40% of coastal primary productivity.[185,186] Kelp forests and seagrass meadows are important nursery grounds for a wide variety of mammals, reptiles, birds, amphibians, and commercially harvestable fish and crustaceans. Giant kelp beds can rival coral reefs for physical magnificence, productivity, and ability to attract numerous vertebrates and invertebrates. Many seaweeds are harvested by humans for fertilizers, mulch, and food (see Chapter 8). Both seagrasses and seaweeds serve to stabilize shores by altering water flow and sediment matrices. Sadly, some of these ecosystems, especially seagrass beds,[187] are disappearing at an alarming rate in both temperate and tropical seas.

4.2   STANDING CROP AND PRIMARY PRODUCTIVITY

Estimates of standing crop and primary productivity of seagrasses and seaweeds are as variable as the methods that have been devised to measure them. The data are so variable as to preclude any meaningful latitudinal patterns (Table 4.1). These variations, even within a genus or species, reflect local variations in limiting factors such as light, nutrients, salinity, and abundance and composition of grazers.

4.2.1   Seagrasses

Seagrasses and seaweeds are usually highly productive autotrophs (Table 4.1). Representative rates of primary productivity of seagrasses range among genera and species from 0.1 to 18.5 g C m^{-2} d^{-1}, with most values falling within a range of 0.4 to 1.5 g C m^{-2} d^{-1}. These rates are somewhat less, on average, than those for marsh grasses, mangroves, and terrestrial plants. Growth rates of seagrasses are often fastest in the subtropics and tropics, but some temperate seagrass beds are as productive as those of lower latitudes. Some of the fastest and slowest rates of seagrass production have been measured in tropical regions.[188] This wide range of growth may reflect various sampling treatments, tagging, and harvesting methods used to measure primary production rather than favorable or unfavorable, growth conditions.[188]

TABLE 4.1

Rates of Net Primary Production for Some Selected Seagrasses
and Seaweeds From Various Locations

Genus/Species	Location	Net Primary Production ($g\ C\ m^{-2}\ d^{-1}$)	Ref.
Kelps			193, 195, 286
Laminaria	North Atlantic	0.3–65.2	
Macrocystis	South America, New Zealand, South Africa	1.0–4.1	
Ecklonia	Australia, South Africa	1.6–3.2	
Rocky intertidal/ subtidal macroalgae			194, 270, 287
Various seaweeds	Europe	0.5–9.0	
Enteromorpha	Hong Kong, U.S.	0.1–2.9	
Ascophyllum	U.S., Europe	1.1	
Distyopteris	Caribbean	0.5–2.5	
Fucus	North America	0.3–12.0	
Sargassum	Caribbean	1.4	
Ulva	Europe	0.6	
Gracilaria	Europe	0.3	
Cladophora	Europe	1.6	
Seagrasses			188, 258, 288, 289
Zostera marina	U.S., Europe, Australia	0.2–8.0	
Thalassia	U.S., Caribbean, Australia, southeast Asia	0.1–6.0	
Halodule	U.S., Caribbean	0.5–2.0	
Cymodecea	Mediterranean, Australia	3.0–18.5	
Posidonia	Mediterranean, Australia	2.0–6.0	
Enhalus	Southeast Asia	0.3–1.6	
Amphibolis	Australia	0.9–1.9	

There are very few estimates of below-ground production of
seagrasses. The sparse data available indicate that production of roots
and rhizomes can rival that of leaf and shoot production.[185,188,189] Below-
ground production constitutes, on average, from 2 to 36% of total plant
production and from 10 to 40% in dense, mature meadows.[188,189] Below-
ground components comprise a greater fraction of total plant biomass,
ranging from 10 to 75% (Table 4.2) and most frequently from 30 to
70%.[189] These variations likely mirror differences in composition and
age of the plants. Recent data from the Philippines (Table 4.2) illustrate

TABLE 4.2

Partitioning of Above- and Below-Ground Biomass and Production of Several Seagrasses in a Mixed Meadow in the Philippines

	Cymodocea rotundata	Cymodocea serrulata	Enhalus acoroides	Halophila ovalis	Halodule uninervis	Syringodium isoetifolium	Thalassia hemprichii
Biomass							
Above-ground	22.9 (58.6)	2.4 (88.9)	53.0 (22.7)	0.1 (50.0)	4.5 (61.6)	12.5 (87.4)	250.2 (76.4)
Below-ground	16.2 (41.4)	0.3 (11.1)	180.2 (77.3)	0.1 (50.0)	2.8 (38.4)	1.8 (12.6)	77.3 (23.6)
Production							
Leaves	199.0 (97.4)	13.0 (87.2)	139.1 (64.5)	9.5 (70.9)	32.8 (78.5)	53.6	1511.5 (94.8)
Vertical internodes	2.2 (1.1)	1.6 (10.7)	0	0	5.4 (12.9)	3.6	61.0 (3.8)
Rhizomes	3.2 (1.5)	0.3 (2.0)	76.5 (35.5)	3.9 (29.1)	3.6 (8.6)	—	22.8 (1.4)

Note: Biomass units are g dry weight m^{-2}. Production units are g dry weight m^{-2} yr^{-1}. Values in parentheses are percentages of total biomass and production.

Source: Adapted from Vermaat, J. E. et al., *Mar. Ecol. Prog. Ser.*, 124, 215, 1995.

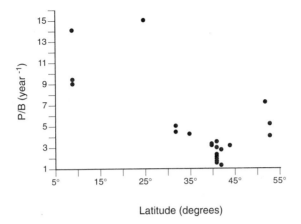

FIGURE 4.1

Changes in annual production-to-biomass (P:B) ratio for seagrasses with latitude. (From Duarte, C. M., *Mar. Ecol. Prog. Ser.*, 51, 269, 1989. With permission.)

variations that can be found in the proportion of below- to aboveground productivity and biomass among seagrass beds dominated by different species.

Rates of seagrass production vary seasonally, even in the tropics. Seasonal variability is not as pronounced in tropical seagrass beds as in temperate meadows, and the seasonal cues are different.[188,190] In temperate regions, seagrasses follow seasonal changes in temperature, growing fastest in spring and early summer. Tropical seagrass growth and productivity are related to:

- Temporal changes in the solar cycle
- Synchrony of tidal exposure
- Occurrence of the wet and dry seasons

Total standing crops vary greatly among seagrasses and with age of meadows from 50 to 850 g dry weight m⁻², with a grand mean[191] of 315 g dry weight m⁻². In his review of standing crops and production-to-biomass (P:B) ratios of seagrass meadows, Duarte[191] found that biomass varies seasonally; there are no clear patterns of biomass with latitude, but it appears that P:B ratios decrease with increasing latitude (Figure 4.1). A positive correlation between production and biomass was attributed to the confounding influence of latitudinal variations. This positive relationship, however, is in agreement with earlier compilations of seagrass biomass and production estimates.[188]

TABLE 4.3

Contribution of Various Primary Producers to Total Net Primary
Production in Selected Seagrass Meadows and Kelp Beds[a]

Habitat	Location	Macrophyte	Epiphyte	Phytoplankton	Other	Ref.
Halodule	Indonesia	0.64	0.23	—	—	290
Syringodium	Fiji	12.0	11.5	—	11.4	291
Halodule	Texas	7.2–96.0	0–98.4[b]	4.8[b]	—	196
Zostera	North Carolina	0.9	0.2	—	—	196
Zostera	Massachusetts	0.4–1.0	0.2–0.8	—	—	196
Seagrass spp.	Denmark	0.24	0.04	0.14	—	288
Seagrass spp.	Philippines	1.4[b]	1.8[b]	—	—	242
Laminaria	Nova Scotia	1.7	—	0.5	—	193
Ecklonia, Laminaria	South Africa	6.5	0.7	6.9	3.3	195

[a] Units are g C m^{-2} d^{-1}, except where noted.
[b] Units are g AFDW m^{-2} d^{-1}.

4.2.2   Seaweeds

Biomass and production estimates of kelps and other seaweeds are as
numerous as those for seagrasses and equally variable, doubtless for
the same reasons. Under favorable conditions, maximum rates of primary production of seaweeds (Table 4.1) can exceed rates of production
of all other macrophytes, including mangroves and marsh grasses.
Their rapid growth potential is evidenced in mariculture farm ponds,
particularly those in Asia.[192] Natural subtidal forests of the kelps, *Laminaria*, *Macrocystis*, and *Ecklonia* can attain rates of primary production of
1000 g C m^{-2} yr^{-1}, with a range of 400 to 1900 g C m^{-2} yr^{-1}. Considering
their individual size and densities, it is not surprising that the biomass
of kelps can exceed several metric tons per square meter of shoreline,
particularly for the giant kelp *Macrocystis*. Standing crop varies drastically for these kelps owing to large variations in grazing pressure.[193]
The very large kelps, such as *Macrocystis*, shunt a large fraction of their
fixed carbon into biomass maintenance. This results in a lower P:B ratio
(~1) than that of the smaller kelps, such as *Laminaria*, which has a faster
turnover time and higher P:B ratio of ~2 to 7.[193]

Other macroalgae, such as *Gracilaria*, *Ulva*, *Fucus*, and *Enteromorpha*,
are also very productive (Table 4.1) and very patchy in their distribution. Biomass estimates have been difficult to obtain owing to difficulties in sampling (such as on rocky coasts) and to their temporal and
spatial patchiness; however, in some locations, their biomass can exceed 500 g dry weight m^{-2} depending upon time of year.[194] Their contribution to total primary production can exceed 10 to 30% in some
estuaries, coral reefs, and lagoons (see Chapters 5 and 6).

In seagrass meadows and in kelp forests, other autotrophs may
contribute significantly to total primary production (Table 4.3). In the

Benguela kelp beds, phytoplankton normally constitute at least 25 to 50% of total production, with lesser contributions from epiphytes and understory algae.[195] In seagrass meadows, epiphytes on leaf blades are often among the largest primary producers other than the grasses themselves (Table 4.3), with their contribution to total primary production ranging from 2 to 60%, with most sites averaging between 20 to 50%.[196]

4.3 PHOTOSYNTHESIS AND WHOLE-PLANT CARBON BALANCE

Seagrasses and seaweeds differ in some aspects of photosynthetic potential, such as their ability to harvest light (Figure 4.2). Light adsorption (as a function of chlorophyll density) is higher in seagrasses than for most seaweeds, likely reflecting the need to compete for attenuating light as submerged, rooted plants (Figures 4.2A and B). Seagrasses, therefore, tend to have greater light compensation points for photosynthesis and growth (Figures 4.2C and D) than do most seaweeds. The light requirements are greater to support seagrass growth because a greater proportion of gross photosynthesis is vested in respiration of nonphotosynthetic parts (15 to 50% of total plant respiration[197]). The high light requirements result in lower photosynthetic efficiency and ability to transform absorbed light energy, leading, in turn, to slower growth rates for seagrasses than for macroalgae (Figure 4.2E). Duarte[197] differentiated between thin and thick seaweeds, suggesting that thick macroalgal species, such as *Laminaria*, perform similarly to seagrasses, whereas thin macroalgal species, such as *Ulva* or *Gracilaria*, perform photosynthetically more similarly to phytoplankton.[198]

The growth and productivity of seagrasses and seaweeds are determined by the way in which individual plants balance their carbon requirements. Models of carbon balance for whole plants exist for several genera, including *Zostera, Thalassia, Fucus*, and *Laminaria*. In Florida seagrass beds, the biomass of individual *Thalassia testudinum*[199] is proportioned mostly into shoots (40.3%) and rhizomes (34.8%), with lesser biomass vested in leaves (15%) and fine roots (9.9%); leaves account for nearly 43% of respiratory demand of the whole plant, followed by shoots (27.6%), roots (17.7%), and rhizomes (12.1%). Below-ground structures thus account for nearly 30% of total respiratory carbon demand with the carbon budget indicating a compensation depth (photosynthesis = respiration) for *T. testudinum* of 4 m. More extensive data for *Zostera marina* similarly indicates that most plant biomass is tied up in shoots and rhizomes, with a smaller fraction (10 to 15%) of total plant respiration contributed by subterranean components. Kraemer and Alberte[200] modeled carbon balance in this species

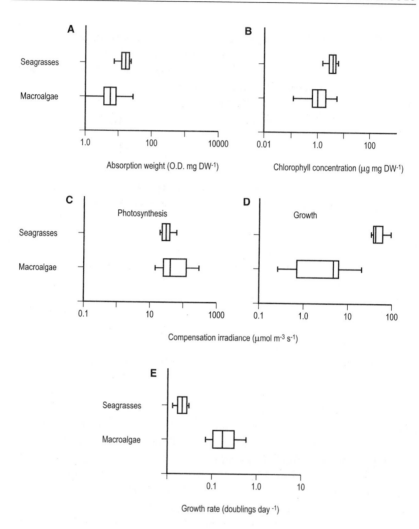

FIGURE 4.2
Comparisons of some photosynthetic characteristics between seagrasses and macroalgae. Each box plot represents grand mean and range of literature values. (Adapted from Duarte, C. M., *Ophelia,* 41, 87, 1995.)

from a subtidal meadow in California and predicted a compensation depth of 4.2 to 11.6 m, with the proportion of subsurface tissue respiration to total respiration increasing from 15 to 25%. The model further predicts that small variations in the production-to-respiration ratio (P:R) and light availability can greatly affect new shoot production.

Carbon budgets of some seaweeds[198,201-203] show significant differences in production and utilization among species, but they do show some common traits:[204]

- Nearly half of assimilated carbon appears as new blade production at growth rates commonly exceeding 1 to 2% per day.
- A small fraction is stored in mature tissues.
- The remainder is presumably lost as dissolved organic carbon (DOC).

The proportion of fixed carbon leached as DOC appears to decrease from temperate to boreal latitudes.

Many seaweeds undergo seasonal carbon imbalances when plants utilize stored photosynthate for maintenance and respiration. Most seaweeds are geared for rapid growth when the opportunity permits, resulting in comparatively large losses of assimilated carbon during these periods. Many seaweeds, however, exhibit distinct seasonal growth cycles that are controlled by endogenous, circannual clocks synchronized to changes in day length. The role of environmental cues is, therefore, linked to seasonal changes in seaweed growth via endogenous rhythms rather than direct and more immediate impacts.[204]

4.4   LIMITING FACTORS

Like other aquatic plants, photosynthesis by seagrasses and seaweeds is physiologically limited by sufficient light, nutrients, and diffusion of inorganic carbon sources. Other factors such as temperature, salinity, and water motion also play an important role for most macrophytes. Seagrasses and seaweeds vary greatly in their physiological tolerances to these factors which determines to what extent these plants fix carbon that will ultimately fuel grazing and detrital food webs.

4.4.1   Carbon Sources and Metabolism

The principle source of carbon is inorganic, either as CO_2 or HCO_3^-. The supply of carbon dioxide and bicarbonate is virtually limitless in seawater, but their availability to plants is limited by diffusion across plant cell membranes, particularly under quiescent and light-saturated conditions.[205] Diel patterns of water-column O_2 and CO_2 reflect the diel cycle of photosynthesis which tends to follow the daily solar light cycle. Plant photosynthesis peaks at late morning and early afternoon. By mid-afternoon, photosynthesis may be depressed by short-term limitation of nutrients or inorganic carbon, as reflected in lower pH and CO_2 concentrations brought about by photosynthesis earlier in the day. How these plants actually transport and utilize inorganic carbon for photosynthesis is poorly understood. In theory, uncharged molecules such as CO_2 can passively permeate plant cell membranes, but charged molecules such as bicarbonate require a special protein channel. Several species of seagrasses display an ability to utilize both HCO_3^- and CO_2, suggesting that such a channel exists.[205] There is evidence[198] that some macroalgae possess two uptake mechanisms: (1) an active bicarbonate transport system linked to a special protein channel (in *Gracilaria* and *Ulva*) and (2) passive uptake of CO_2 facilitated by the external enzyme carbonic anhydrase (in *Chondrus*). Several other species of macroalgae possess this enzyme (in *Soliera, Gracilaria, Fucus,* and *Laminaria*).[198]

Most seagrasses can acquire both compounds, but some species display a high affinity for CO_2.[205] For some species such as *Zostera*, HCO_3^- is assimilated and likely linked to ion fluxes across the cell membrane. Larkum and James[205] propose that the passive diffusion of CO_2 across the membrane (plasmalemma) following dehydration of the bicarbonate ion — facilitated by the enzyme carbonic anhydrase — is the principal or sole assimilation mechanism of inorganic carbon in seagrasses. Their work further indicates that the rate-limiting step in seagrass photosynthesis is the CO_2 concentration within the cell wall which, in turn, is dependent upon the rate of OH^- release from the cell. Ultimately, the factors determining the equilibrium CO_2 concentration limiting seagrass photosynthesis are pH within the cell wall and the thickness of the boundary layer surrounding the leaves. Carbon limitation is thus more likely under low-flow conditions. Limitation is also likely in low-salinity habitats where inorganic carbon concentrations are low and pH is high due to reduced carbonate buffering capacity. Further research is necessary to determine if this cellular mechanism is valid for other species of seagrass and for seaweeds.

Seaweeds are also capable of light-independent carbon fixation. Red and green macroalgae fix CO_2 along with NH_4^+ in the dark providing a mechanism to take up nitrogen while storing carbon. Brown algae have higher rates of dark carbon fixation, although it occurs mainly in young kelp and fucoid plants where it can account for ~20% of total fixed carbon.[198] The ability to fix carbon in the dark can be an ecological advantage in an environment such as kelp beds, where light levels are low and nitrogen is limiting. The principal products of carbon fixation in seaweeds are sucrose and starch (green algae), mannitol and laminaran (brown algae), and low-molecular-weight compounds and floridean starch (red algae).[198] Various secondary metabolites are produced by carbon fixation in seaweeds, namely phenolics and halogenated compounds. These are active antibiotics and ultraviolet protectors and can serve to deter grazers and the colonization and growth of epiphytes.

4.4.2   Light and Temperature

The level of light necessary to saturate a given species shows some relationship to habitat. Most seagrasses[188] saturate at 200 μmol m^{-2} sec^{-1} or less, whereas intertidal seaweeds[198] saturate at 400 to 600 μmol m^{-2} sec^{-1}. Both seaweeds and seagrasses inhabiting shallow subtidal areas saturate at 150 to 250 μmol m^{-2} sec^{-1}; deep-water species saturate at less than 100 μmol m^{-2} sec^{-1}. Whether or not a given habitat reaches light saturation usually depends upon turbidity. In light-saturated conditions, many seagrasses have temperature optima for photosynthesis of between 25 to 35°C; seaweeds appear to have a wider temperature optima.[198] Tropical forms, such as *Sargassum*, can grow optimally at temperatures greater than 35°C, whereas cold-water forms, such as *Fucus*, can grow at near optimal rates at temperatures less than 15°C. Temperature may indirectly affect photosynthesis because some metabolic processes such as respiration and nutrient uptake are temperature dependent. Macrophytes growing at near compensation light levels may attain optimum growth at low temperatures, but at higher temperatures, more light is needed to overcome the effects of respiration to maintain carbon balance. This implies that the growth of seagrasses and seaweeds may be more affected by low light in summer than in winter. Warmer temperatures may also make a plant more susceptible to disease and desiccation, or other stresses.

In situ rates of photosynthesis for seagrasses and some kelps are often limited by self-shading or shading due to epiphyte growth on blades.[188,196,198] Self-shading is largely a function of plant density, but inhibition due to epiphytes is more complicated. Seagrasses can be fouled by epiphytes as a result of nutrient enrichment, causing dieback of meadows in many areas. The distribution and growth of epiphytes are dependent upon plant age, age of different parts of an individual plant, orientation of the blades with respect to the shoot, and age of different blades. Generally, the oldest leaves support the most abundant and diverse epiphytic communities.

Epiphytes reduce macrophyte photosynthesis by limiting available light and the rate of inorganic carbon diffusion. Rates of photosynthesis of many apparently healthy seagrasses are reduced by epiphytes as much as 35 to 60%.[196] The major adaptive strategy to best deal with epiphytes appears to be sloughing off of colonized leaves. Much research remains to be done on epiphyte-host relationships, as epiphytes do offer benefits to their host, such as acquiring limiting nutrients.

4.4.3   Salinity

As with light and temperature, tolerances to salinity vary among seagrass and seaweed species. Seagrasses growing in estuaries are more eurytolerant than stenohaline species inhabiting fully marine and hypersaline water bodies. However, many seagrasses grow well in salinities ranging from 15 to 55 and can survive from a range of 5 to 140.[188] Some notable exceptions are *Halophila ovalis*, which prefers low salinities, and some species of *Halodule* and *Zostera*, which appear to have very wide salinity tolerances.

For seaweeds, most subtidal species tolerate a similar range (18 to 52) of salinity as for many seagrasses. Seaweeds that live intertidally are much more halo-tolerant (0 to 100).[198]

4.4.4   Water Movement

The growth and productivity of floating and anchored plants is greatly influenced by water movement, particularly in relation to surface boundary layers.[206] Many plants living in still water quickly experience growth limitation because of nutrient and gas depletion at the boundary layer. Movement of water is necessary to replenish these pools and to remove metabolites in order to sustain growth. Plants also respond phenotypically to long-term changes in water movement. Conversely, plants can affect hydrodynamics in a number of ways, but only relatively recently has this phenomenon been examined and considered as a key factor in the growth and sustainability of macrophytes.[207]

In seagrass meadows, currents enhance rates of primary productivity by mixing and distributing nutrients and gases and removing wastes, but measurements of seagrass production and boundary layer thicknesses in relation to changes in current velocity are scarce.[208-210] Recent studies have demonstrated that enhancement effects of water flow are attained at a narrow window of current speed; increased turbulence above a given threshold results in suspended material reducing light availability and lowering rates of production. The thickness of the boundary layer surrounding leaf blades varies over space and time.[208-210] Koch[208] exposed blades of *Thalassia testudinum* and *Cymodocea nodosa* to different current velocities in microcosms and found enhanced rates of photosynthesis only at low current velocities (Figures 4.3A and B). Peak photosynthetic rates for *T. testudinum* were measured at a speed of 0.25 cm sec^{-1}. For *C. nodosa* taken from a surf zone, maximal photosynthetic rates occur at 0.64 cm sec^{-1}. There was no further enhancement of photosynthesis for either species at increased flow rates (Figure 4.3), suggesting limitation of carbon diffusion below, and limitation of carbon fixation by enzymes above, the critical flow velocities. The thicknesses of the boundary layers were not uniform owing to the presence of epiphytes causing smooth turbulent flow over the leaf surfaces.

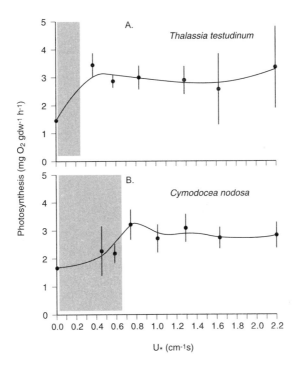

FIGURE 4.3
Variations in photosynthetic rates of blades of the seagrasses *Thalassia testudinum* (A) and *Cymodocea nodosa* (B) with increasing friction velocities (U.) at saturated light levels in the laboratory. Boxed areas depict boundary-layer photosynthesis limited by diffusion. (From Koch, E. W., *Mar. Biol.*, 118, 767, 1994. With permission.)

More frequent measurements of the effects of water flow have been made in kelp beds, particularly those of *Lamineria* and *Macrocystis*.[206,211] The morphology of kelps changes according to current and wave exposure, with kelps in wave-exposed locations characterized by long stipes and thick blades. In more sheltered areas, the stipes may be as long, but the lamina are wider and thinner.[206,211] The functional advantage of modifying blade morphology in response to water motion is still unclear. For instance, Hurd et al.[211] examined the impact of water movement on blade morphology and uptake of inorganic nitrogen for *Macrocystis integrifolia*. In microcosms, blades of *M. integrifolia* taken from both sheltered and exposed sites were subjected to increasing water velocity. They found that for both sheltered and exposed plants, rates of nitrate and ammonium uptake increased with increasing water velocity, suggesting that blade morphology does not enhance nutrient uptake by this species. This is contrary to earlier studies[22] suggesting an energetic advantage to a shift in morphology.

Both seagrass meadows[209,210] and kelp beds[212,213] can modify ambient currents and waves, leading to important ecological effects on:

- Distribution of organisms and their food supply within the understory
- Fluxes of nutrients and gases
- Dispersal of gametes, spores, and larvae

On average, friction between water and plant reduces current speed within the canopy which leads to an increase in the deposition of suspended particles. In beds of the kelp *Nereocystis* on the Pacific coast of North America, Koehl and Alberte[213] continuously measured current velocity within the kelp bed. They recorded maximum velocities on the edge of the beds with a noticeable decline by the middle of the bed. However, turbulent velocity (a measure of small-scale water motion) was highest in the middle of the bed at slack tide. They further found that blade morphology had a critical role to play in water motion, with wide, flapping blades enhancing photosynthesis and turbulent motion but increasing drag on the whole plant, and narrow blades minimizing drag and water motion but also lowering rates of photosynthesis.

In seagrass meadows, water advection and turbulent mixing are reduced among the plants, but significantly increased above the seagrass canopy.[209,210] The extent to which the presence of seagrasses affects local water movements depends not only upon local hydrodynamic conditions but also shoot densities and areal extent of the meadow.[210] Small, discrete beds show little or no alteration of water flow above the canopy; however, turbulent mixing does not appear to be affected by seagrass beds, regardless of size. Our understanding of water movement in and around macrophyte beds is still in its infancy, as evidenced by the disparate results in flume studies and under natural conditions.[206,210,213]

4.4.5   Nutrients

Most seagrasses and seaweeds are nutrient limited. Seagrasses take up dissolved nutrients through the roots and leaves with the dominant route depending upon the nutrient species and its concentration. If water-column concentrations are high, uptake via leaves (or even epiphytes) may dominate. Conversely, if ambient levels in the water column are low, roots may be the main uptake pathway.

Short[214] indicated that seagrasses growing in terrigenous sediments and in temperate areas are nitrogen limited, as phosphorus is readily available in such environments. However, in carbonate sediments where phosphorus is strongly bound to iron oxyhydroxides and in the tropics where phosphorus levels are generally low, seagrasses appear to be phosphorus limited. Evidence collected since Short's review[214] tends to support the notion of phosphorus limitation for subtropical and tropical seagrasses,[215-218] especially in carbonate-rich sediments. Fertilization experiments and examination of available porewater nutrient pools reveal, for instance, that Mediterranean meadows of *Cymodocea nodosa*[216] and beds of *Thalassia testudinum* and *Halodule wrightii* in Florida Bay[217,218] are phosphorus, but not nitrogen, limited. In the Bahamas, fertilization with phosphorus of *Syringodium filiforme* beds inhabiting carbonate sediments resulted in enhanced seagrass growth, biomass, and tissue phosphorus concentration, as well as increased rates of nitrogen fixation in the rhizosphere.[215]

An extensive series of studies performed by Erftemeijer and colleagues[219-221] in seagrass beds in Indonesia suggest that phosphorus limitation is not a universal phenomenon. Fertilization experiments and a comparison of the relative availability of nitrogen and phosphorus in terrigenous and carbonate sediments reveal that neither nitrogen nor phosphorus are limiting to these seagrasses. Dissolved phosphorus concentrations were high, which can be attributed to the limited adsorption capacity of the coarse-grained sediments that these seagrasses inhabit. This suggests that phosphorus limitation for tropical seagrasses is a function of sediment grain size, with phosphorus becoming limiting with decreasing sediment particle size (and increased adsorption). Conversely, adsorption of phosphorus to iron oxyhydroxides in carbonate sediments may also result in iron limitation. Duarte[222] found that fertilizing a mixed Caribbean seagrass bed with soluble iron stimulated growth of *Thalassia testudinum* and increased the chlorophyll *a* concentration in *T. testudinum* and *Syringodium filiforme* tissues. This finding was attributed to the very low, natural concentrations of dissolved iron in plant tissues, carbonate sediments, and in overlying waters. Whether or not iron limits growth and productivity of seagrasses and other algae in other tropical habitats is intriguing and requires further research.

Macroalgae appear to show a similar temperate-tropical dichotomy in nutrient limitation. In Florida Bay, where evidence exists for phosphorus limitation in seagrasses, Lapointe[223] found that the rhodophytes, *Gracilaria tikvahiae* and *Laurencia poitei*, and the phaeophytes, *Sargassum polyceratium* and *Sargassum pteropleuron*, are also phosphorus limited, supporting earlier evidence of phosphorus limitation for macroalgae from the other areas in the Caribbean.[224] In cool temperate waters, macroalgae appear to be nitrogen limited, particularly for nitrate. Nitrate limitation appears to be seasonal for many communities, such as the giant kelp *Macrocystis,* in which both nitrate concentrations and *Macrocystis* growth peak in winter and early spring.[225] Intertidal seaweeds show enhanced nitrate uptake when mildly desiccated, with nitrate uptake positively related to tidal height. Most seaweeds such as *Laminaria, Macrocystis,* and *Gelidium* can take up ammonium and nitrate at equal rates,[198] but *Macrocystis* is unable to store nitrogen nearly as well as *Laminaria.* This has important ecological consequences. In Nova Scotia, for instance, *Lamineria* takes up and stores nitrogen during the winter when water-column concentrations are high. When the spring phytoplankton bloom strips nitrogen out of the water, the kelp is able to live on the stored nitrogen.[226]

Young plants and younger tissues in mature plants take up nutrients faster than older plants and tissues, presumably because of greater metabolic requirements and less storage of nutrients.[198,204] In some seaweeds, such as *Fucus,* the stipe takes up little nitrogen compared to young fronds, which appear to require both ammonium and nitrate. The opposite pattern occurs in the kelps *Macrocystis* and *Laminaria.*[198]

Until recently, seagrasses were thought to take up nutrients almost solely via their roots. A recent review[227] suggests that uptake by seagrass leaves may be underestimated. In a series of laboratory experiments carried out under close-to-natural conditions, it was found that uptake of ammonium by *Zostera marina* leaves accounts for 68 to 92% of total nitrogen uptake.[228] Other studies[198] on European *Zostera marina* populations similarly indicate uptake mostly via leaves. Given that these European communities have a very high above- to below-ground biomass ratio, other *Zostera* populations with proportionally more root and rhizome biomass may behave differently.

The relative concentrations of phosphorus in the water and sediment determine whether or not the roots or leaves of seagrass are the main sites for phosphorus uptake. Recent work[229] suggests that, at least for *Zostera,* most phosphorus assimilated during growth and metabolism is incorporated into leaves, with lesser amounts retained in the roots and rhizomes; only a small fraction is released into the ambient water.

The effects of nutrient enrichment on coastal macrophytes is well known.[230] Excess nutrients supplied to seagrass beds usually results in proportionally greater growth and production of epiphytes and seaweeds, to the extent that the seagrasses die back. Some species are more resistant to eutrophication than others, and sheltered habitats with restricted circulation are more drastically affected than those inhabiting semi-restricted or open coastal areas.[230]

The effects of excess nutrients on most macrophytes are often clear, but indirect effects as a result of complex interactions with other controlling factors are not well understood. Only recently, for instance, have the effects of nutrient availability and grazing been simultaneously examined in seagrass meadows.[231,232] In *Zostera* beds, Neckles et al.[231] found that epiphyte grazers had a greater regulatory impact than nutrients. In spring and summer, grazer removal and nutrient additions increased epiphyte biomass on *Zostera marina* blades, but in autumn, there was little response to treatments, indicating a seasonal factor controlling the interactions. In field and laboratory trials, Williams and Ruckelshaus[232] observed that complex interactions among nutrients, epiphytes, and grazers control the growth of intertidal *Zostera marina* beds in the Pacific Northwest of North America. They predict that growth of the eelgrass can be limited by several conditions:

- Temperatures < 10°C
- Porewater ammonium concentrations < 500 μM
- Light intensity < 100 μmol m^{-2} sec^{-1} (when ambient light is low or epiphyte biomass is high)

Epiphyte growth is high when water-column dissolved inorganic nitrogen (DIN) is > 15 μM and grazing intensity is low. The effects of nutrients and epiphytes on macrophyte growth cannot be adequately understood without considering the impact of grazers. Grazers have a pivotal role to play in the growth and sustainability of seaweeds and seagrasses.

4.5 THE ROLE OF GRAZERS

Few other marine ecosystems are as rich in fauna as are kelp beds and seagrass meadows. A highly diverse range of organisms from microbes to mammals are dependent upon these macrophytes for sustenance and shelter. A large number of predators are exploited by humans. The earliest trophic studies of a marine ecosystem were conducted in Danish seagrass meadows early in this century by the great biologist Carl Peterson. Since his inaugural work[233] which showed that seagrass ecosystems are highly productive nursery grounds for coastal invertebrates and fish, subsequent work has focused on how and how much of the carbon fixed by these plants is transferred up the food chain or lost from the system. This is still not clear, but many workers have documented the:

- Role of macro- and meso-consumers grazing directly on plant tissue and associated epiphytes
- Channeling of carbon derived from decaying seaweeds and seagrasses into detrital food
- Ultimate fate of carbon in the coastal environment

Individual seaweed and seagrass habitats function differently, but there is some commonality with respect to the main pathways of energy transfer and nutrient flow.

4.5.1 Consumers of Living Macrophytes

Direct grazing of seaweeds and seagrasses by herbivores can represent a significant transfer of carbon and energy to food chains.[193] Much evidence exists of the decimation of large expanses of kelp beds by sea urchins and the intense cropping of seaweeds on rocky shores by limpets, chitons, littorinids, sea urchins, and fish.[198] These dramatic effects have been well documented for many rocky shores, such as in the Caribbean, and for kelp beds, such as off Nova Scotia.[193] This section focuses instead on the energetic impacts of grazing as well as less conspicuous consumers. The actual amounts of carbon consumed by large herbivores and the fate of this material are poorly understood, as most research has concentrated on the striking predator-prey cycles and environmental influences affecting these ecosystems, many of which may be driven by anthropogenic disturbances.[234] The energetics of food webs in tropical seagrass beds and seaweed mats are especially in need of attention considering the importance of these macrophytes as fishing grounds for many commercially viable organisms, such as penaeid prawns.

Equally dramatic in directly removing plant biomass are fish,[235] dugongs,[236] turtles,[237] and birds,[238] but the factors involved in herbivory can be complex. For instance, Neighbors and Horn[235] examined the dietary choice of herbivorous fishes off a rocky intertidal habitat in California to determine why these fish eat only a small proportion of macrophytes available to them. Using data on chemical composition, they tested the hypothesis that nutritional quality is the prime factor regulating the choice of macrophyte consumed. They found that there was some overlap in the composition of the dietary and nondietary plant items, but found that no one factor determined the type of food eaten. Brown algae were of higher nutritional quality than red algae but were not eaten, presumably because of noxious secondary metabolites and indigestible structural carbohydrates.[239] Food choice, therefore, cannot be presumed solely on the basis of nutritional quality. This is particularly true in physically dynamic habitats, such as on rocky shores, where food choice and niche partitioning among consumers are very complex and determined by a multiplicity of factors.

Actual estimates of the amounts of living macrophyte biomass consumed relative to production are few. In *Zostera marina* beds in the Netherlands, it was found that the isopod *Idotea chelipes* and eight species of birds consume an amount of living eelgrass tissue equivalent to

7.5% (3.7% by the birds and 3.8% by the isopod) of annual eelgrass production.[238] This is similar to the 3% of eelgrass production estimated to be consumed by birds in Chesapeake Bay.[240]

It is a common assumption that grazing is more intense in subtropical and tropical than in temperate macrophyte systems, yet little data are available to support this supposition. Stomach content analyses indicate that dugongs feed extensively on tropical seagrasses. For instance, in Indonesian *Halodule uninervis* meadows, dugongs can remove as much as 75% of the rhizome-root biomass, preferring to feed on meadows with sparse above-ground, but high below-ground, biomass. The impact of dugong feeding is somewhat complicated by the fact that seasonal decline of *Halodule* in the wet season can also be attributed partly to the synchrony of low tide with daylight, which causes overheating and desiccation.[236] Selectivity has been documented for dugong populations in Australian waters, although estimates of the percentage of total plant biomass or production consumed are not known.[237] The extent of grazing on seagrass material undoubtedly depends upon dugong and plant densities and age structure of their respective populations.

More quantitative information exists for sea turtles, of which only the green (*Chelonia*) and hawksbill (*Eretmochelys*) turtles regularly consume macrophytes.[237] The dietary habits of the green turtle are best known. Green turtles appear to have low daily ingestion rates (0.24 to 0.33% of their body weight) of seagrass and other algae, but they can digest plant material as efficiently as a ruminant — cellulose provides ~15% of their daily energy needs. Like ruminants, gut microflora process ~90% of this material. Thayer et al.[240] estimate in Caribbean seagrass beds that nitrogen in turtle feces returns nearly 3 g N d^{-1} to the whole beds compared to detrital nitrogen release of ~0.04 g d^{-1}, suggesting that turtles can contribute greatly to nutrient recycling in seagrass meadows. Obviously, this is true only when the animals are present in abundance. Annual migrations of herbivores can also drain nutrients from these seagrass beds to other coastal systems. Such transfers of energy and nutrients have never been quantified.

4.5.2   Consumers of Periphyton

Only a small fraction (~10%) of living macrophyte tissue is consumed directly, presumably due partly to poor nutritional quality and digestibility, and strong chemical deterrents. Benthic epifauna rarely consume living macrophytes, as they normally prefer either epiphytes or periphyton.[241] Epifaunal amphipods, shrimps, gastropods, isopods, nematodes, and copepods readily graze down complexes of periphyton consisting of diatoms, chlorophytes, encrusting algae, fungi, protozoa, bacteria, and sedimented material lying on leaf blades and colonizing understory surfaces. In turn, a variety of predators feed on these benthic grazers and may themselves feed, to a considerable extent, on the same foods. Many of these predators, such as crabs, lobsters, and fish, are commercially exploited and can exert considerable control over the abundance of their prey.[198,241]

Most grazers prefer to eat algae rather than detritus or living macrophyte tissue. Feeding experiments have shown that epiphytic algae are grazed heavily and can sustain rapid growth in response to intense grazing pressure.[241] Klumpp et al.[241] recognized four major feeding categories for seagrass epifauna:

- Direct consumers of living tissue (mostly isopods and some amphipods)
- Algal croppers (radula feeding of small gastropods)
- Nonselective feeders (e.g., small amphipods that remove loosely attached diatoms and detritus)
- Selective feeders

Good examples of selective feeders are nematodes that crack open diatom and chlorophyte cells and shrimps that pluck epiphytes from leaf blades.

Little data are available for the extent of grazing on periphyton in tropical seagrass meadows. In mixed seagrass beds in the Philippines, epifaunal grazers remove periphyton biomass equivalent to 20 to 62% of net production, with the major consumers being the gastropods *Strombus mutabilis* and *Cerithium tenellum*.[242] A followup study at the same location found that sea urchins are important grazers, but their impact is extremely variable, as the abundance of sea urchins changes at different times of the year.

In most habitats, food selectivity changes over time and space, so it is often difficult to separate consumption of periphyton from consumption of macrophytes. A good example is the trophic study of sea urchins in the Philippines seagrass meadows (Figure 4.4). A trophic analysis of the sea urchins *Tripneustes gratilla* and *Salmacis sphaeroides* shows that these species exhibit different feeding behavior, with the diet of *T. gratilla* comprised mainly of live seagrasses and that of *S. sphaeroides* being less specialized. Moreover, the behavior of each species varies with location, particularly *S. sphaeroides*.[243]

Many consumers of live and dead algae and seagrass show high rates of assimilation efficiency (Table 4.4). These high rates have been attributed mainly to digestive adaptation of consumers rather than to differences in nutritional quality. Most invertebrates possess gut microflora capable of digesting cellulose and hemi-cellulose components,

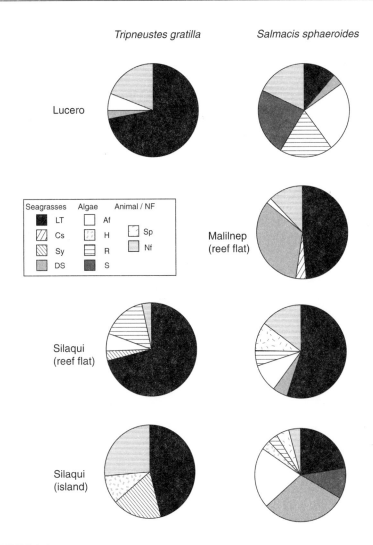

FIGURE 4.4

Differences in feeding behavior of natural populations of the sea urchins *Tripneustes gratilla* and *Salmacis sphaeroides* at different seagrass locations in the Philippines. Pie charts depict mean percentage frequency of feeding upon particular foods. Key: LT, live *Thalassia hemprichii* fronds; CS, live *Cymodocea serrulata* fronds; Sy, live *Syringodium isoetifolium* fronds; DS, dead seagrass fronds; Af, live *Amphiroa fragilissima* (red alga); H, live *Halimeda*; R, coral rubble; S, sediment; Sp, sponges; NF, not feeding. (From Klumpp, D. W., Salita-Espinosa, J. T., and Fortes, M. D., *Aquat. Bot.*, 45, 205, 1993. With permission.)

or themselves possess the necessary digestive enzymes, such as cellulase. Regardless of cause, grazing organisms are relatively efficient at incorporating bacteria, algae, and detritus derived from macrophytes.

TABLE 4.4

Assimilation Efficiencies (%) of Consumers in Seagrass Beds

Consumer	Method	Seagrass Detritus	Algae	Bacteria
Amphipoda	[14]C	11–54	5–67,[a] 41–75,[b] 11–55[c]	60–82
Amphipoda	Gravimetric	—	76–81[c]	—
Isopoda	Gravimetric	60	58–86[c]	—
Caridea	[14]C	73–91	84[b]	—
Gastropoda	[14]C, [51]Cr	46–48	60[b]	76
Penaeidea	Total organics, [14]C	—	63,[a] 68-87,[b] 85[d]	84–93
Polychaeta	[14]C	—	—	57
Portunidae	Lignin marker	61	—	—
Mytilids	[14]C, [51]Cr	—	—	50
Sea urchins	Total organics	6–52	—	—
Parrotfish	Total organics	7–53	50[b]	—
Garfish	Lignin marker	38–51	—	—

[a] Blue-green algae.

[b] Microalgae/epiphytes.

[c] Macroalgae.

[d] Macroalgae detritus.

Source: Adapted and updated from Klumpp, D. W. et al., in *Biology of Seagrasses,* Larkum, A. W. D., McComb, A. J., and Shepard, S. A., Eds, Elsevier, Amsterdam, 1989, chap. 13.

This reveals a close coupling between fixed carbon and secondary consumers but does not obviate the fact that little energy and materials flow through grazing pathways in seaweed and seagrass ecosystems.

4.6   DETRITUS AND MINERALIZATION PROCESSES

Microbial-detrital complexes — the crux of detritus-based food chains — use most of the energy and material fluxing through coastal food webs. Seagrass and seaweed-based ecosystems are no exception. The proportion of fixed carbon channeled through detrital pathways is mostly a function of the composition and abundance of detritus consumers and detritus decomposability, as determined by the chemical composition of the plant tissue.

4.6.1   Detritus Composition and Decomposition

The nutrient composition of detritus varies with age but also reflects the composition of the living tissue from which it is derived. An extensive compilation[244] of nutrient content data of living leaves from

TABLE 4.5

Mean Carbon, Nitrogen, and Phosphorus
Content (% dry weight) of Dead and Living
Parts of the Seagrass *Posidonia oceanica*

Component	Carbon	Nitrogen	Phosphorus
Young leaves	33.8	1.5	0.13
Old leaves	30.6	1.3	0.09
Living rhizome	35.7	0.6	0.03
Dead rhizome	34.1	0.4	0.02
Living roots	38.4	0.5	0.02
Dead roots	36.6	0.4	0.02

Source: Adapted from Romero, J. et al., *P.S.Z.N.I. Mar. Ecol.*, 13, 69, 1992.

46 species of macroalgae and 27 species of seagrass shows that, on average, seagrasses have relatively more carbon (33.5 ± 4.4% dry weight) and phosphorus (0.24 ± 0.13% dry weight) than macroalgae, but equivalent nitrogen (1.9 ± 0.7% dry weight) content. Carbon and phosphorus content for the macroalgae averages 24.8 ± 6.3% and 0.10 ± 0.07% dry weight, respectively. This translates into mean atomic C:N:P ratios of 360:18:1 for seagrass and 647:43:1 for macroalgae. Naturally, these results mask the often considerable variations in nutrient content among and within species and plant parts. Table 4.5 illustrates the variability in carbon, nitrogen, and phosphorus content among living and dead parts of the Mediterranean seagrass *Posidonia oceanica*. Living parts have more nutrients than dead tissues, but the variation in nutrient content is greater for nitrogen and phosphorus than for carbon.[245]

The higher carbon content relative to nitrogen for seagrass reflects the greater amounts of structural tissue required than for seaweeds.[137] Structural compounds such as lignin, cellulose, and hemi-cellulose are more resistant to microbial decomposition than simple carbohydrates. The methods used to examine detritus decomposition are as numerous as the number of workers who have studied the problem, so it is very difficult to compare actual rates and stages of decomposition. Nevertheless, most studies show that seagrass detritus decays more slowly than macroalgal detritus but faster than salt marsh and mangrove detritus.[137] The litterbag study by Buchsbaum et al.[246] of vascular plant and macroalgal decomposition (Figure 4.5) illustrates how most macroalgae decay more rapidly than do most seagrasses. Decomposition of species such as *Gracilaria tikvahiae* and *Ulva lactuca* is very fast, with no material left in litterbags after one month. *Z. marina* detritus decays more rapidly in summer than in winter, but, after several months, ~20% of the original organic matter still remains. The detritus derived from macroalgae and seagrass lost nitrogen at nearly the same

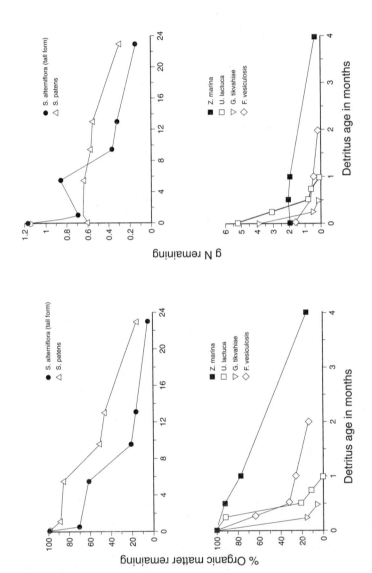

FIGURE 4.5

Comparison of decay of various salt marsh, seagrass, and seaweed detritus in litter bags incubated in the Great Sippewissett Marsh, expressed both as percentage of organic matter (left graphs) and as nitrogen remaining (right graphs). (Adapted from Buchsbaum, R. et al., *Mar. Ecol. Prog. Ser.*, 72, 131, 1991.)

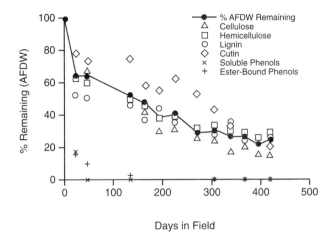

Days in Field

FIGURE 4.6
Changes in various biochemical components of decaying seagrass *Halodule wrightii* expressed as the ash-free dry weight percentage of each component remaining and incubated for over 400 days in a subtropical lagoon. (From Ophsahl, S. and Benner, R., *Mar. Ecol. Prog. Ser.*, 94, 191, 1993. With permission.)

rate as losses of their original mass (Figure 4.6, right). *Spartina* detritus, in contrast, decays very slowly, persisting for 1 to 2 years.

Like detritus derived from their salt marsh and mangrove counterparts (Section 3.4.2.2), macroalgal and seagrass detritus decays in identifiable stages, undergoing initial leaching of soluble compounds, followed by slower decomposition of structural components. In a review of the role of detritus in seagrass ecosystems, Harrison[247] concluded that the decomposition of seagrass detritus exhibits some features different than the decomposition of other vascular plant detritus:

- No net immobilization of nitrogen
- Slow rates of microbial activity (considering the high initial concentration of nitrogen)
- Slow decay rates despite low phenolic content

As we saw earlier for marsh and mangrove detritus, the relative and absolute amounts of nitrogen usually increase over time as microbial colonization increases, adding to the nitrogen pool by contributing cells, extracellular enzymes, and mucopolysaccharides, which complex with phenolic and other refractory nitrogen compounds.

More recent studies[245-246,248-249] tend to support Harrison's conclusion[247] of a lack of nitrogen immobilization as seagrass decay progresses. Changes in the various nitrogen components of decaying seagrass detritus have recently been examined by Buchsbaum et al.[246] and Opsahl and Benner.[248] Buchsbaum et al.[246] observed rapid initial losses of amino acids in *Spartina alterniflora* and macroalgal detritus; significant leaching did not occur from *S. patens* or *Z. marina* detritus, but there was an initial increase in the total amino acid pool before declining. *Z. marina* initially lost little of its soluble phenolics and then, after one month, lost these compounds rapidly. The initial retention of the ortho-dihydroxyphenolics that are bioactive may inhibit microbial activity and thus limit the capacity of microbes to assist in nitrogen immobilization. Opsahl and Benner[248] examined the decomposition kinetics and biochemical composition of the seagrass *Halodule wrightii* and found little selective decomposition in aging detritus compared with the biochemical composition of freshly senescent material. The long-term behavior of the biochemical pools (Figure 4.6) shows that the major constituents follow the ash-free dry weight (AFDW) losses, particularly after 100 days. From these results, they concluded that these constituents vary from most to least stable as follows: hemicellulose ≈ lignin ≈ AFDW > cutin > cellulose >> ester-bound phenols > soluble phenols. Little change in relative abundance of these pools suggests that, from a biochemical perspective, seagrasses have little capacity to immobilize nitrogen.[247]

Microbial activity on aging seagrass detritus may be lower than expected based on total nitrogen content, but the biochemical composition of the total detrital nitrogen pool[246,248] suggests that only a small proportion of this pool is labile, with most being lost during the initial period of decay. In early stages of decomposition, microbial activity can be rapid. Experimental decomposition of *Zostera marina* (Figure 4.7) indicates that bacterial productivity and respiration peak in the first 1 to 2 days, during which time the detritus has lost ~20% of its initial mass; bacterial growth rates peak somewhat later (5 to 10 days), but bacterial biomass peaks during the intermediate and later stages of decay.[250] Detrital carbon was initially assimilated by bacteria at an average efficiency of 8%, increasing during the intermediate stage to an average of 20%, than declining to <5% efficiency thereafter. These patterns of bacterial activity mirror those of the nutritional quality of the detrital material. Less than 8% of the detrital carbon lost in the first 2 days is metabolized by bacteria, but ~53% is metabolized over the remaining period of incubation. Of the total carbon metabolized, 80% is lost as CO_2. Similar patterns of microbial activity and assimilation efficiencies have been found during decay of other seagrasses, but the rates of decay vary,[65,66] probably as a result of different experimental techniques. These differences may be real considering the differences in nutrient content and structural composition among species of seagrass.

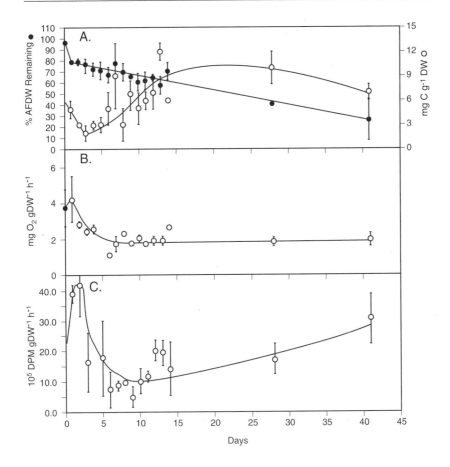

FIGURE 4.7

Changes in (A) decomposition kinetics (closed circles) and bacterial carbon biomass (open circles), (B) respiration, and (C) bacterial productivity in relation to decay of the seagrass *Zostera marina* incubated in litter bags. (Adapted from Blum, L. K. and Mills, A. L., *Mar. Ecol. Prog. Ser.*, 70, 73, 1991.)

Seaweed detritus degrades more rapidly and is converted more efficiently into bacterial biomass than vascular plant detritus. The laboratory work of Newell[253] and Mann[254] and their colleagues on the kelps *Laminaria* and *Ecklonia* have given us considerable insight into the role of microbes in the decomposition of seaweed detritus. In laboratory incubation experiments lasting from 2 to 36 days, Robinson et al.[254] traced the fate of detrital *Laminaria longicruris* carbon and estimated bacterial conversion efficiencies. The efficiencies were highest (43%) in short incubations and declined to a mean of 22% for the 36-day period. Of the initial detrital carbon, 54% leached out as DOC during initial washing, with the remainder split roughly equal into refractory and labile material. Bacterial decomposition of the labile particulate debris (20 to 25% of the initial mass) resulted in ~60% being respired and the rest (roughly 9% of the original material) being incorporated into bacterial biomass. In more extensive experiments, Newell[253] and his colleagues followed the decomposition of the dissolved and particulate fractions of *Ecklonia* and *Laminaria* detritus separately. The dissolved fraction, composed of D-mannitol, hexose sugars, alginates, and laminarins, was used by bacteria in hours (for D-mannitol) to days (for some sugars). In contrast, it took more than 340 days for one half of the particulate fraction to decompose.

The influence of bacterial processes on seaweed detritus can best be considered in relation to annual carbon production.[254] Of the annual carbon production of the kelp *L. pallida*, 15% is released into the water column as DOC and converted by bacteria with an efficiency of 15%, resulting in bacterial carbon production of 26 g C m^{-2} yr^{-1}. Fifteen percent of the kelp production is also released as DOC during fragmentation, but this material is converted by bacteria with higher efficiency (33%), resulting in bacterial carbon production of ~58 g C m^{-2} yr^{-1}. The remaining particulate detritus is converted at a low efficiency of 5.5%, leading to a bacterial carbon production of ~45 g C m^{-2} yr^{-1}. Thus, out of an estimated kelp production of 1172 g C m^{-2} yr^{-1}, bacterial production derived from decomposition of this material is 129 g C m^{-2} yr^{-1} or 11% of the primary production. Nearly all of this bacterial carbon is consumed and dissipated within the microbial loop within a few days.[254]

Bacterial consumption of carbon fixed by seaweed and seagrasses may represent a trophic sink, dissipating energy and recycling nutrients within microbial food chains, rather than bridging a link with metazoans. This is especially true for sediments, where much of the detritus derived from these macrophytes deposits.

4.6.2   Benthic Mineralization and Plant-Microbe Relations

High rates of macrophyte production and biomass, a rich fauna, and the curtailment of water flow leading to enhanced settlement of fine particles suggest that sufficient organic matter deposits within the understory and onto sediments to fuel a highly abundant microbial flora which recycles nutrients back to the plants. The mineralization and recycling of sediment organic matter in macrophyte beds is complicated by the presence of subsurface roots and rhizomes and by various geochemical and biological reworking processes, resulting in most of the below-ground material being decomposed anaerobically. A number of studies have assessed early diagenesis in seagrass sediments,[255] but there are few data of this kind for kelp beds.

TABLE 4.6

Total Organic Carbon, Nitrogen, and Phosphorus Stocks in a
Posidonia oceanica Meadow

Component	Carbon (g m^{-2})	Nitrogen (g m^{-2})	Phosphorus (g m^{-2})
Living leaves	124	5	0.4
Living rhizomes + roots	1522	25	1.8
Dead rhizomes + roots	47,522	539	19.1
Dead fine roots	13,123	195	9.3
Sediment organics	9000	—	—
Total	71,291	764	30.6

Source: From Romero, J. et al., *Aquat. Bot.*, 47, 13, 1994. With permission.

Accumulation of organic matter in seagrass and kelp beds is poorly understood, but some data suggest nutrient storage and low turnover of below-ground biomass in some seagrass systems,[188,256] with net accretion rates ranging from 0.1 to 1 cm yr^{-1}. The *Posidonia oceanica* meadows in the Mediterranean are an extraordinary example, where large quantities of living and dead plant parts are stored, leading to the formation of matte structures similar to peat beds. These matte structures, consisting mostly of dead roots and rhizomes, are rich in nutrients (Table 4.6). Romero et al.[256] calculate that *Posidonia* meadows occupy only 2% of the seabed in the Mediterranean, but this nutrient storage equates to ~10% of inorganic nitrogen discharge from land and ~20% of nitrogen fixation. The storage of carbon equates to 2 to 20% of yearly phytoplankton production. Only a handful of seagrass systems are presently known to have these matte accumulations, so it is unreasonable to conclude that most seagrass beds are net storage sites for nutrients. Some systems are, some are not, depending on the turnover rates of buried detritus. Nevertheless, compared to adjacent unvegetated sediments, most seagrass systems do have large pools of dissolved and particulate nutrients, suggesting intense recycling among the plants, microbes, and sediments.[255]

There is a close interdependence between seagrasses and sediment microbes. The relationship between seagrasses and microbes is fundamentally identical to that for mangrove trees and marsh grasses. The intense coupling of seagrasses with sediments and microbes has been reviewed by Moriarty and Boon,[255] so only more recent aspects will be discussed below.

By releasing gases transported to the roots through the lacunae, oxygen diffusing from the roots causes an oxic zone to develop around the subsurface tissues. It is within these root and rhizome layers that most bacterially mediated nutrient transformations and

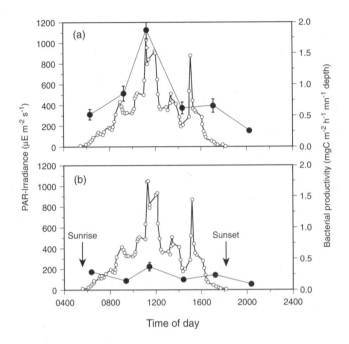

FIGURE 4.8
Diel variations in bacterial productivity (closed circles with error bars) in surface sediments of a (A) tropical seagrass bed and (B) adjacent mudflat in Fij, in relation to incident light. (Adapted from Pollard, P. C. and Kogure, K., *Aust. J. Mar. Freshwater Res.*, 44, 155, 1993.)

recycling take place. During growth, seagrasses also release DOM via their roots. This material is usually rapidly and completely utilized by the sediment bacteria for cell maintenance, growth, and division.

The activities of sediment bacteria are thus cued to temporal variations in plant activity. Figure 4.8 illustrates how rates of bacterial productivity in surface sediments in seagrass beds mirror diel changes in light intensity and, presumably, in rates of seagrass production. Measuring bacterial productivity by uptake of tritiated thymidine, Pollard and Kogure[257] found the highest rates of production in the top centimeter of sediment in *Syringodium isoetifolium* meadows off Fiji. They concluded that bacterial productivity equated to only 4 to 6% and 6% of net primary productivity of microalgae and seagrass, respectively. However, rates of bacterial productivity measured using thymidine seriously underestimate total rates of bacterial carbon mineralization because sulfate-reducers generally do not take up exogenous thymidine, so bacterial carbon production is likely to be a greater proportion of net seagrass production once production by all types of anaerobic bacteria is included.

In tropical seagrass beds in the Gulf of Carpentaria, northern Australia, Pollard and Moriarty[258] found some agreement between rates of sulfate reduction and thymidine uptake. Sulfate reduction ranged from 1.7 to 2.2 g C m^{-2} d^{-1} among four seagrass beds with much lower (0.3 g C m^{-2} d^{-1}) sulfate reduction rates in an open mudflat. Rates of bacterial productivity measured using thymidine were significantly higher among the seagrass and mudflat sites (3.1 to 13.0 g C m^{-2} d^{-1}), indicating that bacteria other than sulfate reducers were active. Variations with sediment depth for both sets of measurements related to subsurface biomass, suggesting that a consortium of bacterial types are fueled by exudates from roots and rhizomes. Both techniques indicate high rates of bacterial activity, with sulfate reduction equating to 27 to 78% and thymidine uptake equating to 75 to 200% of net primary production.

Rates of subsurface bacterial activity appear to vary depending upon the species of dominant seagrass. In carbonate sediments in the Bahamas, Hines[259] measured fastest rates of bacterial sulfate reduction in beds of *Syringodium*, with nearly equivalent rates in *Thalassia* meadows and significantly slower rates in *Halodule* beds. These rates were considerably slower than those measured in a *Spartina alterniflora* marsh in New Hampshire.[259] It is difficult to reconcile such habitat differences considering the small amount of data. It is best to suggest that such differences may relate to the quality and quantity of sediment organic matter among sites, rather than attributing them to some unique, species-specific plant characteristics that are, as yet, unknown.

What we do know of nutrient cycles and transformations by bacteria in seagrass and kelp beds is limited to a relative handful of field studies in which data were collected during only one season. Nitrogen transformations in seagrass sediments have received the most attention, but little work has been done to deduce the phosphorus cycle in either of these ecosystems.[260] Not surprisingly, the link between these elements is very poorly understood.

Recently, Pedersen and Borum[260] evaluated recycling of carbon, nitrogen, and phosphorus in Danish meadows of the seagrass, *Zostera marina*. Their budget (Table 4.7) shows that nutrients required for eelgrass growth must be met from uptake from surrounding water and sediment porewater, as losses do not balance with estimated incorporation rates. Some further insight into the connection between carbon oxidation and nitrogen utilization in seagrass meadows has been provide by Blackburn and his colleagues in studies of seagrass beds in Thailand[172] and in Jamaica.[261] In sparse *Halophila* beds in Thailand, peak rates of sulfate reduction and ammonium turnover co-occur at a 4- to

TABLE 4.7

Estimates of the Annual Budget for Carbon,
Nitrogen, and Phosphorus for a Danish *Zostera
marina* Meadow

	Carbon (g m^{-2} d^{-1})	Nitrogen (g m^{-2} d^{-1})	Phosphorus (g m^{-2} d^{-1})
Incorporation	852	34.5	3.2
Uptake	852	24.7	1.4
Recycling	—	9.0	1.7
Total losses	943	26.5	1.6
Above-ground	—	19.2	1.2
Below-ground	—	7.8	0.5

Source: Adapted from Pedersen, M. F. and Borum, J., *in
Biology and Ecology of Shallow Coastal Waters*, Eleftheriou,
A., Ansell, A. D., and Smith, C. J., Eds., Olsen & Olsen,
Fredensborg, 1995, 45.

6-cm depth, suggesting that either both bacterial types are fueled by
root exudates or carbon oxidized by sulfate reducers is linked to mineralization of organic nitrogen. In more extensive measurements in a
Jamaican *Halodule beaudetti* meadow, Blackburn et al.[261] found that
both carbon and nitrogen are actively cycled, with high rates of sulfate
reduction and rapid turnover of the sediment ammonium pool, probably stimulated by root exudates. Diel changes in the ammonium pool
were ascribed to concomitant changes in plant uptake. No buildup of
sulfides was observed despite the high sulfate reduction rates, suggesting that translocation of oxygen to the roots and subsequent release may have resulted in the oxidation of sulfides. This mechanism
has similarly been proposed for mangroves and salt marshes (Chapter
3). Blackburn et al.[261] proposed that rapid rates of nitrogen mineralization are associated with high rates of carbon oxidation via sulfate
reduction. Surprisingly, the rates of nitrogen mineralization in these
sediments did not balance with the measured rates of nitrogen fixation, which were less than the measured rates of denitrification.[261]
Short-term imbalances may occur in such systems, but it is more likely
that the N_2 fixation and denitrification measurements were made under suboptimal conditions.

In *Zostera noltii* meadows in southwestern France, Welsh et al.[262]
concurrently measured nitrogen fixation and sulfate reduction in the
rhizospheres and found some evidence for the occurrence of fixation
by the sulfate-reducing bacteria. The data indicate that nitrogen fixation activity in the rhizosphere is regulated by organic carbon availability from the roots, as well as low ammonium concentrations caused
by efficient plant uptake in the growing season.

Under steady-state conditions, losses of nitrogen are balanced by inputs. In seagrass meadows and kelp beds, there are three major sources of nitrogen:

- Nitrogen fixation
- Sedimentation
- Nutrient uptake by leaves and fronds

Other sources may include advected water, immigrant organisms, rain, groundwater, and terrestrial runoff.

Losses of nitrogen occur by:

- Denitrification
- Diffusion from sediments
- Consumption by migrating animals
- Export of sloughed leaves or fragments
- Leaching from plant tissue

A few recent attempts have been made to budget nitrogen losses and gains from individual seagrasses.[263-265] In Danish meadows, leaves and roots of *Zostera marina* were found to take up exogenous nitrogen equally. This nitrogen supply accounts for 73% of annual incorporation, with the balance supplied from internal regeneration.[263] Nitrogen gains and losses in beds of *Potamogeton perfoliatus* and *Zostera marina* have also been traced in North American seagrass beds[264,265] during the growing season. Caffrey and Kemp[264] observed faster rates of nitrogen transformation in seagrass deposits compared to bare sediments. In the *Zostera* beds, rates of ammonification exceeded plant requirements but potential rates of nitrification were not tightly coupled to denitrification rates. Ammonification and nitrification rates were correlated in beds of *P. perfoliatus*, coinciding with peak plant biomass and productivity.

Microbes regulating these transformations are likely to be enhanced by the plants. For example, ammonification is fueled by organic nitrogen inputs and nitrification is stimulated by translocation of oxygen and release from the roots. Using ^{15}N as a tracer in sediments and *Potamogeton perfoliatus*, Caffrey and Kemp[265] calculated a mass balance of nitrogen flow through the plant. Root and rhizome uptake accounted for ~90% of total nitrogen uptake in May, but only 20% in July; however, 70 to 75% of this nitrogen was subsequently translocated to the shoots. Roughly 75% of the ^{15}N lost from sediments was denitrified and 25% was taken up by the plant. Denitrification was tightly coupled to nitrification. The difference in ^{15}N uptake between spring (May) and summer (July) was attributed to the greater proportion of new shoots available to take up nitrogen after peak shoot growth in the spring. Variations in nitrogen uptake and translocation may therefore be partly a function of root-to-shoot ratio as the plant grows and ages. These results demonstrate that seagrasses directly take up ammonium and nitrate and reduce the concentrations of these dissolved nitrogen species in sediments and overlying water.

4.6.3 Benthic Detritivory

A consortium of benthic organisms, from protists to large epifauna, feed on sediment bacteria and bacterial-detrital complexes. Considerable research over the past two decades has attempted to clarify the suite of feeding modes and interactions that exist within detritus-based food chains in sediments, including within seagrass and kelp beds. As in intertidal flats, mangroves, and salt marshes, only a small proportion of bacterial biomass is probably consumed within seagrass and kelp bed sediments.

Grazing of bacterial-detrital aggregates is intense, however, on and within oxidized surface sediments, on surfaces of leaves and fronds, on shoots, and on natural hard substrata. Surface-feeding macrodetritivores consume both seaweed detritus and associated bacteria and diatoms that accumulate on the sea bed[266,267] off rocky coasts and other temperate subtidal areas. Involvement of benthic organisms further stimulates detrital aging and microbial growth. Meiobenthic organisms such as harpacticoid copepods and nematodes are particularly voracious consumers of bacteria, chlorophytes, diatoms, fungi, and some protozoa growing epiphytically on plant parts. Combined with microbes associated with oxic habitats, this diverse group of detritivores serves as food for larger predators.[241,253,255] Anaerobic bacteria, in contrast, are not directly linked to higher trophic groups, but mineralize and recycle nutrients to sustain the primary producers upon which all higher animal life ultimately depends.

4.6.4 Detritus Mineralization in Overlying Water

The bulk of detritus mineralization occurs in sediments, but rapid microbial activity and consumption also occurs in waters bathing seagrasses and seaweeds. Living and dead macrophytes exude or leach considerable amounts of dissolved organic matter (DOM) and slough off a substantial fraction of dead and decaying leaf parts and fronds at the tips, stimulating pelagic bacteria and other members of the microbial loop.

Pelagic microbes will respond not only to biological processes occurring within macrophyte beds, but are also sensitive to changes in the environment. Bacterioplankton production ranges from 12 to 96 µg C l^{-1} d^{-1} in some seagrass and kelp beds[253,255] and varies seasonally, particularly in temperate meadows. In *Posidonia oceanica* meadows in the Mediterranean,[268,269] changes in bacterioplankton production and growth rates were attributed to concomitant changes in the availability of phytoplankton exudates and to time lags in phytoplankton biomass, but the data strongly indicate some concordance with seasonal changes in water temperature (Figure 4.9). Phytoplankton production, however, may not at times be sufficient to provide the carbon and nitrogen required to sustain the observed rates of bacterial production.[268,269] Some utilization directly from the plants may often be necessary.

Of course, not all macrophytes are anchored to the seabed. Many seaweeds, such as *Sargassum* and *Cladophora*, trap gases to create floating algal mats which bloom and cover large expanses of shallow coastal waters.[198] Mass blooms of floating macroalgae are becoming increasingly common in coastal waters,[192,198] coincident with the dieback of seagrass beds, probably as a result of eutrophication. These mats are highly productive, often outpacing phytoplankton growth, and can take up large amounts of dissolved nutrients and gases from the surrounding water to the extent that nutrient and oxygen deficits occur below them. A nitrogen budget (Figure 4.10) constructed for *Cladophora sericea* mats in Danish coastal waters shows that the algal mats, utilizing inorganic nitrogen (mainly ammonium), are virtually closed systems, with 63% of its nitrogen requirements supplied by internal recycling. The remaining nitrogen comes from the atmosphere, lateral transport, and sediment efflux. Consumption by the mat system accounts for nearly 95% of available nitrogen and 85% of available phosphorus, resulting in clarification of the water beneath the mats. These mats sink or are exported from local waters, where their decomposition can result in a local decline in water quality, fish and shellfish harvests, and in seagrass abundance.[270]

4.6.5   Consumption of Pelagic Detrital-Microbial Aggregates

The often high rates of production and fragmentation of seaweeds and, to a lesser extent, of seagrasses infer that a substantial proportion of their fixed carbon becomes available for pelagic and suspension-feeding consumers. This is especially so in kelp beds as kelp detritus is more readily assimilated and of generally greater food value than seagrass detritus (Section 4.6.1) This phenomenon is best seen in the South African work on detritus consumption by ascidian and bivalve filter-feeders (see review of Branch and Griffiths[195]). Kelp beds also receive intrusions of upwelled, nutrient-rich water that

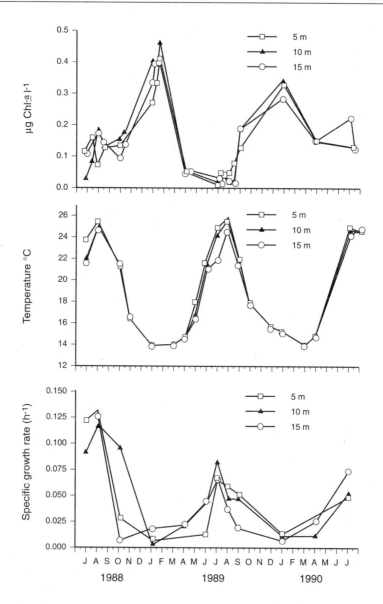

FIGURE 4.9
Seasonal changes in bacterial growth rates (bottom) and phytoplankton biomass (top) in relation to temperature (middle graph) in the water column at three depths above a Mediterranean seagrass bed. (From Velimirov, B. and Walenta-Simon, M., *Mar. Ecol. Prog. Ser.*, 80, 237, 1992. With permission.)

stimulate phytoplankton production. Many kelp bed consumers are, therefore, sustained by food resources in the water column rather than from strictly benthic sources.[195,267,271]

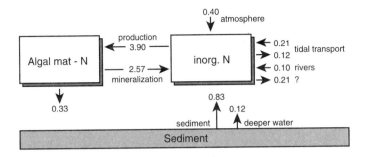

FIGURE 4.10
Nitrogen budget for mats of the macroalga *Cladophora sericea* during early summer in shallow Danish waters. (From Thybo-Christensen, M. et al., *Mar. Ecol. Prog. Ser.,* 100, 273, 1993. With permission.)

Manipulative studies show that the high densities of benthic suspension feeders within kelp forests are not simply the result of rich, kelp-derived food but also depend upon local hydrodynamic forces. Local water motion redistributes suspended food particles and affects larval recruitment. In kelp beds off the Aleutian Islands, Duggins et al.[267] transplanted populations of the mussel *Mytilus edulis* and the barnacle *Balanus glandula* across sites of differing kelp abundance and measured growth rates of mussels naturally occurring at the same sites. After one year, the recruitment and biomass of the transplanted suspension-feeding populations correlated positively with kelp abundance. The growth rates of naturally occurring mussels were greater in the sites of highest kelp biomass. Duggins et al.[267] further used $\delta\ ^{13}C$ analyses to distinguish between kelp- and phytoplankton-derived food resources. Finding a clear differentiation in $\delta\ ^{13}C$ signatures between kelp detritus (–17.7 per million ± 2.3) and phytoplankton (–24.0 per million ± 1.0), they found that 10 of 11 consumers fed mostly on kelp-derived food. At the kelp-dominated sites, consumers averaged ~58% kelp-derived carbon, whereas consumers from sites without kelps averaged 32%.

Farther south off Washington state, Duggins and Eckman[271] transplanted suspension feeders to real and artificial kelp forests to differentiate between nutritional and physical factors that may affect consumer growth. Contrary to the Aleutian experiments, they found no enhancement effect of suspension-feeder growth between treatments. Duggins and Eckman[271] attributed these results to the fact that in the subtidal environment off Washington, the waters are well mixed, leading to kelp-derived particles being suspended in the water column for considerable periods of time, long enough to change their nutritional quality and to be transported long distances. This is in contrast to the Aleutian sites, which are dominated by a different species of kelp. These results show that kelp beds in different physical settings and dominated by different species are unique. Consumers may thus live and thrive in proximity to a given habitat for a variety of reasons ultimately related to hydrodynamic forces.

The trophic importance of pelagic and suspension-feeding consumers in seagrass ecosystems are similarly regulated by hydrodynamics. For instance, in *Posidonia oceanica* meadows off Naples,[272] suspended particles show no similarity to seagrass particles except during periods of strong water movement. Pelagic and suspension-feeding consumers may not always be strongly linked trophically to these seagrasses. In contrast, suspended particles in waters overlying *P. oceanica* beds off Corsica — where water movements are not as strong — were found to be mostly of seagrass origin.[273] Such hydrodynamic differences underscore the uniqueness of seemingly identical ecosystems, even over comparatively short distances.

4.7   ECOSYSTEM BUDGETS

The uniqueness of individual seagrass and kelp ecosystems is clear, but this does not obviate the need for systems-level budgets of energy and nutrient flow. Only a few attempts have been made to quantify energy, carbon, or nitrogen flows in seagrass[243,274] and kelp[275,276] ecosystems.

Considering their importance to humans, it is surprising that no complete energy or nutrient budget exists for an entire seagrass ecosystem. The energy flow estimates by Thayer et al.[274] for a California seagrass community remain one of the few early and successful budgeting attempts. Recent information from other seagrass communities suggests that this model (and earlier ones) underestimate the role of bacteria in detritus aging and availability to consumers and the trophic role of epiphytes. However, the California model is still likely to be correct in depicting the macrofauna, particularly polychaetes and bivalves, as being among the major consumers of energy and respiration being the cause of the major loss (48%) of fixed carbon. Roughly 6% of available carbon may be exported, but this figure may be an underestimate considering potential additional carbon input from epiphytes.

A more recent model[243] of carbon and nitrogen flux within a tropical seagrass community highlights the energetic role of periphyton compared to carbon fixed by the seagrasses (Figure 4.11). Nearly all of the periphyton production in this Philippine seagrass bed, which accounts for 63% of the total carbon and 64% of the total nitrogen produced, is shunted through small- to intermediate-sized grazers, with much lesser amounts consumed by sea urchins and juvenile fish. In contrast, only ~23% of living seagrass (as carbon) is consumed directly, with the fate of the remaining carbon being unknown, presumably entering detrital food chains or exported or both. A similar proportion of seagrass nitrogen is unaccounted for, suggesting either that nitrogen is not conserved in this system or that it is conserved within unquantified pathways. This model is preliminary but demonstrates a dichotomy in the nutrient flow pathways of periphyton and seagrass (Figure 4.11).

Similarly, only two preliminary budgets exist for entire macrophyte-dominated, rocky shore ecosystems. Field[277] pieced together a preliminary carbon model for an exposed rocky intertidal community near Cape Peninsula in the Benguela upwelling system, and Hawkins et al.[278] deduced the major carbon flows through exposed, semi-exposed, and sheltered rocky shores on the Isle of Man in the U.K. (Table 4.8).

The South African model indicates that phytoplankton, advected from offshore by tides, longshore currents, and waves, contributes substantially with seaweed detritus to the nutrition of filter feeders, whereas grazers feed mainly on the thin veneer of microalgae and sporelings covering rock surfaces. The filter feeders dominate the biota on this exposed rocky shore, accounting for, on average, 56% of dry-weight biomass; herbivores/detritus feeders and carnivores/scavengers/omnivores account for only 5% and 2% of total biomass, respectively. Phytoplankton contribute 37% to this figure. The intertidal algae (*Ulva, Gelidium, Gigartina,* and *Bifurcaria*) common to this shore are estimated to produce 1100 g C m^{-2} yr^{-1}. The relative amounts of carbon assimilated from seaweed and phytoplankton by filter feeders are not known.

In the Isle of Man model (Table 4.8), the major flux of carbon from producer to macroconsumer shifts with degree of exposure. In sheltered rocky shores, the main route is from fucoids to the gastropod, *Littorina,* and further on to other consumers; microalgae and phytoplankton contribute little compared to macroalgae. From semi-exposed to fully exposed shores, phytoplankton, and to a lesser extent, microalgae, become the major primary producers supplying carbon to higher consumers. These autotrophs are more abundant and productive than macroalgae on hard surfaces pounded by ocean swell.[278]

The flow of energy and carbon through the rocky intertidal is speculative at best, yet it is clear that the rate of water transport is crucial in determining the productivity of these exposed rocky shores. Clearly, much more work is necessary to quantify the energetics of exposed and sheltered, rocky shore communities, including the role of mucus, which has recently been found to be a significant source of carbon for microbes on rocky substrata.[279]

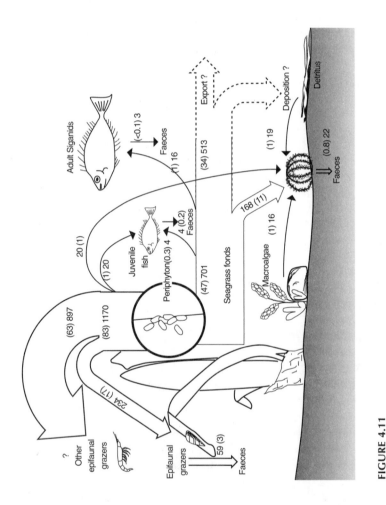

FIGURE 4.11

Carbon and nitrogen flow model for the reef-flat seagrass meadows at Bolinao, The Philippines. Units are mg C m^{-2} d^{-1}; units in parentheses are mg N m^{-2} d^{-1}. (From Klumpp, D. W., Salita-Espinosa, J. T., and Fortes, M. D., *Aquat. Bot.*, 45, 205, 1993. With permission.)

TABLE 4.8

Carbon Flow (g C m^{-2} d^{-1}) Through Food Chains on Sheltered, Semi-Exposed, and Exposed Rocky Intertidal on The Isle of Man, U.K.

	Trophic Transfer		
	Sheltered	Semi-Exposed	Exposed
Level 1			
Fucoids ⇒ *Littorina*	125	8	<0.01
Microalgae ⇒ *Patella*	30	62	75
Phytoplankton ⇒ *Semibalanus*	29	116	460
Level 2			
Littorina ⇒ other carnivores	3	8	<0.01
Patella ⇒ other carnivores	0.2	—	0.4
Semibalanus ⇒ *Nucella*	7.5	2.5	7.5

Source: Data adapted from Hawkins, S. J. et al., *in Plant-Animal Interactions in the Marine Benthos,* John, D. M., Hawkins, S. J., and Price, J. H., Eds., Clarendon Press, Oxford, 1992, 1.

Off the Cape Peninsula rocky shores, extensive kelp beds exist on rocky subtidal reefs along the west coast of southern Africa. These kelp beds within the Benguela system have been examined in sufficient detail to provide us with some of the most complete energy and nutrient budgets yet available for any macrophyte ecosystem.[195,275,276] These budget exercises have led to considerable insights into the relative importance of phytoplankton, macrophytes, and microbes in carbon and nitrogen flow through coastal ecosystems. The major conclusion was that phytoplankton and particulate detritus are the major carbon resources for the dominant filter-feeding consumers because bacteria are inefficient at incorporating carbon derived from particulate kelp detritus and feces.

Figure 4.12 summarizes nitrogen and carbon cycling within a "typical" South African kelp bed. The primary producers are macrophytes, mostly *Laminaria pallida* and *Ecklonia maxima,* and phytoplankton, which are strongly influenced by upwelling and downwelling events. These primary producers exude dissolved organic carbon (step 1) equivalent to 30% of their net production (425 g C m^{-2} yr^{-1}), which is converted into microbial biomass at 65% efficiency. A portion of the kelps fragment into particles which leach DOC at a rate of 318 g C m^{-2} yr^{-1} (step 2) but are converted into microbial biomass at a much lower efficiency. Filter feeders assimilate particles at 50% carbon conversion efficiency (step 3) and expel mucus and excreta (step 4), which are also utilized by bacteria. The total carbon incorporated into bacteria from both the detrital and DOC pathways is 392 g C m^{-2} yr^{-1} (= 99 g N m^{-2} yr^{-1}). Microbial carbon amounts to ~27% of that available for direct consumption and

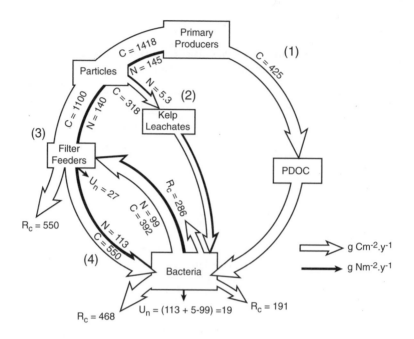

FIGURE 4.12

Relative flux of carbon and nitrogen through a kelp bed community off the Cape Peninsula, South Africa. Numbers in parentheses correspond to those in text to explain stages. R_c = respired carbon; U_n = nitrogen excreted but unaccounted for by bacteria. (Reprinted from Newell, R. C. and Field, J. G., *Mar. Biol. Lett.*, 4, 23, 1983. With permission.)

only 30% of the 1297 g C m^{-2} yr^{-1} required to sustain the macroconsumers, indicating that the major flow of carbon is directly from primary producers to consumers. However, nitrogen is conserved and converted by bacteria at much greater efficiency than carbon. Total microbial incorporation of nitrogen amounts to as much as 59 to 73% of the nitrogen required by the consumers, equating to ~70% of the nitrogen flowing directly from the primary producers.

The differing contributions that phytoplankton (C:N = 5.8) and macrophytes (C:N = 15.6) make to carbon flow are more discernible from Table 4.9. Macrophytes make up 65% of net primary production and also exude a similar proportion of photosynthate as DOC. Bacteria incorporate nearly three times more macrophyte carbon than phytoplankton carbon by detrital pathways; twice as much bacterial carbon originates from macrophytes than from phytoplankton. The carbon contributed from both primary producers and incorporated into microbial biomass equates to 392 g C m^{-2} yr^{-1}.

Carbon and nitrogen appear to be loosely coupled, as nitrogen is conserved by microbes but carbon is readily available to consumers

TABLE 4.9

Differences in Carbon Flow from Phytoplankton and Macrophytes
Through Bacteria in a South African Kelp Bed Community

	Phytoplankton (g C m^{-2} yr^{-1})	Macrophytes (g C m^{-2} yr^{-1})	Total (g C m^{-2} yr^{-1})
Primary production	501	917	1418
DOC at 30% of primary production[a]	150	275	425
DOC incorporation into bacteria[b]	98	179	277
Detritus incorporation into bacteria	32	83	115
Total carbon incorporated into bacteria	130	262	392

[a] Photosynthetically produced dissolved organic carbon averaging 30% of primary production.

[b] Photosynthetically produced dissolved organic carbon (DOC) was incorporated at 65% carbon conversion efficiency based on empirical measurements.

Source: From Newell, R. C. and Field, J. G., *Mar. Biol. Lett.*, 4, 249, 1983. With permission.

directly from both phytoplankton and kelp. Bacteria, therefore, are a relatively more significant source of nitrogen than of carbon owing to their higher efficiency in sequestering nitrogen. Unlike for carbon, nitrogen remineralization is limited in this system, so nitrogen from external sources (e.g., upwelled water) is probably required to sustain primary production.

4.8 CARBON BALANCE: EXPORT AND LINKS TO ADJACENT SYSTEMS

Most mass balance models treat communities as closed systems, ignoring the fact that there may be considerable exchange between adjacent ecosystems. Seagrass meadows and seaweed-dominated rocky shores and subtidal reefs are no exception. The data in Table 4.10 — by no means exhaustive — indicate that many systems show some net export of particulate materials, with the absolute (0.01 to 1.3 g dry weight m^{-2} d^{-1}) and relative (1 to 100% of net plant production) amounts of exported material differing greatly among locations. Few data are available for seaweed-based systems, but sufficient evidence, such as from beach wrack deposits (Chapter 2), exists to suggest that export of seaweeds does commonly occur along many coasts.

The proportion of organic material exported is highly variable, but the data suggest net autotrophy. Because the absolute exchange of dissolved nutrients is unknown for these systems, it cannot be stated with any certainty that they are net autotrophic or net heterotrophic.

TABLE 4.10

Net Exchange of Particulate Materials Between Representative Seagrass
and Seaweed Systems and Adjacent Coastal Areas

Habitat/ Location	Net Exchange	Rate/ Volume (g DW m^{-2} d^{-1})	% of Net Primary Production (NPP)	Ref.
Seagrass/ Virgin Islands	Export	0.02 0.18–0.36	1% (*Thalassia*) 60–100% (*Syringodium*)	292
Seagrass/Fiji	Export	6.96	88% (*Syringodium*)	293
Kelp/South Africa	Export	0.32[a]	10% (mixed kelps)	195
Seagrass/ North Carolina	Export	0.01–0.26[b] 0.23–0.57[b]	1–8% (sheltered *Zostera*) 10–30% (exposed *Zostera*)	294
Seagrass/Florida	Export	0.14–0.38	8–22% (*Syringodium*)	295
Seagrass/ Mediterranean	Export	0.05–0.36[a]	35–55% (exposed *Posidonia*)	245
Embayment/ Philippines	Import Export	0.45 0.76–1.3	8% (mangroves) 6.5–6.7% (mixed seagrasses)	283
Coral Island/ Indonesia	Export	0.08	35.5% (*Cymodocea*) 1.6% (*Enhalus*) 22.9% (*Thalassia*)	293

[a] Units are g C m^{-2} d^{-1}.

[b] Units are g AFDW m^{-2} d^{-1}.

It is likely that most of these systems are roughly in balance between
autotrophy and heterotrophy, with considerable oscillation over time.

As discussed for mangrove forests and salt marshes, net exchange
of material depends upon many factors, such as hydrography, geomorphology, climate, and species-specific characteristics (e.g., growth
rates, buoyancy). There is some indication of enhanced export from
more exposed habitats. A good example is the *Zostera* beds in the
Beaufort estuary in North Carolina, where rates of export are greater
from the exposed (0.23 to 0.57 g AFDW m^{-2} d^{-1}) than the sheltered (0.01
to 0.26 g AFDW m^{-2} d^{-1}) seagrass beds; in relative terms, a greater
percentage of net plant production is also lost from the exposed habitats (Table 4.9).

The extent of export losses may also be genus or species specific.
Some genera, such as *Syringodium* and *Posidonia*, lose a greater proportion of total net primary production than *Zostera*, although it is not
clear why this is so. Species do vary in leaf buoyancy. For instance,
species with leaves with large lacunal spaces, such as *Syringodium
isoetifolium*, are known for their buoyancy. It is likely that the transport
of such relatively buoyant material is affected by climate and tidal
height. Wet tropical habitats, dominated by *S. isoetifolium* and large
tides, would likely export considerably more particulate material than,

for instance, *Zostera* meadows within a microtidal, sheltered lagoon receiving little freshwater input. This scenario is speculative, merely suggesting that differences in physical setting would undoubtedly affect the rate and direction of macrophyte transport with adjacent ecosystems.

In a mass balance model of an Indonesian seagrass community, Erftemeijer and Middelburg[280] suggested that the direction and quantity of net material transfer also depend upon nutrient stoichiometry. An increase in the N:P mineralization ratio greater than the N:P ratio of the seagrass leaf litter implies that an import of dissolved nutrients is required and that the system is net autotrophic, exporting particulate material. Conversely, a decrease in the N:P mineralization ratio below that of the seagrass material suggests that the system requires an import of particulate nutrients and, being net heterotrophic, exports dissolved materials. It is unlikely that such a modeling exercise is applicable to other systems, but it does highlight the importance of the balance between net autotrophy and heterotrophy in relation to nutrient stoichiometry. Their model indicates a rough balance between gross primary production and respiration in the Indonesian seagrass meadows.

Other seagrass and seaweed ecosystems may or may not be in balance, varying with season and the age and composition of autotrophs and heterotrophs. Frankignoulle and Bouquegneau[281] demonstrate, for instance, that *Posidonia oceanica* meadows in the Mediterranean are a net source of carbon from August to November but are a carbon sink from December to July. Some systems may not even be in steady state.

The export of particulate material from seagrass and seaweed ecosystems may be small, on average, but several studies report that the trophic impact of this material is disproportionate to the actual amount exported. Many ecological links do not show up in mass balance calculations. For instance, in coastal food chains off the island of Tasmania, seagrass detritus rather than phytoplankton is the main carbon source for pelagic food webs. Storms transport buoyant rafts of seagrass detritus offshore, where microbes break down the material which serves as food for larvae of the region's major finfish predator, *Macruronus novaezelandiae* (blue grenadier).[282] Also, Fortes[283] provides significant evidence of close trophic and material links between seagrass beds and adjacent mangroves in the Philippines. In Calancan Bay, detritus is exchanged between mangroves and seagrass beds. Mangrove detritus is imported into the seagrass beds, at a rate equivalent to 18% of net seagrass production. Seagrasses, in turn, export detritus to the mangroves, equal to 7% of production. There is also high overlap between mangroves and seagrass beds in the similarity of fish (30%), crustacean (51%), and epiphytic algal (32%) communities, implying close trophic links between these adjacent systems.[283]

The Gazi Bay system in southern Kenya is another tropical ecosystem which shows a broad interdependence among seagrass beds, mangrove forests, and coral reefs.[284,285] Gazi Bay is a semi-enclosed, shallow embayment, 10 km² in area. Mangrove forests, an additional 5 km² in area, line the shore. Seagrass beds are at the center of the bay and cover 7 km². Coral reefs are located ~1 km seaward away from the mouth of the bay. Stable isotope analyses[284] indicate that mangrove-derived particulate organic matter (POM) is exported to the adjacent subtidal seabed. The signature for this material decreases rapidly with distance from the forests, indicating that the seagrass beds most proximate to the forests receive most of the material. Moreover, changes in the $\delta^{13}C$ signature of the seagrass *Thalassodendron ciliatum* parallel the inputs of mangrove material, suggesting that mangrove carbon is assimilated by the seagrass, perhaps by uptake of the CO_2 derived from mineralization of the outwelled mangrove carbon. This transport of organic matter is not unidirectional; on flood tides, the flux reverses with some seagrass-derived carbon flowing back into the forest. Much of the organic material in the seagrass bed is derived from *in situ* production and allochthonous sources. The exchange of materials between mangroves and seagrass beds is driven by tidal currents which, in concert with onshore winds and alongshore currents, promotes the trapping of brackish, turbid water inundating the mangrove forests and seagrass beds but not the coral reefs. The connection between mangroves and seagrasses is through tides and river plumes in the wet season. The seagrass beds and coral reefs are weakly linked, mostly by tidal currents.[284,285]

Mangroves, seagrass beds, and coral reefs co-occur in many other tropical coastal areas and undoubtedly are linked both hydrodynamically and trophically, like those of Gazi Bay. These interconnections need to be examined further, as nearly all coastal habitats are linked to one another to some degree. Coral reefs — often considered to be isolated from other coastal and shelf ecosystems — are discussed in the next chapter.

Chapter 5

CORAL REEFS

5.1 INTRODUCTION

Corals reefs are truly masterpieces of nature, often described as oases in an oceanic desert. This description, while accurate visually, does not reflect the true status of coral reefs with respect to their surrounding environment. Commonly thought to be nearly devoid of plankton and nutrients, the translucent, azure waters bathing coral reefs are, in fact, teeming with microbial life. The concentrations of nutrients are low, but they are processed and recycled rapidly through the microbial food chains. By complex water movements, these microbial communities are swept onto the framework and into food web processes of the coral reef.

The supply of essential nutrients supporting productivity on coral reefs comes from a variety of sources:[296]

- Advection
- Upwelling and endo-upwelling
- Groundwater
- Rain
- Terrestrial runoff
- Guano
- Immigrant organisms

The external origin of these sources attests to the importance of the connections between coral reefs and their immediate environment and their proximity to other coastal and oceanic habitats.

Connections between the various zones of a reef are also vital to the sustainability of an entire reef. Seawater impinging upon and moving within reefs ensures that food webs and material flow within the different zones of a coral reef are interlinked, helping to maintain a balance between production and consumption of energy. Reef geomorphology undoubtedly plays a role in determining the residence time of water and organisms and thus the flux of energy and material transfer within a coral reef. The global distribution of coral reefs reflects the confluence of physical, chemical, and geological conditions in subtropical and tropical seas that are most suitable for reef growth.

Much debate has focused on the contribution of coral reefs to the world's ocean processes and resources, but it is clear that they contribute in a manner disproportionate to their total area. Coral reefs compose only ~0.17% of the total ocean area and roughly 15% of the world's sea floor shallower than 30 m,[297] but they are important habitats for a variety of tropical marine communities — including humans. Coral reefs are fished heavily by local populations that are increasing rapidly, putting further pressure on these fragile ecosystems. In truth, coral reefs are being destroyed at an alarming rate,[14] and many may even disappear early within the next century, ironically obviating the need for so many speculations already discoursed about the possible consequences of the Greenhouse scenario (see Chapter 8).

5.2 SOURCES OF CARBON PRODUCTION

The traditional notion that coral reefs sustain high rates of gross primary productivity is valid, but the contribution of various sources (including net inputs of allochthonous carbon) and variations in productivity among reef zones are still emerging into a coherent picture. Part of the reason for this uncertainty is the difficulty in contending with such a high degree of spatial and organismal diversity. A rich variety of symbiotic zooxanthellae, phytoplankton, seagrasses, and micro- and macroalgae living in and on hard substrata and unconsolidated sediments contribute to production of fixed carbon on a coral reef, aptly described recently by Hatcher[298] as "a beggar's banquet".

A larger contribution is made by hermatypic corals and other animals, such as foraminifera, sponges, bryozoans, molluscs, and echinoderms, to fix inorganic carbon as calcium carbonate (calcification) with help from their photosynthesizing symbiotic zooxanthellae. As the nature of this relationship is currently being debated — yet unquestionably vital for reef construction — this topic will be discussed first, followed by the role of other carbon producers.

5.2.1 Coral Photosynthesis and Calcification

5.2.1.1 Rates and Mechanisms

Current debate rages over the precise relationship between corals and symbiotic dinoflagellates.[299,300] The long-held view — currently accepted

TABLE 5.1

Variations in Gross Photosynthesis (P_G) and Respiration (R)
Among Some Species of Corals at Optimum Light Intensity

Coral Species	P_G (μg C cm^{-2} d^{-1})	R (μg C cm^{-2} d^{-1})	P_G /R (μg C cm^{-2} d^{-1})
Acropora cervicornis	119	72	1.6
Acropora diversa	375	115	3.3
Acropora palmata	231	267	0.9
Millepora tenera	81	113	0.7
Montastrea annularis	152	139	1.1
Montastrea cavernosa	58	73	0.8
Pocillopora damicornis	52	63	0.8
Pocillopora eydouxi	165	79	2.1
Porites porites	375	103	3.6
Stylophora pistillata	137	118	1.2
	105	94	1.1
	137	180	0.8
Turbinaria reniformis	112	163	0.7

Source: Adapted from Sorokin, Y. I., *Coral Reef Ecology*, Springer-Verlag,
Berlin, 1995.

by most reef scientists and for which substantial data exist — is that
coral calcification is dependent upon the metabolic processes of photosynthesis by the endosymbionts.[300] The corals acquire carbon fixed
by algal photosynthesis (glycerol and lipids) for metabolic needs, and
the symbionts take up inorganic nutrients released as metabolic waste
products by the coral. Densities of zooxanthellae in coral are normally
within a narrow range of 1.0 to 5.0 $\times$ 10^6 cells cm^{-2} of coral surface.[296]
Maximal rates of gross primary production and respiration for
zooxanthellate corals at optimal illumination are very variable among
species (Table 5.1). The wide range of production to respiration ratios
indicates that some corals are net autotrophic, while others are net
heterotrophic. A mean rate of gross primary production or respiration
must therefore be considered very cautiously. Both Smith[301] and
Kinsey[302] suggest a mid-range value for gross primary production of
25 g C m^{-2} d^{-1}, a value considerably greater than the range (0.5 to 14 g
C m^{-2} d^{-1}) suggested by Muscatine,[303] but lower than rates reported for
some corals, such as for *Pocillopora damicornis*.[304]

A few studies have shown that calcification rates increase coincident with increases in photosynthesis in relation to available light.
Estimates of gross carbonate production usually range from 3 to 6 kg
CaCO$_3$ m^{-2} yr^{-1}, with some workers reporting extreme rates as slow as
1 kg CaCO$_3$ m^{-2} yr^{-1} and as fast as 35 kg CaCO$_3$ m^{-2} yr^{-1}, respectively.[300]
These calcification rates translate into accretion rates of from 1 to 7 mm
yr^{-1}, with a mean of 3 mm yr^{-1}.

Photosynthesis can enhance calcification by:

- Providing an energy source or substrate
- Taking up CO_2

Calcification is a process that requires energy, but attempts to stimulate calcification by supplying exogenous photosynthate have failed. Uptake of carbon dioxide during photosynthesis would shift the carbonate equilibrium to favor precipitation. However, the nature of the reaction

$$Ca^{2+} + H_2O + CO_2 \Leftrightarrow CaCO_3 + 2\ H^+ \qquad (5.1)$$

indicates that removal of CO_2 leads to calcification but no further carbonate precipitation, unless protons created by the reaction are either removed or neutralized. The protons can be neutralized by ionic compounds, perhaps accumulated as a metabolic waste product (e.g., ammonium). The problem with this mechanism is that uptake of carbon dioxide during photosynthesis favors calcification but resists further CO_2 uptake as pH increases shift the carbonate equilibrium. As we have seen for seagrasses (Section 4.4.1), photosynthesizing algae, including zooxanthellae, may catalyze bicarbonate using carbonic anhydrase in the reaction:

$$2\ HCO_3^- + H_2O + CO_2 \Leftrightarrow CaCO_3 + 2\ H^+ \qquad (5.2)$$

which requires removal or neutralization of the hydroxide formed when carbon dioxide is released or uptake of a proton; such a balance may be achieved by input of another bicarbonate ion into the calcification reaction.

The ratio of carbon incorporated into skeleton to that fixed into photosynthate varies among corals and calcareous algae but usually favors photosynthesis by a factor of 4 to 8. Calcification, therefore, is not stoichiometrically linked to photosynthesis. There is some evidence to suggest that hermatypic corals can calcify in the dark, although very slowly; the evidence for dark calcification in coralline algae is ambiguous.[300]

5.2.1.2 Limitations

Calcification and photosynthesis in zooxanthellate corals cannot be considered in isolation from the effects of changes in light, nutrient concentrations and coral heterotrophy and may be limited by such factors. Dubinsky and Jokiel[305] offer a scenario to explain the main interactions and feedback mechanisms among light, nutrients, and

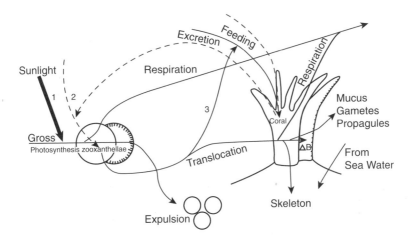

FIGURE 5.1
Graphic of interactions between carbon and nitrogen fluxes in zooxanthellate corals.
Solid arrows represent carbon flow; dashed lines represent nitrogen flow. Pathways 1
and 2 are the main external forcing functions; Pathway 3 is an internal feedback loop.
Shaded areas represent growth in biomass. (From Dubinsky, Z. and Jokiel, P. L., *Pac.
Sci.*, 48, 313, 1994. With permission.)

feeding in zooxanthellate corals (Figure 5.1). In a milieu of constant
nutrient levels, light intensity regulates nutrient limitation (as light
intensity increases, C:N ratios surpass the Redfield ratio of 6.6 to 1).
Nitrogen availability determines the fate of photo-assimilated carbon;
under high C:N ratios, most carbon is respired, excreted as mucus,
and/or is shunted into calcification. At low C:N ratios, translocation
and calcification are reduced, and density of zooxanthellae increases.
Under low light, feeding on zooplankton supplies carbon for metabolism, and under high light, coral feeding provides nitrogen to both
algae and animal. Under normal environmental conditions, zooxanthellae are closed systems in terms of nitrogen, and their growth is not
balanced with respect to fixed carbon; excess carbon is translocated to
the host. However, if the nitrogen supply is increased, the flow of
translocated carbon is reduced as the zooxanthella growth rates increase. The Dubinsky and Jokiel scenario[305] suggests that a low nutrient milieu is required to maintain a close, balanced association between algal symbiont and animal host.
    Zooxanthellate corals respond differently, however, to nitrogen
and phosphorus enrichment. Several studies have shown that additions
of dissolved inorganic nitrogen (DIN) increase zooxanthella densities,
but the effects of dissolved inorganic phosphate (DIP) enrichment are
more complicated. In a series of enrichment studies on the reef coral
Pocillopora damicornis, Snidvongs and Kinzie[306] found that additions of

DIN increased zooxanthella densities and nitrogen content, but not carbon content. Increased CO_2 demand by the larger zooxanthellae populations may have created CO_2 limitation for the corals. The addition of DIP resulted in a decline in calcification and a decrease in carbon and phosphorus cell content of the zooxanthellae. This would in turn lead to reduced availability of carbon dioxide in coral tissue. Why phosphorus would result in decreased calcification is not known.

It is generally accepted that carbon is not limiting to zooxanthellate corals considering the large amounts of dissolved inorganic carbon in seawater. However, recent studies[307,308] on the boundary layers surrounding coral tissues and their micro-environment indicate that availability of dissolved inorganic carbon may at times be diffusion limited and greatly affected by water motion and coral morphology. In a series of microcosm experiments,[307] individuals of the coral *Pocillopora damicornis* were subjected to different flow velocities. The experiments showed that the coral exhibited significantly increased enzyme activity, photosynthesis, and respiration with increasing water velocity. Under different low-flow regimes, the coral skeleton changed morphologically in such a way as to minimize boundary layer thickness and to maximize carbon delivery. Coral with fixed skeletal morphology, when exposed by different flow velocities, exhibited physiological and biochemical plasticity to facilitate carbon transport for assimilation.

Micro-electrode measurements[308] have shown that boundary layer thicknesses can vary and may result in O_2 limitation and depletion affecting photosynthesis and respiration. Gradients in gases and dissolved materials may therefore exist within a scale of millimeters to several centimeters. Gladfelter et al.[309] observed significant variations in primary production, respiration, and zooxanthellae cell growth rates within different parts of a colony. Along a branch of the reef coral *Acropora palmata*, they found significantly higher rates of respiration and lower rates of primary production proximal to the branch tip compared to distal portions; cell division of zooxanthellae was more rapid closer to the tip, but densities were greater with distance from the tip, underscoring the high spatial heterogeneity even at micro- and mesoscales.

Recent experiments with hermatypic and ahermatypic corals suggest that instead of enhancing coral growth, the symbiotic algae may be suppressing calcification.[310] Working on the accepted notion that the cause-and-effect relationship between corals and their endosymbiots is not clear, Marshall[310] compared the calcification rates of polyps of the hermatypic coral *Galaxea fascicularis* and the non-reef-building coral *Tubastrea faulkneri* by measuring uptake of ^{45}Ca over a 4-hr period. He found that under normal daytime conditions, both corals incorporated the labeled calcium at equivalent rates of ~0.6 μmol g^{-1} hr^{-1}. A second series of experiments carried out for 6.5 hr and at higher temperature, with and without the addition of photosynthetic inhibitor, found that calcification rate declined significantly in the hermatypic coral with the inhibitor, but there were no treatment effects for the ahermatypic coral. Marshall[310] found no difference in uptake of labeled Ca by *T. faulkneri* in the light or dark; dark calcification rates for *G. fascicularis* were significantly slower compared to calcification in the daytime. In a final series of experiments, dark incorporation of ^{45}Ca by the hermatypic coral was confined to tissues devoid of symbionts.

Marshall[310] hypothesized that during the day the symbiotic algae manipulate calcification by increasing the proton supply needed to convert bicarbonate into CO_2 for photosynthesis. At night, the algae shut down to conserve the coral's metabolic energy. How the algae do this is unknown, and more experiments with other species are necessary to confirm these controversial findings. Moreover, the rates of calcification need to be compared on both a weight-specific and surface area-specific basis to fully delimit differences between reef-building and ahermatypic corals. These results are contrary to the established beliefs regarding the coral host-algal symbiont relationship and beg the question: why are hermatypic corals successful, and ahermatypic corals unsuccessful, at building reefs?

5.2.2   Free-Living Primary Producers

Carbon is fixed on coral reefs by a variety of benthic autotrophs on hard and soft substrata and by pelagic primary producers. Table 5.2 summarizes the rates of gross and net primary productivity and respiration, by zones and functional groups, on coral reefs. Epilithic algal communities — composed mostly of small filamentous green and red turf algae, cyanobacteria, crustose coralline algae, and macroalgae — grow rapidly and cover large areas of hard substrates, often accounting for most of the primary productivity on many reef flats. Their biomass is normally low because of intense grazing pressure, but their net production rates are high (Table 5.2), varying with season and location.

It is a common misconception that the structure and function of communities on coral reefs vary greatly over space but comparatively little over time, an idea perhaps misconstrued from the obvious spatial heterogeneity and apparent constancy of sea-surface temperatures, solar insolation, and oceanic nutrient concentrations in low latitudes. That this notion is false is demonstrated by the epilithic algal community (EAC). On Davies Reef, the Great Barrier Reef, EAC composition varies both among reef zones (being most abundant on the reef flat) and over

TABLE 5.2

Ranges of Gross (P_G) and Net (P_N) Primary Productivity ($g\ C\ m^{-2}\ d^{-1}$) and Production-to-Respiration (P:R) Ratios

	P_G	P_N	P:R Ratio
Major Reef Zones and Entire Reefs			
Ocean water (n = 8)[a]	?	0.01–0.65	?
Outer reef slope			
Fore-reef (n = 4)	2.0–7.0	–1.0–5.1	0.5–5.5
Reef flat			
Reef crest (n = 2)	2.0–7.0	0.3–1.5	1.0–4.0
Back reef (n = 25)	2.6–40.0	–1.7–27.0	0.7–3.2
Lagoon			
Sand, shallow patch reefs (n = 8)	0.9–12.9	–0.5–3.4	0.7–1.4
Water (n = 21)	0.01–2.0	–1.3–1.4	0.1–1.4
Entire reef (n = 5)	2.3–6.0	–0.01–0.17	1.0

	P_G	P_{Nsp}	P:R Ratio
Producer Group[b]			
Symbionts			
Corals (n = 10)	0.77–10.2	8.0–40.0	0.5–5.0
Endo- and epilithic algae			
Corallines (n = 7)	0.8–2.8	0.06–11.7	1.0–5.4
Turfs (n = 13)	0.9–12.1	17.0–280.0	1.2–6.7
Macroalgae (n = 6)	2.3–39.4	2.5–118.0	1.2–6.3
Epipelic and rhizobenthic assemblages			
Microalgae (n = 9)	0.08–3.7	?–363.0	1.1–10.3
Seagrasses (n =5)	3.0–16.0	4.0–8.8	1.5–2.5
Macroalgae (n = 5)	?– 4.0	0.2– 40.0	1.9–2.8

[a] n = number of empirical measurements.

[b] P_{Nsp} is weight-specific primary productivity in mg C dry weight d^{-1}.

Source: Data from Hatcher, B. G., *Trends Ecol. Evol.*, 3, 106, 1988.

time[311] with turf algae accounting for 60 to 80% of total algal cover on the reef flat and slopes. Crustose coralline algae dominate the fore- and back-reef crests and decrease away from the crest; encrusting brown algae and cyanobacterial mats are minor contributors to total algal cover. There were no clear seasonal patterns of algal biomass, but EAC productivity was >1.5 times greater in summer than in winter, with maximum net productivity of 3 g C m^{-2} d^{-1}. On average, the epilithic algal community contributed 28% to total reef flat productivity and accounting for all net production. Other studies[312] have similarly found that the EAC contributes greatly to reef productivity, with differences among locations attributed to factors such as grazing pressure.

The contribution of benthic microalgal communities can be significant in reef zones in which soft sediment accumulates, such as on the leeward side of coral substrata and in lagoons.[313] Rates of gross primary production by benthic microflora in reef sediments can range from 0.08 to 3.7 g C m^{-2} d^{-1}, with respiration rates ranging from 0.1 to 3.8 g C m^{-2} d^{-1}. These communities, consisting of diatoms, chlorophytes, phytoflagellates, cyanobacteria, and some zooxanthellae living in symbiotic foraminiferans, attain maximum biomass as algal mats covering unconsolidated lagoonal sediments.

A variety of floating and rooted macrophytes also contribute to reef primary productivity, although their presence is highly variable from zone to zone and from reef to reef, controlled by hydrodynamic forces, temperature, and nutrient availability. Seagrasses can form thick mats in shallow, sandy lagoon bottoms or on reef flats, whereas some seaweeds can attach to hard outcrops on reef flats, including on dead coral heads.[313] In terms of areal coverage macrophytes, on average, can cover from 20 to 30% of lagoon floor, 20 to 80% of reef flats, and 2 to 20% of outer slopes; biomass and productivity are in direct proportion with these percentages.[298] Rates of gross primary productivity and respiration of these larger plants (Table 5.2) are identical to those measured in other habitats (Chapters 3 and 4), so their contribution to the productivity of an individual reef will depend greatly upon their colonization success and areal coverage. They tend to be more prevalent on high-latitude reefs, where higher nutrient concentrations and more variable temperatures and light intensities competitively favor the growth of macrophytes.

Early estimates of phytoplankton biomass and production in coral reef waters indicated little contribution to reef primary productivity, which is in agreement with other studies showing that phytoplankton communities in tropical waters are of low biomass and unproductive.[298] Improvements in methodology and a paradigm shift emphasizing the importance of the microbial loop have resulted in a more accurate depiction of the role of phytoplankton in coral reef waters. As in most other tropical waters, size fractionation has shown that pico- and nanoplankton usually dominate the phytoplankton community in coral reef lagoons.[314] In several reefs within the Great Barrier Reef, picoplankton (0.2 to 2.0 μm in cell size) account for 34 to 65% and 47 to 69% of the total phytoplankton standing crop and primary productivity, respectively, whereas nanoplankton contribute 13 to 32% of standing crop and 15 to 38% to phytoplankton production.[314] Net plankton increase in relative importance in summer. Total phytoplankton production ranges from 0.2 to 1.6 g C m^{-2} d^{-1}, correlating inversely with flushing rates and residence times of water in the lagoons,[315] and is greater within semi-enclosed lagoons than in surrounding open-ocean water, suggesting enhanced uptake and utilization of reef-derived nutrients by pelagic autotrophs.

Blooms of phytoplankton occur during calm periods[314] and after the passage of cyclones. Delésalle et al.[316] observed that after the passage of a cyclone over the barrier and fringing reefs of Moorea Island in French Polynesia, phytoplankton production and biomass was stimulated either by nutrient enrichment from resuspended sediment or terrestrial runoff. The composition of the phytoplankton community shifted from the dominance of small-size classes to diatoms, which are competitively superior to smaller groups under high nutrient loads. Moreoever, the diatom bloom was followed by a noticeable increase in the growth of benthic green macroalgae.

It is clear that primary productivity of coral reefs varies greatly over time and space.[298] Improved methods and additional measurements in reef zones other than on reef flats have shown that differences among reef slopes, crest, fore- and back-reef areas, lagoons and overlying water are significant (Table 5.2). This led Kinsey[302] to identify coral/algal assemblages, algal pavement, and sediment autotrophs as the three main centers of production, with grand mean rates of gross primary production of 20, 5, and 1 g C m^{-2} d^{-1}, for each of these respective functional groups. Hatcher[298] was unable to delimit differences in total reef productivity with latitude and proximity to land but noted that the contribution of coral/algal production decreases and production from phytoplankton and macrophytes increases with increasing latitude and increasing proximity to land. Rates of gross primary productivity tend to remain within the range of 5 to 20 g C m^{-2} d^{-1}, despite the fact that productivity can be highly seasonal and follow daily changes in light.

Major controls on reef primary production and biomass include:

- Grazing
- Light
- Structural complexity and physical stability of the substrata
- Nitrogen and phosphorus concentrations (and their availability as affected by water residence times)
- Local hydrodynamics
- Whole gamut of interactions/factors that determine the areal extent and composition of coral/algal cover

The regulatory control each of these factors exert individually on reef production are well understood, but synergistic-antagonistic relationships among them are not. Physical disturbance has more recently

TABLE 5.3

Macro-Herbivores on Coral Reefs Categorized
as Functional Groups

Functional Groups	Specific Groups
Scraping (excavating)	Scarids, echinoids, limpets, chitons
Denuding	Acanthurids, signids, pomacentrids (some), blennies, kyphosids
Non-denuding	Pomacentrids, gastropods (excluding true limpets), amphipods, polychaetes

Source: Adapted from Steneck, R. S., Proc. *Sixth Conf. Coral Reef Symp.*, 1, 37, 1988.

been recognized as playing a regulatory role. More inter- and multi-disciplinary studies are needed to better understand how these factors operate simultaneously, or in sequence, to regulate primary productivity. Variations in primary productivity subsequently lead to spatial and temporal heterogeneity in consumers and the fate of this fixed carbon.

5.3 THE FATE OF ORGANIC MATTER

Various mass balance and energy flow models (Sections 5.5 and 5.7) indicate intense consumption and recycling of dissolved and particulate organic matter within reefs, with small (but very variable) amounts of material available for net export. Early food chain work focused on the more conspicuous swimming and epibenthic metazoans, particularly herbivores. More recent emphasis on detritivory and microbial ecology demonstrates that detritus-based food webs are well developed on coral reefs, as little, if any, organic matter remains unconsumed for long.

5.3.1 Herbivory, Carnivory, and Mixotrophy

Three functional groups of herbivores on coral reefs have been identified by Steneck (Table 5.3):[317]

- Scrapers
- Denuders
- Non-denuders

Scrapers, particularly deep grazers such as sea urchins and parrotfishes, have the greatest impact on reef algae, feeding on a wide array of algal

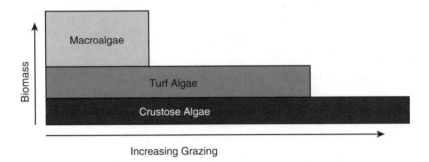

FIGURE 5.2
Idealized view of shift in algal community composition induced by grazing herbivores.
(Adapted from Steneck, R. S., *Proc. Sixth Coral Reef Symp.*, 1, 37, 1988.)

groups. Denuding herbivores — mainly other fishes and some gastropods — can significantly reduce fleshy algae but do not feed on crustose coralline algae or leathery or articulated forms. Non-denuders
include some pomacentrid fishes, gastropods, amphipods, and polychaetes, and these organisms generally do not severely graze down
algal biomass. Steneck's analysis[317] includes the more obvious herbivores, not smaller organisms, such as algal-feeding meiobenthos, zooplankton, nonselective suspension feeders, and some protozoans, all of
which, on a much smaller or local scale, can significantly graze down
small and larger autotrophs. Nevertheless, most of these smaller organisms can be considered as non-denuders.

Algal communities are heavily grazed by one or more of these
groups on most shallow (≤10 m) reefs. Parrotfishes, surgeonfishes, and
sea urchins are usually the most abundant herbivores and are capable
of severely grazing down reef algal biomass and shifting community
composition.[318,319] Herbivory tends to be most intense just below mean
low water where strong wave action is ameliorated to permit feeding.
Under intense grazing, algal communities may be characterized by
low biomass algal turfs and some crustose algae. Zonation patterns of
herbivory have been identified, but consistent patterns are lacking or
obscured on reefs with large tides, those in proximity to land, or reefs
that are nutrient enriched.[320]

Manipulative experiments consistently reveal that the composition of reef algal communities shifts from macroalgae to turf algae and
from turf algae to crustose algae, with increases in intensity of grazing
(Figure 5.2). Increased grazing from scraping and denuding herbivores
— regardless of species— also results in decreases in algal biomass and
diversity, but usually increases in rates of production. The impact of
grazing is related to intensity and frequency of feeding.

Herbivory on coral reefs is often studied from a population or
community ecological perspective, in which competitive effects among

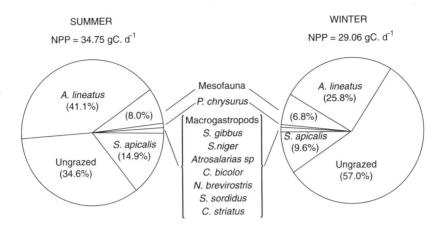

FIGURE 5.3
Pie-charts of the percentage of benthic algal net primary productivity (as carbon) consumed by different grazers from the *Acanthurus lineatus* zone in summer and winter, Davies Reef, Great Barrier Reef. (From Polunin, N. V. C. and Klumpp, D. W., *J. Exp. Mar. Biol. Ecol.*, 164, 1, 1992. With permission.)

herbivores are often stressed and grazing assessed using predator-prey models.[321] Shifts in algal community structure, biomass, or productivity resulting from grazing pressure are often measured,[322,323] but the impact of herbivory on energy or nutrient flow or the energetics (ingestion and assimilation rates) of individual herbivores have rarely been assessed. This is unfortunate given the impressive impact herbivores have on reef algal communities. Such intensely grazing organisms must have an ensuing impact on other food web members, such as detritivores, and the bacteria which undoubtedly consume and recycle substantial amounts of excreta derived from these algal consumers.

The significance of large grazers to energy and nutrient flow on coral reefs has only recently been examined.[322–325] Polunin and Klumpp[325] estimated the grazing demand by herbivores in terms of their consumption of benthic algae on the windward crest of Davies Reef, Great Barrier Reef. They found that grazing was intense in winter and more so in summer. Net primary production increased with the change in season from 29.06 to 34.75 g C d^{-1}, but grazing increased by a much greater proportion, accounting for 65% of net primary production in summer and 43% in winter. It is clear that on the windward reef crest, the surgeonfish *Acanthurus lineatus* and the damselfish *Stegastes apicalis* dominate the herbivorous grazing community; algal consumption by populations of both species together account for ~56% of the net primary production in summer and ~35% in winter (Figure 5.3). Mesograzers such as amphipods, copepods, molluscs, polychaetes, and other crustaceans plus large gastropods consume algal biomass equivalent to ~9% and ~8% of net algal production in summer and winter, respectively. There was significant export of unconsumed algae to other reef zones, implying that this material is either consumed in the water column or incorporated into detrital food chains. It was suggested that secretion of dissolved organic matter (DOM), microbial decay, and export of particulate algal material account for an unknown proportion of the unconsumed algal crop.

It is difficult to quantify the significance of mobile epibenthic and infaunal grazers as consumers of autotrophic standing stocks because of their omnivorous dietary habitats, methodological problems of separating true grazing from disturbance effects, and the often stimulatory effects of benthic feeding activities which often results in no discernible grazing impact. How much algal carbon is consumed by benthic epifauna and infauna is also a function of their abundance which may be greatly affected, in turn, by interactions with other organisms.[313] For example, Alongi[326] found very low densities of algal- and bacterial-feeding nematodes in sediments of the Davies Reef lagoon, ascribed to deleterious impact by the intense bioturbation activities of thalassidean shrimps. In an earlier review, Alongi[313] concluded that these shrimps, common in carbonate sediments, are the major determinants of autotrophic and heterotrophic benthic community organization and abundance in many reef lagoons.

Herbivory among other reef invertebrates is well established.[317-320] Sessile invertebrates such as sponges and soft corals (particularly those that are asymbiotic) derive most of their nutritional needs by straining food particles from the overlying water. Zooplankton, detritus, and associated bacteria are food items, but, until recently, phytoplankton as a food source for asymbiotic sessile organisms has been ignored. A recent study[327] has found that some asymbiotic soft corals found in the Red Sea feed almost exclusively on phytoplankton. Fabricius et al.[327] found that the common soft coral *Dendronephthya hemprichi* ingested phytoplankton cells in response to different rates of water flow (Figure 5.4). These results demonstrate how often-stated suppositions are unsupported and frequently simplistic. The often-observed decline in phytoplankton abundance across coral reefs may signify more intense herbivory by a greater diversity of organisms than currently believed.

Reef organisms display a variety of feeding modes, including uptake and assimilation of dissolved organic matter. As we have observed for other marine ecosystems, most pelagic and benthic animals eat a variety of foods in order to maintain a balanced diet. A given organism can, at most times, be a herbivore but may supplement its diet by turning carnivorous or using DOM to obtain vital essential elements lacking in algal tissue. Cnidarians and poriferans are particularly adept at defying classification of their trophic status (hard corals

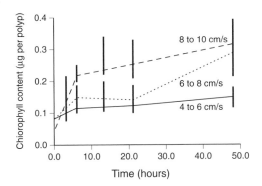

FIGURE 5.4
Influence of water-flow rates on the soft coral *Dendronephthya hemprichi* feeding on phytoplankton in the laboratory after a period of starvation. (From Fabricius, K. E., Benayahu, Y., and Genin, A., *Science*, 268, 90, 1995. With permission.)

are discussed in Section 5.3.3). Depending on location within a coral reef, some soft corals exhibit mixotrophy, depending upon both phototrophy and heterotrophy to supply their nutritional require-ments.[328,329] The energetic costs of growth, reproduction, and mucus production for soft corals are virtually unknown, but it is clear that under normal light conditions, phototrophy is insufficient to meet their respiratory requirements. Clearly, more information is needed in regard to the role of mixotrophy in coral reefs.

The Like herbivory and mixotrophy, the effect of carnivory on energy and nutrient flow in coral reefs is not clear. Large and small predators, especially fishes, exist in great abundance on most reefs and are un-doubtedly important to the trophic organization of reef food webs. Conversely, the rich diversity and abundance of life on coral reefs attracts and sustains many large and solitary oceanic predators that may not survive without them.[330]

The trophic importance of top predators to the functioning of coral reefs is likely to be disproportionate to their energetic significance, considering the attenuation of energy from lower to higher trophic levels; however, as observed for soft corals, such assumptions remain untested. For instance, the carbon flux model of Johnson et al.[331] im-plies that carnivory accounts for ~10% of total carbon throughput on Davies Reef. Moreover, while top predators feed at higher trophic levels, they also feed at many lower trophic levels such that the "aver-age" trophic level of feeding can be low. Hamner et al.[332] proposed, for instance, that planktivorous fishes on the windward face of a reef form a "wall of mouths" that removes most of the zooplankton from the water prior to its impingement onto the reef. The point of both of these studies is that, until recently, carnivory has been underestimated on coral reefs and may be responsible for a significant proportion of energy and carbon flux through food webs.

5.3.2   Detritus and Detritivory

5.3.2.1 Detritus Sources and Fluxes

The role of detritus and detritus-based food chains on coral reefs was virtually ignored until the early 1980s.[333] Much of the delay in recognizing the energetic importance of detritus in coral reef ecosystems stems from our belated ignorance of microbial food webs and biogeochemistry and the fact that most of the efforts to construct mass balance models have centered on the shallower and highly productive reef crest and reef flat zones, rather than on the deeper, more quiescent lagoons.

The bulk of detritus on coral reefs is produced *in situ*, derived mostly from:[334]

- Epilithic algal communities
- Sediment microalgae and cyanobacterial mats

and secondarily from:

- Benthic and floating macrophytes
- Filamentous endolithic algae
- Encrusting coralline algae
- Phytoplankton
- Symbiotic zooxanthellae
- Mucus secreted by corals and other reef invertebrates

On some coastal reefs, there may be allochthonous inputs via groundwater seepage and river runoff. According to Hatcher,[333] from 10 to 80% of net primary production may remain ungrazed by herbivores, implying a significant supply of detritus within most coral reefs. The considerably wide range in Hatcher's estimate[333] reflects variations not only in rates of primary production and consumption among reefs, but also variations as influenced by reef geomorphology and hydrography. These differences induce changes in water motion and residence time, in synchrony with local and regional differences in sea level, tides, currents, and other hydrographical characteristics.

Despite the obvious ecological implications of physical processes, rarely have temporal and spatial heterogeneity in nutrient and detritus standing stocks and fluxes been examined concurrently with hydrographic conditions on a coral reef.[334] Water motion within a coral reef can be exceedingly complex. For instance, Wolanski[335] observed boundary mixing at lagoonal surfaces within Davies Reef lagoon. This mixing may account for all vertical transport of water and suspended matter, and a sharp pycnocline separating water layers was detected which may inhibit settling of particulate material to the lagoon bottom. This phenomenon may result in longer residence time of detritus in the water column, facilitating decomposition. Furthermore, the availability and potential limitation of nutrients on coral reefs appear to be a partial function of their residence times as determined by relative rates of water exchange.

The distribution and abundance of benthic and pelagic communities inevitably mirror the distribution and availability of essential dissolved and particulate nutrients. Other factors such as light and temperature play a role, but zonation patterns of organisms have been primarily attributed to:

- Gradations in sediment granulometry
- Aggregations of essential nutrients
- Rates of water turbulence and current flow
- Detritus deposition

Variations in water motion have complicated attempts to measure rates of detritus transport and deposition within coral reefs. The data indicate very high spatial and temporal heterogeneity in rates of detritus flux into reef lagoons.[333,336-339] At One Tree Reef in the southern Great Barrier Reef, Koop and Larkum[336] measured an average rate of detritus deposition into the lagoon of 1500 mg C m^{-2} d^{-1}. This rate is high, attributable to the shape and shallowness of the lagoon (~6 m). The lagoon is fully enclosed, which may result in efficient retention and longer turnover time of detritus. High rates of deposition and suspended POC concentrations have also been measured in enclosed lagoons in a Tuamotu atoll[337] (mean = 350 mg C m^{-2} d^{-1}) and in the southwest lagoon of New Caledonia[338] (mean = 844 mg C m^{-2} d^{-1}).

Rates of detritus deposition appear to be lower in more open lagoons. In the semi-enclosed lagoon of Davies Reef, Hansen et al.[339] measured detritus deposition at shallow sites immediately behind the reef flat and at deeper sites further within the back lagoon. Large day-to-day variations were observed (Figure 5.5) to the extent that no significant differences could be observed among sites, but there were seasonal differences. Rates of deposition were slow in summer (mean = 9 mg C m^{-2} d^{-1}) and rapid (mean = 141 mg C m^{-2} d^{-1}) in winter, in agreement with earlier studies of the epilithic algal communities that

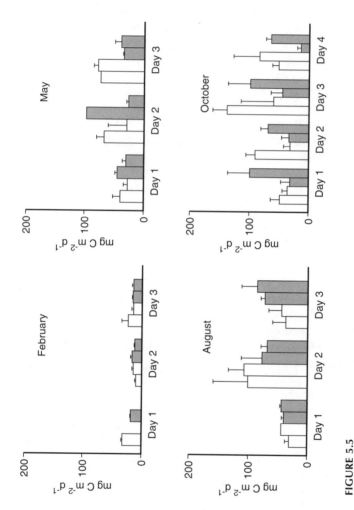

FIGURE 5.5

Day-to-day variations in organic carbon deposition into Davies Reef lagoon at shallow (open bars) and deep (shaded bars) sites in summer (February), autumn (May), winter (August), and spring (October). (From Hansen, J. A. et al., *Mar. Biol.*, 113, 363, 1992. With permission.)

TABLE 5.4

Proportion of Reef Flat Net Primary Production Reaching the Lagoon
as Detritus in Winter and Summer at Davies Reef, Central Great
Barrier Reef

	Winter	Summer
Algal net primary production (mg C m^{-2} d^{-1})	1157	1851
Detritus deposition rate (mg C m^{-2} d^{-1})	67.5	14
Algal net primary production (mg C m^{-1} d^{-1})	521	833
Detritus deposition rate (mg C m^{-1} d^{-1})	204	42
Percent algal net primary production deposited in lagoon (d^{-1})	39	5

Note: The lower two sets of values are expressed as lineal rates.

Source: From Hansen, J. A. et al., *Mar. Biol.*, 113, 363, 1992. With permission.

indicated more intense grazing and less algae available for export in
summer (Figure 5.3). These rates were much less than those measured
using similar type sediment traps at One Tree Reef, indicating the
potential effects of differences in water residence times between enclosed and semi-enclosed lagoonal systems.

Much of the material deposited into reef lagoons is likely to be
derived from turf algae on the reef flat as a result of fragmentation,
erosion, grazing, and export as fecal material. Most zooplankton advected onto reefs are consumed at the reef front,[332] and coral mucus is
likely to be a minor contributor given the generally slow rates of
mucus production by corals.[340] Carbon may also be imported from
waters impinging on the reef, but no data are available to quantify this
potential route.

The proportion of net primary production on the reef front transported as detritus to the lagoon varies seasonally — on Davies Reef, for
example, from 5 to 39% (Table 5.4). In summer, net primary productivity is higher than in winter, but grazing is more intense, resulting in
less algal material available for export. Estimates on a lineal basis
indicate that only 5% of the daily reef flat net primary production
deposits in the lagoon. The pattern is different in winter, as proportionally much more algal-derived carbon settles into the lagoon (Table
5.4). It is not known to what extent this pattern and rates of material
flux are applicable for other reefs as such measurements are lacking.

5.3.2.2 The Role of Sediment Bacteria

How important are bacteria in consuming detritus on coral reefs?
Again, the Davies Reef studies[339,341] provide a glimpse of the extent to
which detrital and fresh algal carbon is utilized by microbes in reef
lagoons. The sum of carbon input from detritus deposition and gross

TABLE 5.5

Comparison of Rates (mg C m^{-2} d^{-1}) of Detritus Deposition, Benthic
Gross Primary Production (GPP), Bacterial (P$_B$) Production, and
Sediment Respiration (R$_S$) over Four Seasons at Two Lagoonal
Habitats, Davies Reef

Season	Habitat	Detritus Deposition	GPP	P$_B$	R$_S$
Summer	Shallow	16 ± 3	373 ± 20	105 ± 9	458 ± 23
	Deep	12 ± 1	320 ± 37	177 ± 16	331 ± 36
Autumn	Shallow	58 ± 12	365 ± 27	121 ± 8	389 ± 29
	Deep	73 ± 34	193 ± 9	96 ± 3	208 ± 22
Winter	Shallow	69 ± 15	98 ± 17	81 ± 3	133 ± 14
	Deep	66 ± 7	32 ± 6	46 ± 3	84 ± 13
Spring	Shallow	78 ± 17	161 ± 16	240 ± 19	214 ± 35
	Deep	72 ± 19	89 ± 11	238 ± 32	119 ± 15

Note: Each value is a mean ± 1 standard error.

Source: From Hansen, J. A. et al., *Mar. Biol.*, 113, 363, 1992. With permission.

primary production virtually balance respiration (Table 5.5). This indicates that most of the organic material reaching the lagoon is mineralized rapidly. Based on estimates of surface-sediment bacterial productivity and assuming a carbon conversion efficiency of 50%, bacteria could consume from 54 to 100% of the available carbon. Boucher and Clavier[343] came to a similar conclusion for microbes dominating energy flow within the lagoon sediments in New Caledonia.

These estimates for Davies Reef are, however, at odds with the modeling exercises of Johnson et al.,[331] which imply that the proportion of the total carbon pool fluxing through bacteria is small. Further, their Davies Reef model suggests that detritivores and filter feeders account for much more detrital carbon flux than do bacteria. These divergent views may be partly reconciled if one considers (1) that bacterial productivity was estimated by the tritiated thymidine method which has numerous drawbacks; and (2) neither study took into account the large temporal and spatial variations in rates of detritus flux, bacterial activity, and distribution and abundance of higher consumers within the reef. Bacteria may be relatively unimportant trophically for brief periods of time. Even if metazoans consume a large share of the detritus pool, ultimately, waste and most unassimilated material will be either respired or recycled by microbes.

In any case, the fate of sediment bacteria and detritus are interlinked on coral reefs. As in other marine ecosystems, the fate of bacteria, microalgae, and particulate detritus in sediments cannot be easily separated owing to the dichotomy between oxic and anoxic environments and the lack of selectivity of most benthic detritivores. Benthic deposit feeders, such as holothurians, acquire food by swallowing large volumes of sediment relative to their body size and weight, processing both microbial and nonliving carbon for sustenance.[334] In reality, detritivores are omnivores, feeding on living and dead heterotrophic and autotrophic biomass, the relative contributions of which depend upon availability, caloric value and nutritional quality.

The rate of detrital utilization by benthos also depends on the rate of detritus supply and source. As most reef detritus is algal-derived, a large proportion of this material is probably utilized without the need for significant microbial enrichment. As discussed in the previous chapter, algal detritus is more readily assimilated than vascular plant detritus. On Davies Reef, Hansen et al.[339] found that the C:N ratio of the settled detritus ranged from 3:1 to 12:1 (mean = 8:1) and that the C:N ratio in the lagoon sediments averaged 3:1, suggesting that this material may be of fairly high nutritive value. Hansen et al.[342] observed that thalassinid shrimps (*Callianassa* spp.) in Davies Reef lagoon lower organic content on their mounds to the extent that bacterial growth rates are lowered.

Bacterial, algal, and detrital foods are utilized rapidly, fitting the common observation that epibenthos and surface infaunal populations are very abundant in most reef lagoons. Few data exist on actual ingestion rates and grazing of microbes and detritus by reef benthos.[334,350-352] Nearly all such studies have focused on the conspicuous epifauna such as holothurians and gastropods. For example, on Lizard Island, Great Barrier Reef, Moriarty et al.[350] found that grazing by the holothurian *Holothuria atra* depressed both microalgal and bacterial productivity, consuming roughly 10 to 40% of bacterial carbon per day, but not their biomass. On Heron Island, Great Barrier Reef, several species of deposit-feeding holothurians show preferential consumption of detrital matter and associated microbes, avoiding niche overlap.[351]

Field manipulative experiments, more often than not, are equivocal and the mechanism(s) responsible for a positive or negative effect cannot be discerned. For instance, on One Tree Reef, the gastropod *Rhinoclavis aspera* feeds on microalgae and detritus, attaining peak abundance on the back-reef sandflat.[352] Caging experiments revealed no consistent effects of the gastropod on either microalgal biomass or rates of bacterial growth, but bacterial numbers were depressed in cages with high densities of gastropods. Physical disturbance was found to produce a similar effect (although it was concluded that the gastropods were indirectly responsible) by stimulating meiofaunal grazers which, in turn, consumed bacteria. Clear demonstration of consumption is extremely difficult under natural conditions, as finely balanced geochemical gradients, to which different suites of microbial groups are adapted, are nearly always disrupted.

Bacteria may or may not be tightly linked to higher consumers, yet the fact remains that they are highly abundant and productive in unconsolidated carbonate sands and in hard substrata, including hard coral matrices (see Chapter 4 in Sorokin[296]). Standing stocks of bacteria and microbes can constitute most of the non-detrital organic carbon in reef sediments, as shown for the Mururoa atoll (Figure 5.6). Villiers et al.[344] calculated that 97 to 98% of total sedimentary organic carbon was detrital, but of the tiny fraction of organic carbon derived from benthic organisms, bacteria and other microbes made up from 54 to 81% of the living non-detrital fraction, varying with water depth and grain size. Benthic microalgae made up most of the remainder of the living fraction.

Biogeochemical studies indicate that most bacterial activity in reef sediments is aerobic. Rates of sulfate reduction in reef carbonates are low,[345,346] usually ranging from 0.6 to 6.0 mmol S m^{-2} d^{-1}. The proportion of total carbon metabolism attributed to anaerobic processes tends to increase with increasing organic carbon content (and finer sediment) and terrigenous inputs. In the southwest lagoon of New Caledonia, the proportion of total benthic metabolism that is due to anaerobic bacteria varies from a low of 35% for "white-sand" bottoms to a high of 49% for muddy sediments.[347] At Davies Reef, where organic carbon concentrations and terrestrial influences are low, sulfate reduction accounts for only 5% of total organic matter decomposition in the lagoon.[346] Like other sediments, reef carbonates are devoid of free oxygen below the upper few millimeters,[348] supporting some evidence[349] that other diagenetic pathways (e.g., denitrification, methanogenesis) are operative in reef carbonates, but the extent to which they contribute to total benthic carbon decomposition is not known. It is clear that bacteria mineralize most of the organic carbon in reef sediments, although how much of the remaining organic matter is buried is unknown.

5.3.2.3 *The Role of Pelagic Bacteria*

Benthic-pelagic coupling appears to be tight on coral reefs.[298,334] Benthic algal fragments, algal and coral exudates, amorphous aggregates, and associated microflora are continually produced and transferred from unconsolidated and hard substrata to the overlying water, whereas fecal pellets, dead and living plankton, marine flocs, and other suspended particles (including microbes) either settle passively to the bottom or are actively filtered by a variety of suspension-feeding benthic animals, including corals.[333,334] A seemingly endless variety of pelagic organisms feed on living cells and detrital material advected onto reefs from surrounding waters or produced *in situ*. The activity of pico- and nanoplanktonic communities is enhanced by the release of dissolved

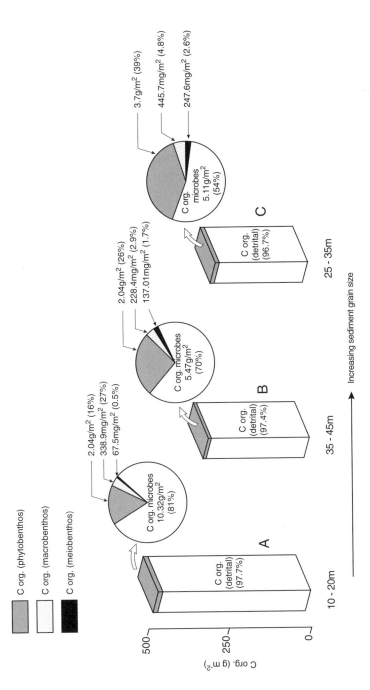

FIGURE 5.6
Proportion of living- and nonliving detrital carbon fractions in sediments at different water depths in Tuamotu Atoll, French Polynesia. (Adapted from Villiers, L., Christien, D., and Severe, A., *ORSTOM Oceanographic Notes Documents No. 36*, 1987.)

materials from corals and other benthos. Coral reef waters are no different to other seawater parcels in cycling energy and nutrients through the microbial loop. Estimates of the flux of energy and nutrients through the microbial loop are hampered by a lack of information about autotrophic and heterotrophic, pico- and nanoplankton in reef waters.

The composition of reef plankton is dominated by coccoid cyanobacteria less than 1 μm in diameter.[353] On Davies Reef, Ducklow[353] found that the plankton carbon biomass was composed of autotrophic picoplankton (46%), nanophytoplankton (32%), net phytoplankton (13%), bacteria and bacteriovores (7%), and ciliates and zooplankton (3%). The few data available suggest that they vary in abundance across reef zones and between reefs and surrounding tropical waters.[353,354] Ayukai[354] observed a consistent pattern of lower densities of phytoplankton, bacteria, nanoflagellates, and ciliates over leeward reef flats in open ocean water or on reef fronts at several reefs within the Great Barrier Reef. The decline in numbers constitutes a retention of these communities equivalent to a maximum of 90 mg C m^{-2} d^{-1}, nearly equal to net community production. The decline in total microbial numbers may reflect not only consumption in the water column, but also consumption by benthic filter feeders. The retention of pelagic organisms may be one mechanism for coral reefs to conserve limited nutrients.

Consumption of bacteria by protozooplankton and zooplankton has rarely been quantified in reef waters, but presumably their trophic importance depends upon their growth rates and whether or not the cells are free living or attached to particles. On average, a greater proportion of bacteria are associated with particles in reef waters compared with bacteria in surrounding oceanic waters. However, Ducklow[353] concluded that consumption of bacteria in the water column appears to be a minor sink for reef bacterial production; far more pelagic bacteria are consumed by benthic filter feeders. Zooplankton seem to be differentially removed at the reef front by the "wall of mouths" (see Section 5.3.1), resulting in the dominance of phytoplankton and bacteria in the water column.

Rates of bacterioplankton production measured using the tritiated thymidine technique range from 0.1 to 64 μg C l^{-1} d^{-1} in coral reef waters, within the range of values measured in other productive ecosystems. Some calculations suggest that phytoplankton and bacteria are not effectively removed but are exported from coral reefs.[357] Excess bacterial biomass is caused not by high rates of production, but by a large influx of bacterial cells with inflowing water. This conclusion is puzzling, as other workers have consistently observed what appears to be retention of bacteria and other microbes in reef waters. This disparity may reflect the fact that no data are available for bacterial abundance and productivity over time; most studies have measured bacterial densities and activity at one or two sites and usually for one season.

Better understanding of the availability of dissolved and particulate materials would help to clarify fluxes of energy and material through microbial food chains. Coral mucus is one conspicuous component for which inconsistent findings exist. Hermatypic corals may release from 40 to 60% of their photosynthetically fixed carbon as mucus.[340] This material, rich in wax esters, triglycerides, fatty acids, and other nutrient-rich compounds, serves as food for heterotrophic bacteria and other animals, such as infauna, commensals, and some specialized fish.[296] However, rates of mucus production and nutritive value vary greatly, and earlier speculations of the contribution of mucus to energy flow in reef food webs have recently been questioned.[340]

In an extensive series of biochemical analyses of mucus from several *Porites* spp., Coffroth[340] found that fluid mucus and mucus sheets had mean caloric content of 4.7 and 3.5 cal mg^{-1} ash-free dry weight (AFDW), and 22 and 68% ash content, respectively, suggesting low nutritional value. Bacterial numbers were higher on the mucus than in surrounding water, but bacteria accounted for less than 0.1% of the total carbon and nitrogen content of the mucus. Annual mucus production rates are slow and show lunar periodicity, suggesting also that mucus and mucus sheets make only a small contribution to energy flow. It is likely that much of the DOM and particulate organic matter (POM) on coral reefs is of low quality, which helps to explain why nutrients are limiting and recycled efficiently.

5.4 NITROGEN AND PHOSPHORUS: CYCLES AND LIMITATION

Coral reefs are often considered to be limited by the availability of nutrients, but little consensus exists as to whether nitrogen or phosphorus are limiting and under what conditions such limitation occurs.[355,356] Nutrient limitation has not been directly tested on coral reefs but has been inferred from the fact that reefs lie in oligotrophic waters and demonstrate very high rates of nitrogen fixation. Such inferences may be incorrect. High rates of nitrogen fixation do not preclude further increases in productivity, and Smith[357] suggests that coral reefs have nutrient requirements similar to those communities in the surrounding seas. Part of the problem is that studies of nutrient limitation have utilized physiological, bioassay, mathematical, and stoichiometric approaches that are all different in emphasis and scale.

5.4.1   Nitrogen

Some aspects of the nitrogen cycle are better understood than others. Rarely has the entire cycle been deduced for entire benthic and pelagic components of a coral reef. Capone et al.[358] recently measured some nitrogen transformation processes in surface carbonates on reefs within the Great Barrier Reef (Figure 5.7). All three reefs show rapid rates of ammonification in relation to pool size, suggesting highly dynamic nitrogen pools and turnover times of less than a day. Nitrification is also quantitatively important and, as on other reefs, nitrogen fixation accounts for a large fraction of ammonium production and turnover. Denitrification rates were low (methodological problems may have biased these results), but such low rates do suggest that nitrogen is being conserved. Most measured rates of nitrogen cycling processes in reef sediments (see Table 5.6 and references in Capone et al.[358]) show great variability because of different methods used, differences in sediment grain size and organic content, variable rates of particulate organic nitrogen (PON) input, and seasonality. Regardless, the information indicates that these small nitrogen pools turn over rapidly — as found in other tropical marine deposits.[12]

Sediments are normally the prime location for nutrient transformations, but the extreme spatial complexity and diversity of coral reefs make it likely that some reef components, previously thought to play a minor role in nutrient exchange, may play a larger role. Recent discoveries show that photosynthetic symbionts in some hermatypic corals[359-361] and sponges[362] can fix nitrogen and can nitrify.

Nitrogen transformations in reef waters are less understood than in sediments, complicated by very high spatial and temporal heterogeneity of nutrient concentrations owing to complex patterns of water mixing and variable residence times.[363,364] Availability of nutrients may be limited by exchanges with sediments, although this is not always true.[365] Hopkinson et al.[366] found that rates of ammonium regeneration in the water column at Davies Reef were highly variable in winter, ranging from undetectable to rates from 0.0013 to 0.0112 μmol N l^{-1} hr^{-1} — among the lowest values reported in the literature. They did find, however, that regeneration rates were more important in terms of supplying regenerated nitrogen to plankton from the reef front to the back lagoon.

Benthic-pelagic coupling is evidently facilitated, and perhaps partly mediated, by demersal organisms that feed at the sediment-water interface. In several reef lagoons within the Great Barrier Reef, Bishop and Greenwood[367] measured rates of nitrogen excretion by some demersal macrozooplankton. They found that these organisms excrete 12 and 34 μM NH_4 m^{-2} d^{-1} within the water column and sediments, respectively. This equates to 29% of nitrogen regeneration reported for

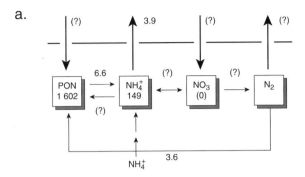

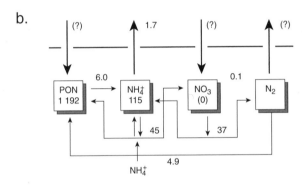

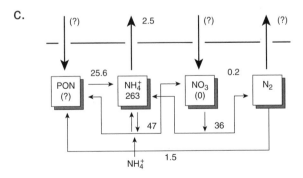

FIGURE 5.7
Nitrogen cycling in surface (0 to 2 cm) sediments of Pandora (a), Hopkinson (b), and Bowl (c) Reefs, Great Barrier Reef. Standing stocks: mmol N m^{-2}; fluxes: mmol N m^{-2} h^{-1}. (Adapted from Capone, D. G. et al., *Mar. Ecol. Prog. Ser.*, 80, 75, 1992.)

TABLE 5.6

Range of Rates of Various Nitrogen Transformation Processes in
Coral Reef Sediments from Various Locations[a]

Transformation Rate	Method and Location
NH$_4^+$ Release	
0.7–1.0	Diffusive modeling; fringing reef, Puerto Rico
0–0.8	Diffusive modeling; Tikehau lagoon, French Polynesia
50–704	Chambers; anoxic sediments, Tikehau lagoon
0.4	Diffusive modeling; Hydrolab, St. Croix
35	Dark chamber; Hydrolab, St. Croix
3	Chambers; back-reef, Tague Bay, St. Croix
0.4–4	Diffusive modeling; various reefs, Great Barrier Reef
0.02	Chambers; Davies Reef lagoon, Great Barrier Reef
Ammonification	
4.5	Direct, tube pack; Tague Bay, St. Croix
5–300	Diffusive modeling; Bermuda
4.4–50	Core incubation; various reefs, Great Barrier Reef
Denitrification	
19	Acetylene blockage; lagoonal sediments, Bermuda
50–100	Acetylene blockage; fringing reefs, Puerto Rico
0.1–13	Acetylene blockage; various reefs, Great Barrier Reef
N$_2$ Fixation	
2.7	Acetylene reduction, $^{15}N_2$; Barbados
5.1	Acetylene reduction; Kanehoe Bay, Hawaii
1.7–15	Acetylene reduction; central Great Barrier Reef
6.5–12.8	Acetylene reduction; muds, Puerto Rico
0.4–3.6	Acetylene reduction; sands, Puerto Rico
3–12	Acetylene reduction; muds, Bermuda
0.4–11.5	Acetylene reduction; sands, Bermuda
6	Acetylene reduction; Hydrolab, St. Croix
4–10	Acetylene reduction, $^{15}N_2$; southern Great Barrier Reef

[a] References can be found in original source.

Note: Units are μmol N m^{-2} hr^{-1}.

Source: From Capone, D. G. et al., *Mar. Ecol. Prog. Ser.*, 80, 75, 1992. With permission.

microheterotrophs and 9% of nitrogen utilization by phytoplankton. Compared to the benthos, these excretion rates are equivalent to 10 to 13% of ammonium released from the sediments and from 5 to 32% and 2 to 13% of benthic ammonification and utilization rates, respectively. Nitrogen excretion by some demersal macrozooplankton may even exceed excretion by some infauna.[367]

5.4.2   Phosphorus

Phosphorus cycling is even less well understood than nitrogen cycling on coral reefs.[355-357] The transformations of phosphorus are straightforward compared to those for nitrogen, as gaseous phases are lacking:

$$POP, DOP \Leftrightarrow HPO_4^{2-} \Leftrightarrow mineral\ P\ (aragonite, calcite) \qquad \textbf{(5.3)}$$

where POP = particulate organic phosphorus, and DOP = dissolved organic phosphorus. The mainly biotically driven processes of assimilation (inorganic P $\rightarrow$ organic P), excretion and hydrolysis (organic P $\rightarrow$ inorganic P), and the abiotic processes of precipitation, sorption, and dissolution between mineral phosphorus and the other phosphorus pools have been not been well studied until recently.

A number of studies indicate rapid regeneration of phosphorus within reef waters and between reef waters and the benthos.[368,369] On reefs within the Capricornia section of the Great Barrier Reef, Sorokin[368,369] used ^{32}P to trace the flow of phosphorus in food chains. He found that, although dissolved phosphorus concentrations were very low, turnover was rapid in the water column, ranging from 2 to 8 days. Bacterioplankton was the major consumer of inorganic phosphorus, accounting for 50 to 90% of total phosphate uptake. Further experiments by Sorokin[369] showed that bacterioplankton transfer phosphorus when consumed by reef benthos and that the benthos releases phosphorus (in most instances roughly equal to the amount consumed), resulting in no or little net uptake. In some cases, there is a small net release of dissolved phosphorus from sediments.

Atkinson and his colleagues[370-373] and the ORSTOM (Office de la Recherche Scientifique et Technique Outre-Mer) program in French Polynesia[374] provide us with a glimpse of the modes and conditions under which coral reefs cycle phosphorus. For the lagoonal atoll of Tikehau, Charpy and Charpy-Roubaud[374] found that phosphorus concentrations in lagoon waters are slightly greater than in the surrounding water; excretion by pelagic heterotrophs supply 75% of phytoplankton phosphorus requirements and all of the phosphorus required for benthic primary production is met by release from the sediments. Their budget suggests that phosphorus is exported from the lagoon in particulate form, but equates to only 2.5% of phytoplankton production.

Atkinson and his co-workers[370,371] have proposed the mass transfer limitation hypothesis, which predicts that the extent to which phosphorus is regulated within a coral reef is independent of concentration but dependent on the rate at which phosphorus is transported onto and into the reef. Phosphorus would be regulated, or even become limiting, if the rate of uptake by reef organisms exceeds its rate of input. Nutrient regulation and limitation on coral reefs, they argue, is therefore dependent upon the rate of water motion into and across a coral reef.

To test their hypothesis, Bilger and Atkinson[372] used an engineering approach to calculate the ratio of the uptake rate to advection rate (known to engineers as the "Stanton number", a dimensionless figure) on a simplified reef flat. Their calculations yielded a Stanton number of 5×10^{-5}. Field measurements yielded a value 40 times greater (1.7×10^{-3}). To test the explanation that this anomalously high mass transfer of phosphorus was due to the difference between a smooth surface and the real, three-dimensional benthic surfaces of a coral reef, Atkinson and Bilge[373] performed a series of flume experiments on communities of several reef organisms. Their experiments confirmed higher than predicted rates of mass transfer. They found that increasing the water-flow velocity increased the rate of phosphorus uptake, indicating that uptake is regulated by diffusive boundary layers near the surfaces of the organisms and that the natural geometry of reef surfaces facilitates transfer of nutrients. In most other respects, the data were consistent with the engineering model of mass transfer, being somewhat influenced by additional variables such as water temperature, solar insolation, and nutrient loading.[373]

Applying the mass-transfer limitation idea to other phosphorus data, Atkinson[371] concluded that:

- There is a maximal rate of phosphorus uptake, under highest water flow conditions, but the residence time of the water over the reef is shortest.

- The amount of phosphorus removed from the water is only ~5% of the amount that passes over the reef; this may explain the difficulties in earlier studies to observe phosphorus uptake over a short distance.

The idea of coral reefs being limited by mass transfer of nutrients is intriguing as it agrees with the strong water-flow dependence observed for growth of many reef organisms. But as it is unlikely that any ecosystem conforms closely to engineering equations, the hypothesis requires further testing in order to determine the scope of its applicability. The view that phosphorus is mass transfer-limited is not necessarily at odds with the view that recycling is an important regulator on coral reefs. It is conceivable that under low-flow conditions, recycling mechanisms are the pre-eminent means of conserving and modulating the phosphorus cycle. Under high-flow conditions, supply rate may outpace the rate of phosphorus recycling on coral reefs. The mass transfer idea is an elegant refinement of Smith's earlier supposition[357] that the rate of hydrodynamic flux determines whether or not nitrogen or phosphorus is limiting to coral reefs. The conditions and the extent to which this is true remain to be investigated. It is highly unlikely that any one factor limits energy and nutrient flow under all climatic conditions.

Uncertainty about the relative importance of recycling and the mass transfer limitation of nutrients does not mean that coral reefs require a large supply of "exotic" nutrients to explain the high rates of nutrient flux, as suggested by the endo-upwelling concept of Rougerie and Wauthy.[375] Their hypothesis states that geothermal heat deep within the limestone and basaltic framework underlying Pacific atolls facilitates upwelling and emergence of nutrient-rich Antarctic intermediate water to the shallow surface waters to fuel high rates of carbon fixation. Endo-upwelling may be a significant geochemical process, particularly on Pacific atolls but is unlikely to be a significant mechanism for nutrient supply to living continental shelf reefs.

5.5   THE CORAL FACTORY: CARBON AND ENERGY BUDGETS

It is reasonable from an energetic perspective to consider an individual reef coral as a mini-ecosystem. Each coral colony consists of many living, individual polyps fueled mostly, if indirectly, by sunlight and supplemented by DOM and POM (the latter usually by carnivory), growing, calcifying, respiring, and serving as home for many animals, and, whether zooxanthellate are present or not, sharing nutrition and energy with a rich assortment of microorganisms. The ecophysiology of corals has obvious relevance to our understanding of coral reefs as ecosystems.

Like other organisms, living hard and soft corals must balance their energy requirements. Various estimates[298,303] have been made of the energy budgets of living corals. The majority of studies indicate that coral energy requirements can be met by the photosynthate produced by their zooxanthellae; however, as light diminishes with increasing depth, photosynthesis decreases, and corals must exhibit obligate heterotrophy. As pointed out by Muscatine,[303] the assimilation efficiency of carbon is very low (~8%), even for corals living in shallow water, because of nitrogen deficiency. Zooxanthellae translocate amino acids and dissolved inorganic nitrogen to their coral hosts, but they still must acquire exogenous nitrogen from the overlying water in order to meet their nitrogen requirements, usually by capturing plankton and detritus but also by actively taking up nanomolar concentrations of dissolved free amino acids.[376] The recent discovery of nitrogen-fixing[359] and nitrifying[360] bacteria within coral skeletons suggests

TABLE 5.7

Carbon Budget for Light- and Shade-Adapted
Colonies of the Coral *Stylophora pistillata*

Flux	Light (μg C cm^{-2} d^{-1})	Shade (μg C cm^{-2} d^{-1})
Gross primary production	140.1	32.2
Symbiont respiration	2.6 (2%)	1.0 (3%)
Net primary production	137.5	31.1
Algal growth	1.2 (1%)	0.7 (2%)
Translocation to coral	136.3 (99%)	30.4 (98%)
Coral respiration	94.8 (70%)	52.2 (173%)
Dissolved organic matter (DOM) release	8.2 (6%)	14.0 (46%)
Storage/assimilation	33.3 (24%)	?

Note: Values in parentheses indicate percentage of either gross primary production (GPP) or net primary production (NPP).

Source: Adapted from Muscatine, L., *Proceedings of the Great Barrier Reef Conference,* Baker, J. T., Carter, R., Sammarco, P. W., and Stark, K. P., Eds., James Cook University and Australia Institute of Marine Science, Townsville, 1983, 341.

translocation of nitrogen from bacteria to the corals, but the magnitude of this pathway is uncertain.

Muscatine[377] constructed a carbon budget for the Red Sea coral *Stylophora pistillata* under light- and shade-adapted conditions (Table 5.7). The budget shows that under light, roughly 99% of the net fixed organic carbon is translocated to the coral host. Approximately 70% of the translocated carbon is respired, 6% is released as dissolved organic carbon (DOC), and the remaining 24% is stored or assimilated for growth. Under shaded conditions, the translocated carbon is less than the daily carbon respired, suggesting an energy imbalance.

The energy budget[378] of *Porites porites* is similar, but most of the translocated carbon (78% of gross primary production, GPP) is lost (45% of net primary production, NPP), presumably as mucus, enzymes, nematocysts, cell debris, etc. The remainder is respired (26.3% of net primary production) and channeled into growth (6.3%) and reproduction (0.4%).

A more comprehensive analysis of the effect of daylight changes on energy flow within corals was made in Hawaii by Davies.[379] He found that under optimal light conditions, all three species (*Pocillopora damicornis*, *Montipora verrucosa*, and *Porites lobata*) fixed excess carbon for growth and respiration, but there was a loss of 12 to 28% of the carbon fixed by zooxanthellae. On a "normal" day, the rate of photosynthesis was lower but fixed carbon was still in excess (12 to 27% of the total).

TABLE 5.8

Mean Annual Carbon and Nitrogen Budget for the Elkhorn Coral
Acropora palmata

Flux	Carbon (g C yr⁻¹)	Nitrogen (g N yr⁻¹)
Inputs		
Gross primary production	16.4 (91%)	0.0
Feeding on particulate organic matter (POM)	1.6 (9%)	0.5 (70%)
Dissolved organic matter (DOM) uptake	0.0	0.2 (30%)
Outputs		
Respiration	11.5 (64%)	0.0
DOM losses	3.0 (17%)	0.35 (50%)
Organic tissue growth	2.1 (12%)	0.30 (40%)
Gamete production	1.6 (9%)	0.10 (11%)

Note: Values in parentheses are percent of nutrient requirements.

Source: Adapted from Bythell, J. C., *Proc. Sixth Coral Reef Symp.*, 2, 535, 1988.

On an "overcast"day, all three species had to draw carbon from their lipid reserves, although to a much lesser extent in the case of *M. verrucosa.* Davies[379] suggested that under suboptimal light conditions, coral catabolize some of their lipid reserves, again replenishing their lipid stores when light conditions are optimal, and secreting excess carbon as mucus when their lipid reserves are full. Other studies have demonstrated the favorable effects of light[380] and water velocity[381] (but see Atkinson et al. [382]) on the energetics of hermatypic corals.

Simultaneous measurements of the flux of nitrogen and carbon in corals have rarely been made, which is unfortunate because the extent of carbon-nitrogen coupling within coral reefs is very poorly understood. The construction of carbon and nitrogen budgets for the elkhorn coral, *Acropora palmata,* by Bythell[383] is a notable exception, providing us with some insight into the mechanisms and extent of linkage between carbon and nitrogen fluxes. The inputs and outputs (Table 5.8) show that both nutrients behave differently within the colony. Most carbon (91%) is assumed to come from symbiont photosynthesis, whereas nitrogen is acquired from heterotrophic feeding on water-column particulates and by taking up ammonium and nitrate. Most carbon (81%) is lost by respiration and DOC release. One half of the nitrogen is excreted as mucus dissolved organic nitrogen (DON), but the rest is incorporated into growth of somatic tissue and reproduction. The low assimilation ratio (3 to 1) of particulate carbon to nitrogen suggests a requirement for high quality foods, such as bacteria or organic aggregates rich in amino acids or proteins. Bythell[383] suggested that the large losses of nitrogen may be from released mucus and that the low release rates of carbon compared to nitrogen indicate that

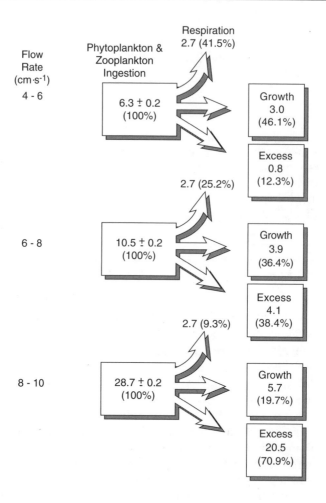

FIGURE 5.8

Effect of water flow on ingestion rates (mg C polyp^{-1} d^{-1}) for the soft coral *D. hemprichi*. Budgets are preliminary, as respiration was measured only at a flow speed of ~16 cm sec^{-1}. "Excess" represents carbon which is not assimilated into somatic growth or respiration. (Adapted from Fabricius, K. E., Genin, A., and Benayahu, Y., *Limnol. Oceanogr.*, 40, 1290, 1995.)

nitrogen is not limiting for growth. There are other equally plausible explanations for the pathways of carbon and nitrogen flux in this species, but it is clear that they do behave differently reflecting the biochemical needs of the organism.

Even less is known about the energetics of soft corals. Fabricius et al.[384] followed up their initial experiments with the herbivorous Red Sea octocoral *Dendronephthya hemprichi* by constructing a preliminary carbon budget (Figure 5.8). Zooplankton contribute <5% to the carbon demand of this coral. Increases in water velocity alter the flow of carbon demand. At a flow rate of 4 to 6 cm sec^{-1}, nearly equal amounts of carbon are respired and channeled into growth. As water flow increases, so does the rate of phytoplankton ingestion, and proportionally less carbon is respired, but more is lost, either as mucus or as leakage of DOC from the gastrovascular cavity (Figure 5.8). Although proportionally less carbon is contributed to growth, the absolute amount of new tissue carbon increases. This preliminary budget shows the influence of water motion on the energetics of soft corals. Soft corals can partition their carbon input in the same way as other invertebrates, with a large expenditure of energy as respiration (and/or "excess") for comparatively little net gain of new tissue.

5.6   SYSTEMS-LEVEL PERSPECTIVES: MODELS AND BUDGETS

Ignorance of the energetics of individual organisms, populations, and communities has not impeded attempts since the early efforts of the Odums[385] to construct energy-flow models of reef food webs, or to estimate the mass balance of nutrients on coral reefs. These models and budgets have clear limitations, but give some insight into how coral reefs function internally and to what extent they interact with the atmosphere and surrounding ocean. More importantly, they provide clues on missing pieces in the reef engine puzzle.

5.6.1   Recent Models of Energy and Carbon Flow: What Do They Tell Us?

Sorokin[386] recently constructed an energy-flow scheme of an idealized coral reef, but such generalizations are misleading, as energy and nutrient flows vary from zone to zone and from reef to reef.[387] Have attempts to model energy and carbon flows since the first model of the Eniwetak atoll[385] shed more light on the functioning of coral reefs? Models have been few, and their reflection of reality corresponds directly to the validity of their assumptions.

The application of the ECOPATH model[388-390] to the French Frigate Shoals in the northwest Hawaiian Islands was an ambitious attempt to estimate the biomass, consumption, and production of organic matter in the ecosystem. The simplified food web model indicates that organic matter decomposition occurs by direct intake, not by diagenetic pathways in sediments or through the microbial loop; however, the model did not include a realistic depiction of the roles of detritus and microbial food webs. The emphasis on using optimal grazing rates in the model led to the assumption that grazing controls system productivity. This reflects poor understanding of detritus and biogeochemical constraints on ecosystem-level nutrient fluxes. Earlier models (reviewed by Sorokin[296]) suffer from similar shortcomings, particularly with regard to fluxes in lagoons.

More recent models of energy and carbon[331,391] flow on coral reefs are somewhat more realistic, attempting to either balance grazing and detritus food webs or considering the role of external forces on flux rates. The carbon flux model constructed by Johnson et al.[331] is a more sophisticated version of earlier carbon flow models[334] of Davies Reef, Great Barrier Reef. They used a large store of empirical data obtained from previous studies on the reef to model the shallow front slope. Assumptions were still made for many aspects of the model, but it is arguably the best synthesis of current knowledge of carbon flow within a coral reef zone. The results of the static model offer several important ideas:

- Carbon flux on the slope is dominated by external inputs and outputs due to high water-flow rates, with *in situ* GPP accounting for ~1% of total organic carbon input.
- Recycling accounts for 26 to 32% of carbon flux, but relatively little of this is via the microbial loop.
- The role of bacteria and other microbes is consequently small.
- Trophic transfer efficiency averages 11.5%.

In modeling an often-observed shift from live coral cover to epilithic algae when reefs are disturbed, Johnson et al.[331] found that proportionally more carbon is recycled and transferred to detrital food chains after such disturbance. Not surprisingly, there is also a shift in trophic dependencies among groups.

The model of normal conditions is greatly different from the model of Sorokin[386] and the earlier models of Davies Reef.[334] There are several reasons for these differences, but the main reason is that Johnson and his colleagues modeled the front reef slope, most of which is hard bottom with small patches of unconsolidated, coarse-grained sediment. The earlier Davies Reef models were of the back lagoon, where fine sediments predominate and conditions are comparatively quiescent. The idealized model of Sorokin[386] is similar to the earlier models of Davies Reef, but both suffer from the fact that bacterial carbon production was estimated using different methods which may not provide an accurate measurement of total bacterial carbon oxidation.

Energy flow and the trophic function of fringing and barrier reefs of Tiahura in French Polynesia have been assessed using the ECOPATH

II program.[391] The modeling exercise shows that most of the net primary production is recycled via the microbial loop, allowing long trophic pathways but low trophic efficiencies. The model indicates two cycles: a main cycle involving detritus and a smaller grazing pathway. These results are not surprising if one compares the front slope of Davies Reef with the Tiahura reef complex. For instance, in contrast to the open front slope of Davies Reef, the Polynesian reef system is semi-enclosed and has a greater proportion of unconsolidated sediments. These differences seem to conform to the view that the proportion of carbon that is recycled within a coral reef depends upon reef zone and the rate of water flow. Low-flow conditions may facilitate internal recycling, whereas on high through-flow reefs or zones, a proportionally small amount of carbon is shunted through recycling pathways.

Carbon budgets of other reef lagoons[338] and reef flats[302] demonstrate that individual zones will behave differently with respect to energy and nutrient flow, with some shallow reef fronts and flats being frequently net autotrophic and some deep lagoons appearing to be net heterotrophic. The carbon budget for the southwest lagoon of New Caledonia (Figure 5.9) is incomplete but clearly demonstrates that most organic carbon flows through detrital pathways in the soft sediments of this lagoon-dominated reef complex. Trophic efficiency is low (6 to 13%) within the detritus food web in this lagoon, but bacterial mineralization rates were not measured.

These few modeling and budget exercises show that each coral reef, and certainly each zone within a coral reef, is different with respect to the proportion of organic carbon entering grazing and detrital food chains. More importantly, they serve as a cautionary reminder of the danger of sweeping generalizations that are often made when working at the systems level. This does not, however, preclude searching for some ecologically meaningful patterns concerning the energetics of coral reef ecosystems.

5.6.2   Mass Balance Estimates: Are Coral Reefs Sources or Sinks for Carbon?

Kinsey[302] reviewed the large number of estimates of carbon production and consumption on various coral reefs (Figure 5.10) and concluded that, on average, the ratio of GPP to R (respiration) approximates 1, suggesting that most coral reefs are in balance with respect to organic carbon with little, if any, net production (see also Table 5.2). Kinsey[302] warned that the deviations of some reef zones from unity are real, although somewhat confounded by the different methods used, seasonality, and spatial heterogeneity. Other factors such as hydrology, tidal amplitude, and proximity to land must play an unquantified role

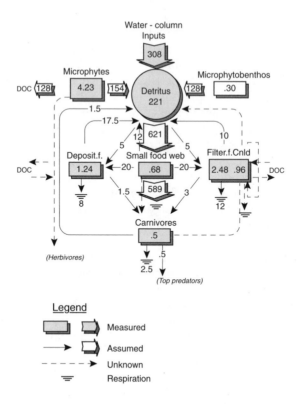

FIGURE 5.9

Carbon flow in sediments of the south west lagoon of New Caledonia. Biomass/standing stocks: g C m^{-2}; fluxes: g C m^{-2} yr^{-1}. (Adapted from Chardy, P. and Clavier, J., *Proc. Sixth Int. Coral Reef Symp.*, 2, 521, 1988.)

in striking a balance (or not) between production and consumption. It is obvious from the data compiled in Table 5.2 and Figure 5.10 that organic carbon fluxes have been more frequently measured within some zones than in others. The majority of studies have focused on the shallower reef front and reef flat zones rather than lagoons. Studies of entire reefs are equally few.

An earlier, unsubstantiated view held that shallow reef fronts and reef flats are the chief zones of carbon fixation and that lagoons are mainly depositories for any excess carbon transported from the shallower, front reef zones. It is now clear from some lagoonal studies discussed earlier[336,339,341-344] that this is oversimplified. Reef lagoons are mosaics in which some areas of the lagoon floor are carpeted with algal mats fixing carbon, while other areas, such as those densely populated by thalassinid shrimps, are zones consuming organic carbon. Depending on water depth, sediment texture, and hydrodynamics (to name a few key factors), some lagoons may be net producers of organic carbon,

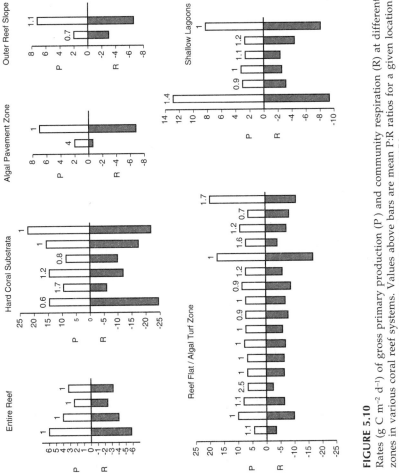

FIGURE 5.10

Rates (g C m^{-2} d^{-1}) of gross primary production (P) and community respiration (R) at different zones in various coral reef systems. Values above bars are mean P:R ratios for a given location. (Data from Kinsey, D. W., *Proc. Fifth Coral Reef Congr.*, 4, 505, 1985.)

some may be roughly in balance, and other lagoons may be zones of net heterotrophy.

Because carbon flow on reefs, as in nearly all other marine ecosystems, is dominated by benthic processes, factors affecting spatial heterogeneity of soft and hard substrata will greatly affect the production-to-respiration (P:R) balance. For instance, on a coral reef flat in the Philippines, Yap et al.[392] measured organic carbon production and consumption on different substrata and found that corals were the most metabolically active with a P:R $\geq$ 1. Coral rubble followed with a P:R $\approx$ 1, and unconsolidated sands were slightly heterotrophic (P:R $\leq$ 1). Similar variations in the organic carbon balance have been found in other reef zones.[342]

How such variations within and across reef zones affect the P:R ratio of an entire coral reef depends not only on relative rates of gross primary production relative to respiration, but also to other factors including reef geomorphology. If we accept the oversimplification that lagoons are net heterotrophic and reef flats are net autotrophic, a coral reef that is mostly shallow reef flat will very likely show an overall P:R $\geq$ 1. Conversely, a coral reef that is mostly lagoonal will have an overall P:R $\leq$ 1. Oversimplification perhaps, but a salient example of how the organic carbon balance of entire coral reefs may deviate from unity.

The study of Nakamori et al.[393] on the Shiraho and Southern Reefs of the Ryukyu Islands is a good example of how spatial heterogeneity and water circulation across a reef can be incorporated into an analysis of organic and inorganic carbon fluxes. By measuring changes in pH, total alkalinity, ΣCO_2, and water flux across the different reef zones and by converting their results to an area-specific basis, they were able to show how carbon fluxes within zones on Shiraho Reef link to the adjacent Southern Reef and contribute to net export of organic carbon from the system (Figure 5.11). The driving force behind the export of organic carbon is the strong unidirectional movement of water from the front edge of Shiraho Reef to Southern Reef caused by tidal oscillations and wave action at the reef front.

The mass balance estimates of Smith and his colleagues[297,395-398] suggest that coral-dominated reefs export little, if any, organic carbon. Taking a "total reef" view of the data on production and consumption in the different reef zones, Smith[394] observed that the few comprehensive studies of the metabolism of a total reef system show a net productivity ranging from –5 to 170 mg C m^{-2} d^{-1}, averaging 130 mg C m^{-2} d^{-1}. Considering the rates of gross primary production and respiration across reef zones (Figure 5.10), variations associated with the methods used, seasonality, and extrapolations to entire reef area, these net values are exceedingly small and well within the boundaries of probable error. Moreover, these estimates pertain to healthy, coral-dominated reefs,

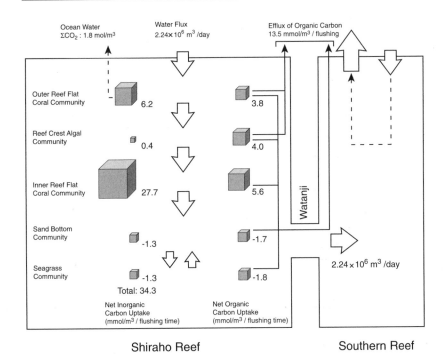

FIGURE 5.11
Model of water (open arrows) and carbon (small arrows) fluxes over Shiraho Reef and
connections to adjacent Southern Reef, Ryukyu Islands. (From Nakamori, T., Suzuki,
A., and Iryu, Y., *Cont. Shelf Res.*, 12, 951, 1992. With permission.)

not those disturbed by anthropogenic inputs, having undergone a trophic
shift in dominance from corals to algae.

Smith[394] observed that coral reef productivity is somewhat enhanced
compared to surrounding oceanic plankton communities and suggested
two reasons why this might be so. First, the stoichiometry of benthic
plants (C:N:P = 550:30:1) indicates that nitrogen and phosphorus are
depleted with respect to carbon, in comparison to the C:N:P ratio
(106:16:1) of oceanic plankton. This difference may constitute a production advantage for coral reefs (which are benthic-carbon dominated) in
that less nitrogen and phosphorus are required by benthic plants to fix
a given amount of carbon compared to plankton. For phosphorus, the
advantage may be a factor of 5, given the above-stated C:P ratios. It is
unlikely that the advantage is so great, because element ratios of tropical benthic plants and plankton deviate greatly in stoichiometry. The
second reason offered was that primary producers may shift from
nitrogen to phosphorus limitation as water flows from the surrounding
ocean onto the coral reef. The idea comes from the observation near

several atolls that ocean waters are deficient in dissolved inorganic nitrogen relative to dissolved inorganic phosphorus, whereas the ratio of reef water DIN:DIP is significantly greater, perhaps due to high rates of nitrogen fixation (Table 5.6). The applicability of this supposition for coral reefs depends upon whether or not this difference in water stoichiometry holds for fringing and barrier reefs as well. Variations in water flow and rates of recycling may ameliorate any such advantages in production.

On a global basis, coral reefs contribute gross CO_2 production on the order of 700×10^{12} g C yr^{-1} to the world's ocean and an excess production of approximately 20×10^{12} g C yr^{-1}. Crossland et al.[395] calculated that 15% of this net production is incorporated into reef structures, 10% is available for human harvest, and the remaining 75% is available for export to adjacent waters. This value, however, is very small compared to gross production, suggesting that most organic material produced on coral reefs is recycled within reef food webs. These calculations are estimates subject to considerable uncertainty and illustrate our ignorance of the fate of organic production, its temporal and spatial variability, and its incorporation into reef structures over long time scales.

Coral reefs appear to be small net producers of organic carbon, but the significance of coral reefs in terms of inorganic carbon, particularly CO_2, is less understood. Kinsey and Hopley[397] showed that coral reefs are net sinks of inorganic carbon, mostly as $CaCO_3$ accretion. Coral reefs globally produce ~900 million metric tons of calcium carbonate per year, thus acting as a sink for ~111 million metric tons of inorganic carbon per year. Production of calcium carbonate has important consequences also for the exchange of CO_2 between coral reefs and the atmosphere. Precipitation of 1 mol of $CaCO_3$ results in the release of 0.6 mol of carbon dioxide;[399,400] therefore, on a global basis, coral reefs contribute from 2.2 to 8.2×10^{13} g C yr^{-1} to the atmosphere, which equates to roughly 0.4 to 1.4% of the current rate of anthropogenic carbon production from combustion of fossil fuels.[398]

Recent studies have successfully clarified the role of CO_2 fluxes from coral reefs.[399-402] Gattuso and his co-workers[399,401,402] provide the first direct confirmation that coral reefs are a source of CO_2 to the atmosphere. Working in coral reefs of Moorea, French Polynesia, they measured community metabolism and found that in waters surrounding the reef, net air-sea gas exchange was negligible, but over the reef front and back, carbon dioxide fluxes mirrored the diel patterns of primary production, respiration, and calcification. CO_2 fluxes were positive at night due to respiration and calcification increasing the partial pressure of CO_2 and favoring release to the atmosphere. As the day progressed, CO_2 fluxes declined to negative values, as the net effect of photosynthesis (CO_2 uptake) dominated. Overall, the reef was a source (1.5 mmol CO_2 m^{-2} d^{-1}) of carbon dioxide.

Not all coral reefs are net sources of CO_2 to the atmosphere. The calculations of Gattuso et al.[402] predict that a coral reef exhibiting net organic carbon production and low $CaCO_3$ precipitation would be a sink of carbon dioxide; only reefs with a balanced metabolism (P:R ≈ 1) would be expected to show a net release of CO_2. Field evidence supports these calculations. Shiraho Reef in the Ryukyu Islands — a net exporter of organic carbon — appears to be a sink for carbon dioxide, as shown by recent field experiments.[400] This qualification suggests that the global calculations of Crossland et al.[395] are overestimates, considering that many, but not all, coral reefs are in metabolic balance.

5.7 THE ROLE OF CORAL REEFS IN THE TROPICAL BIOSPHERE

The energetic role of coral reefs in the global ocean is small, but reefs contribute to the tropical biosphere in ways that do not enter into mass balance considerations. Fishes, crustaceans, molluscs, turtles, and a host of other animals are major sources of sustenance for many human populations dwelling close to coral reefs. The contribution of coral reef fisheries to the world's sustainable catch is small, but for the tropical ocean, their yield can be impressive — roughly six million metric tons per year.[403]

The immense physical structures of reefs, built by corals over vast time spans, have also been used to hindcast past climatic events, as rates of calcification are sensitive to changes in solar insolation, water clarity, and temperature. These environmental changes translate into variations in rates of calcification, displayed within the hard coral matrix as bands resembling growth rings in tree stems.[404] These proxy records are invaluable in providing climatologists with a record of past variations in climate.[405] Most of these coral records show considerable variation in the earth's paleoclimate, demonstrating that forecasts of climate change in the future must be considered against a backdrop of considerably large natural variations.

Perhaps the most subtle and often forgotten role that coral reefs play is in the amelioration of water motion, permitting the development of a quiescent seascape suitable for the colonization of mangroves and seagrass meadows[406] and helping to maintain stable shorelines. The organic connection between coral reefs and adjacent tropical ecosystems is often of localized or regional significance.[406] The inorganic connection, however, may be more widespread, as sedimentary facies of some low-latitude continental shelves are composed of varying mixtures of terrigenous and carbonate materials, with much of the latter originating from adjacent coral reefs and reef-associated organisms. These and other shelf habitats are discussed in the next two chapters.

Chapter **6**

THE COASTAL OCEAN I. THE COASTAL ZONE

6.1 INTRODUCTION

The interface between land and the open sea — the continental margin — occupies only about 8 and 0.5%, respectively, of the surface area and volume of the world's ocean. However, the high productivity of coastal seas equates to nearly 30% of total net oceanic primary production and at least 90% of the global fish catch.[407] This high productivity is driven by nutrient inputs from rivers and groundwater, upwelling, exchanges at the continental shelf edge, and atmospheric inputs. The coastal ocean acts as a highly efficient trap for materials from land. A recent estimate suggests that >90% of river-derived materials, including trace metals and pollutants, are retained within deltaic and inner shelf areas.[408] This figure is an overestimate as it ignores the role of tropical riverine export to the ocean margin, but the tight coupling at the land-sea boundary reveals how crucial the coastal ocean is for estuarine and marine food chains, including fisheries, and for the cycling of elements.

Most of our concepts of how the coastal ocean functions arose from the disproportionate research expended within coastal plain estuaries and the adjacent nearshore zone in temperate seas. More recent efforts to understand the role of the tropical coastal ocean, physical-biological interactions, and the atmosphere have begun to broaden our perspective of how coastal seas contribute to the world ocean.

6.2 THE COASTAL OCEAN DEFINED

The traditional notion of the coastal ocean as being defined from the high-water mark on shore to the shelf break is essentially correct but ignores the fact that a dynamic continuum exists between land and the open sea. Figure 6.1 illustrates the main boundaries of the coastal

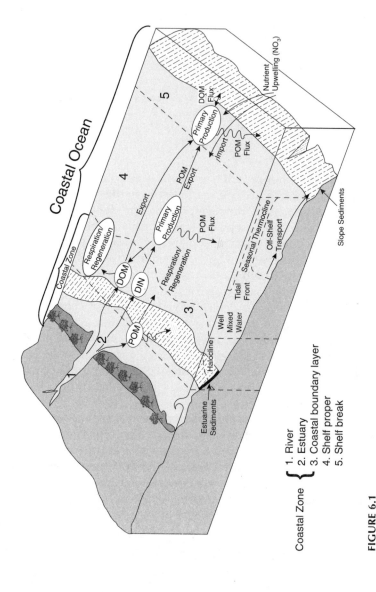

Coastal Zone {
1. River
2. Estuary
3. Coastal boundary layer
4. Shelf proper
5. Shelf break

FIGURE 6.1

Idealized scheme defining the coastal ocean and the coastal zone, with some key biogeochemical fluxes linking land and sea and pelagic and benthic processes. The latter are not to scale.

ocean. The coastal zone encompasses tidal river, estuary, and adjacent inner shelf waters, with the upper limit at the tidal freshwater zone and the seaward limit at the coastal boundary layer, the latter often being delineated by a tidal front. Such boundary layers and tidal fronts are dynamic, oscillating latitudinally and longitudinally over time. Boundary layers are formed when turbid inshore and estuarine waters are mixed and trapped along the coast, particularly in calm weather; there is much slower mixing between coastal and offshore waters. Boundary layers and fronts break down often under strong winds and river discharge.

The coastal zone, therefore, varies in breadth depending on the strength and salient characteristics of local ocean circulation, river discharge, shelf width, climate, and latitudinal position. The coastal zone may not be located close to shore on an arid or semi-arid continental shelf, as such shelves are often macrotidal with mixing of inshore and offshore waters extending to mid-shelf. Conversely, in the wet tropics, the coastal zone often extends beyond the shelf edge in proximity to large rivers — the Amazon being exemplar.

The coastal zone is the subject of this chapter. The mid- and outer-shelf areas, where the coastal ocean interacts with the open sea, are considered in the next chapter.

6.3   WHAT IS AN ESTUARY?

Like the coastal zone of which it is an integral part, the estuary is difficult to define namely because no one definition can encompass all of the diverse combinations of geomorphological settings and water movements that make up the 500,000 to 1,000,000 km of the world's coastlines.[409] Pritchard's classic definition[410] that "an estuary is a semi-enclosed coastal body of water which has a free connection to the open sea and within which sea water is measurably diluted with fresh water derived from land drainage" is overly simplistic. It ignores the role (if any) of tides and excludes water bodies in arid and semi-arid areas which receive freshwater only sporadically (such as in northern Western Australia, Mexico, and parts of Africa) and in wet tropical areas, where mixing of freshwater and seawater occurs on the shelf. Most definitions have therefore not considered the vagaries of tropical estuaries where the mixing boundaries are often not distinct. The circulation in tropical estuaries is often evaporation driven, resulting in what Pritchard called an "inverse estuary". This phenomenon occurs in areas where evaporation exceeds runoff from precipitation, such as in lagoons from Texas to southern Mexico, in Baja California, and in northern Australia.

Here, Kjerfve's definition of an estuary is excerpted:[411]

> An estuarine system is a coastal indentation that has a restricted
> connection to the ocean and remains open at least intermittently.
> The estuarine system can be subdivided into three regions: (a)
> a *tidal river zone* — a fluvial zone characterized by lack of ocean
> salinity but subject to tidal rise and fall of sea level; (b) a *mixing
> zone* (the estuary proper) characterized by water mass mixing
> and existence of strong gradients of physical, chemical, and
> biotic quantities reaching from the tidal river zone to the
> seaward location of a river mouth or ebb-tidal delta; and (c) a
> *nearshore turbid zone* in the open ocean between the mixing zone
> and the seaward edge of the tidal plume at full ebb tide.

Kjerfve[411] not only defines the estuary proper, but also places our
definition of the entire coastal zone on a clear hydrographical foot-
ing.

6.4   HYDROGRAPHIC CLASSIFICATION
OF COASTAL SYSTEMS

Coastal circulation is driven by energy derived from solar heating or
gravity, barometric pressure, and the density of oceanic waters.[411]
Mixing results from tides, wind-driven waves, and buoyancy effects
from river runoff. Water mixing and circulation are greatly affected by
geometry and bathymetry of the coastal zone. Three main types of
estuarine and coastal circulation are recognized:

- Gravitational (due to river runoff)
- Tidal (tidal pumping)
- Wind-driven

Tidal circulation is usually the most important, with interaction by
coastal boundaries generating turbulence, advective mixing, and longitudinal mixing and trapping, the latter setting up coastal boundary
fronts or layers.[411,412] All three circulation patterns may, however, operate simultaneously in a given estuary.

A classification of shallow coastal zones based on the principal
mixing forces (tides, wind, and buoyancy) and related to geomorphology has recently been proposed by Dronkers.[412] No scheme is perfect,
but Dronkers' classification[412] (Table 6.1) is used here because it distinguishes the various types of coastal ecosystems based on water exchange processes (river flow, tides, waves) that greatly affect energy
and material exchanges.

TABLE 6.1

Classification of Coastal Systems Based on Relative Importance
of River Flow, Tides, and Waves

Type	River Flow	Tide	Waves	Description
I	+	−	−	River delta
II	+	−	+	River delta (plus barriers)
III	+	+	−	Tidal river delta
IV	0	+	−	Coastal plain estuary
V	−	+	+	Tidal lagoon
VI	−	+	−	Bay
VII	−	−	+	Coastal lagoon

Note: Plus and minus designations indicate relative impacts; for
example, − + + means that river discharge is very small relative
to tidal and wave energy

Source: Adapted from Dronkers, J., in *Coastal-Offshore Ecosystem
Interactions*, Jansson, B.-O., Ed., Springer-Verlag, Berlin, 1988, 3.

The first three coastal types are river-dominated systems, further
distinguished by the importance of tidal and wave energy. Coastal
plain estuaries represent a transitional phase in which river flow is
moderate and wave energy is minimal compared to the importance of
tides. The last three types (bays and lagoons) are tidal or wave-energy
dominated systems with minimal river input in comparison to tidal or
wave energy. Some coastal systems, such as small, restricted marine
inlets and large, semi-enclosed bays, defy simple classification but are
discussed where most appropriate. Some shallow nontidal fjords, such
as those in Danish waters, may be placed within the coastal lagoon
category.

Deep, silled fjords are also not classified and are not discussed
here. The reader may consult the excellent recent review by Burrell[413]
for an analysis of the energetics of these high latitude systems.

6.5 COASTAL PLAIN ESTUARIES, TIDAL LAGOONS, AND BAYS (TYPES IV, V, AND VI)

The greatest wealth of coastal data exists for coastal plain estuaries and
bays, particularly in North America and in Europe, so we shall examine
these habitats first. Coastal plain estuaries, lagoons, and bays are considered together as the relative importance of tides in these systems is
generally greater than the influence of river inflow and waves (Table 6.1).

Despite two decades of synthesis (see Day et al.[414]), a multi-disciplinary perspective of how coastal plain estuaries function is only now

emerging. Long viewed separately by individual disciplines, scientists are now realizing just how crucial are the interactions among physical, chemical, geological, and biological processes to sustain food webs in these water bodies.

6.5.1 Pelagic Processes

6.5.1.1 Sources of Primary Production and Regulatory Factors

It is often stated that estuaries and adjacent nearshore waters are among the most highly productive ecosystems in the world.[414] This is acknowledged for most North American and European estuaries, but primary productivity varies widely in coastal waters worldwide (7 to 560 g C m^{-2} yr^{-1}) and many estuaries in Asia, Africa, South America, and Australia remain wholly unexplored.[414]

Several factors combine to explain the observed and potential for high estuarine productivity:

- Consortia of phytoplankton, macrophytes, and benthic microalgae that maximize available light and space
- Tidal energy and circulation
- Abundant nutrients
- Conservation, retention, and efficient recycling of nutrients among benthic, wetland, and pelagic habitats (i.e., coupling of sub-systems)

Macrophytes often dominate the primary production of estuaries, depending upon such factors as geomorphology, tidal amplitude, and the relative surface area of wetlands to total estuary area. Many estuaries, however, are dominated by phytoplankton and, to a lesser extent, by microphytobenthos. This can be seen clearly, for instance, in various estuaries along the Dutch coast (Figure 6.2) where rates of primary productivity vary among estuaries, ranging from 160 g C m^{-2} yr^{-1} in the Ems-Dollard to 450 g C m^{-2} yr^{-1} in the Veerse Meer. The contribution of the three main primary producers also varies greatly among the six estuaries, underscoring the wide variation in ecological characteristics even among estuaries in very close proximity. Nevertheless, phytoplankton are the major primary producers in each of the six estuaries.[415]

Phytoplankton biomass and production vary greatly with season and along the longitudinal axis of most estuaries. In temperate estuaries, primary production is not continuous, peaking usually during spring and summer mainly in response to increased solar insolation, water temperature, and nutrient availability.[414] Spatially, phytoplankton production varies from the head to the mouth of most estuaries. In

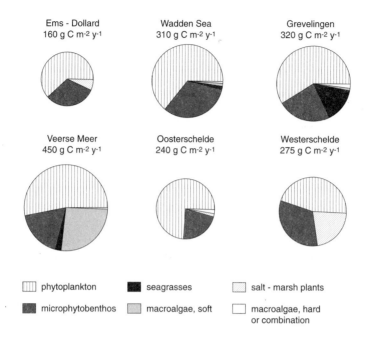

FIGURE 6.2

The relative importance of the different primary producers to carbon production in several Dutch estuaries. (From Nienhuis, P. H., *Hydrobiologia*, 265, 15, 1993. With permission.)

the Ems-Dollard estuary, for instance, phytoplankton production increases seaward, whereas benthic primary production increases in the opposite direction.[415] This pattern is true for most temperate estuaries, but other estuaries show different patterns as to the location and onset of phytoplankton peaks corresponding to temporal and spatial variations in regulatory factors. Year-to-year changes within the same estuary are the norm owing to changes in factors such as river runoff, tidal fronts, coastal water mass movements, grazing, re-suspension, and sedimentation.[416]

In his review of the dynamics of phytoplankton blooms in San Francisco Bay, Cloern[416] noted that blooms are responses to changes in physical characteristics of the coastal ocean, atmosphere, and land. These physical forces vary over time and space such that algal blooms can be episodic, seasonal, or rare events. Phytoplankton production has important biogeochemical consequences, leading to:

• Depletion of inorganic nutrients
• Supersaturation of O_2 and CO_2 removal
• Compositional shifts in the isotopes of carbon and nitrogen

- Production of trace gases (e.g., methyl bromide, dimethylsulfide) that are active climatically
- Changes in toxicity and availability of trace metals (e.g., As, Cd, Ni, Zn)
- Changes in the chemical nature and reactivity of suspended particulate matter
- Transformation and incorporation of inorganic elements into organic matter required for the sustenance of pelagic and benthic heterotrophs

Cloern[416] further emphasizes that evidence is mounting for the alteration of natural bloom cycles by human activities, including inputs of contaminants and nutrients, changes in river discharge, and movements of whole populations. Unfortunately, it is often difficult to distinguish between natural and anthropogenic changes in phytoplankton production; studies of such processes using holistic approaches are required to recognize the complex interactions of phytoplankton with physical forces originating from land, the atmosphere, or within the coastal zone.

In tropical estuaries, physical forces are also important, but the strength and timing of such forces are different: phytoplankton blooms are often cued to the onset of monsoons.[417] Peak production occurs immediately after the monsoons when vertical mixing of seawater and freshwater maximizes the availability of nutrients. Temperature, light, and other abiotic conditions are optimal in the pre-monsoon season, but nutrient levels are depleted. Production is usually low during the monsoon period due to high turbidity and low light levels. For example, in the Mandovi estuary in India,[417] pre-monsoon, monsoon, and post-monsoonal rates of primary productivity average 570, 262, and 1077 mg C m^{-2} d^{-1}. Rates of primary productivity are high in many other Indian estuaries due to high rates of organic pollution.

In unpolluted tropical estuaries, there are normally clear shifts in nutrient concentrations and physicochemical conditions between wet and dry seasons. Taking the Moresby river estuary in northern Australia as an example, the changes in nutrients and water conditions between seasons are evident (Table 6.2). In the wet season, river water is slightly less acidic over the same salinity range. Concentrations of most dissolved and particulate nutrients are greater in the wet season, but the lower dissolved phosphate levels can be attributed to sorption to the increased suspended loads. Biological activity was not measured, but nonconservative mixing curves indicate that nitrogen was biologically removed from the water column.[418]

In some tropical, semi-enclosed bays and lagoons, the effect of monsoons on primary productivity may depend upon factors other

TABLE 6.2

Variations in Water Chemistry and Nutrient
Concentrations Between Wet and Dry Seasons in the
Moresby River in Tropical North Queensland, Australia

Parameter	Dry Season	Wet Season
pH	5.30 ± 0.11	5.60 ± 0.18
PO_4^{3-} (μM)	0.4 ± 0.4	0.2 ± 0.1
NO_3^- (μM)	2.3 ± 2.8	31.2 ± 26.3
Si $(OH)_4^+$ (μM)	186.4 ± 101.5	211.4 ± 54.9
Particulate bioavailable P (μmol g^{-1})	0.9 ± 0.4	6.6 ± 7.8
Particulate total P (μmol g^{-1})	14.9 ± 6.6	79.2 ± 41.5
Particulate N (μmol g^{-1})	13.7 ± 3.7	58.1 ± 55.5

Note: Values are mean ± 1 standard deviation.

Source: Adapted from Eyre, B., *Estuarine Coastal Shelf Sci.,* 39, 15, 1994.

than destratification of the water mass. For instance, in shallow Bietri Bay within Ebrie Lagoon on the Ivory Coast of Africa, primary productivity is not noticeably changed by the onset of the monsoon.[419] This can be attributed to the conservative behavior of nutrients in this eutrophic bay and the dominance of microheterotrophic activity. The influence of monsoons on tropical estuaries generally declines with increasing distance from the equator.

Primary production in estuaries can be estimated from daily light irradiance, photic depth, and biomass[420] using the empirical formula:

$$P_{Np} = 150 + 0.73B\ I_oZ_p \qquad (6.1)$$

where P_{Np} = net primary production; B = biomass; I_o = daily irradiance, and Z_p = depth of the photic zone (1% of ambient light). This relationship does not mean that other parameters such as nutrient concentrations and biological interactions are not important, but that they are accounted for in the biomass number as it integrates the balance of growth from nutrient resources and mortality due to grazing. This equation gives a good fit for primary production in North American estuaries but does not give a best estimate for some turbid European estuaries.[421]

Recent attempts have been made to clarify the problem of how light, nutrients, water mixing, and other environmental factors interact to regulate phytoplankton production in temperate estuaries.[416] Water-column mixing in estuaries and bays has the antagonistic effect of dampening light availability via sediment resuspension and the synergistic effect of exposing phytoplankton cells to a variable light and nutrient regime. This effect was investigated by Mallin and Paerl[422] in a North Carolina coastal plain estuary by use of a rotating light simulator to mimic the effects of a constantly varying light regime in the estuary. They found that constantly varying irradiance stimulated phytoplankton productivity by reducing photo inhibition and ameliorating the effects of light limitation by increased turbidity. These results also suggest that static incubations to measure phytoplankton production underestimate the true rates of production.

Simultaneous or alternating periods of nutrient and light limitation also occur, but these phenomena have not been measured often. In the Delaware estuary, Pennock and Sharp[423] found that in winter, light is limiting for phytoplankton growth due to high turbidity and a well-mixed water column. In late spring, phosphorus becomes limiting as biogeochemical processes remove dissolved phosphate as it transits the estuary until it is stoichiometrically depleted relative to dissolved inorganic nitrogen. During summer, limitation shifts from phosphorus to nitrogen as biological and geochemical desorption processes regenerate phosphorus, resulting in a greatly decreased dissolved inorganic nitrogen-to-phosphate ratio.

This scenario is contrary to the well-established tenet that nitrogen is the limiting nutrient for temperate phytoplankton. Obviously, nutrient limitation is more complicated that currently believed in that physical flushing and water-mixing rates, geochemical equilibria reactions, and biological processes all determine nutrient availability and, in turn, phytoplankton growth in estuaries.

6.5.1.2 Nutrient and Food Web Dynamics

As exemplified by the recent work in the Delaware estuary, nutrient limitation can oscillate over time and space, but recycling of nutrients is normally intense. The flux of ammonium is particularly intense, being the preferred form of nitrogen for most pelagic autotrophs and microheterotrophs,[424] with lesser consumption of urea, nitrite, and nitrate.[425] In the turbid coastal zone, most nutrients (mainly in the reduced forms, NH_4^+ and urea) are supplied to the plankton by regeneration from the microbial loop, from sediments, and from land.[424]

Regional and seasonal hydrographic conditions, however, may result in greater availability and utilization of the oxidized forms of nitrogen. For instance, during spring and autumn in the Chesapeake Bay, nitrite becomes available in the photic zone via nitrification in the deeper waters when the pycnocline breaks down and ammonium-rich water mixes into the oxygenated surface layer.[426] Nitrate from land may also constitute a significant input of new nitrogen, particularly from agriculture land use. The supply and use of autochthonous vs. allochthonous nutrients, however, have rarely been distinguished in inshore waters.

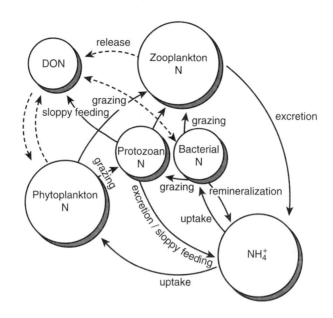

FIGURE 6.3
Idealized depiction of dissolved nitrogen cycling between phytoplankton, zooplankton, and members of the microbial loop.

Figure 6.3 illustrates the cycling of dissolved nitrogen among phytoplankton, bacteria, protozoa, and zooplankton via grazing, excretion, remineralization, and various uptake processes. On average, pico- and nanoplankton (0.2 to 2 μm and 2 to 20 μm cells, respectively) are responsible for the bulk of nitrogen regenerated in coastal waters, with inputs (assimilation) balancing outputs (remineralization). Protozoans in particular regenerate a major portion of the nitrogen pool in most water-column environments, but not all. Bacterioplankton and micrometazoans also contribute to nutrient regeneration depending upon:

- Physiological and nutritional status
- Temperature
- Rates of primary productivity
- Nutrient concentrations and stoichiometry
- Plankton community composition

Regardless of the source of nutrient supply, the microbial loop (dominated numerically by bacterioplankton) is at the center of pelagic food web dynamics in estuaries, bays, and lagoons. In their review of bacterial production, Ducklow and Shiah[163] noted that rates of bacterial production and abundance are higher in the coastal zone than in offshore waters, owing to greater rates of organic input and higher rates of primary productivity as mixed by hydrographic processes. Rates of bacterial productivity range widely from 0.5 to 803 mg C m^{-3} d^{-1} as does cell abundance (0.1 to 38×10^9 cell l^{-1}) and bacterial biomass (2 to 760 mg C m^{-3}). Bacterial cells attached to particles often correlate positively with higher suspended solid loads in turbid coastal waters,[163] and rates of bacterial production can exceed rates of primary productivity. Further seaward, bacterial production rates are less than rates of primary production; bacterial abundance and activity are rarely related to any single factor. Each estuary, bay, and lagoon is different to the extent that any combination of factors can regulate microbial activity and abundance over different spatial and temporal scales.

Nevertheless, ordinarily there is close coupling between primary and bacterial production in coastal waters because pelagic bacteria are largely fueled by dissolved and particulate carbon derived from phytoplankton exudates and phytodetritus. This material is degraded rapidly by the microbial consortia present, but mostly by the heterotrophic bacterial community — up to 60% of carbon fixed by phytoplankton flows through bacteria, either lost as respiration or assimilated into new biomass. Picoplankton dominate the energetics of pelagic food webs as demonstrated by their high rates of respiration, on average accounting for 40 to 90% of pelagic respiration in temperate coastal areas.[163,421] Understanding the role of heterotrophic bacteria is crucial to understanding nutrient flux processes and food web dynamics in coastal waters.

What controls bacterial numbers and productivity in the coastal zone? As with all populations of organisms, coastal bacteria are regulated by availability of resources, environmental cues, and some biological interactions. Ducklow and Carlson[427] argue that bacteria must be considered within the context of the entire plankton community rather than in isolation. This is because recent theoretical work suggests that organisms may be regulated within the food web structure as a "trophic cascade". Derived from the study of food webs in temperate lakes, the concept states that regulation works in two ways. First, there are some "bottom-up" controls in which a change in resources at the base of the food web will propagate up the food web with decreasing level of control. For coastal bacteria, this would mean, for example, that an increase in organic matter input into an estuary would result in a concomitant increase in bacterial biomass and activity but smaller increases in zooplankton and fish biomass. Second, some regulation works by "top-down" control in which changes in a consumer population results in a decline in prey immediately below, but in an increase in the prey population further down the web. For coastal plankton, this would mean that an increase in the carnivorous copepod population would lead to a decrease in protozooplankton numbers but an increase in bacterioplankton biomass caused by a release from immediate predatory control. Bacteria and other coastal plankton are thus regulated by a combination of "bottom-up" and "top-down" controls.

A variety of evidence indicates that pelagic bacteria, being located immediately above the producer level near the base of coastal food chains, are regulated more by changes in environmental factors (light, food, temperature) than by predation.[163] This is seen most clearly from temporal and spatial variations in bacterial numbers and activity within large temperate estuaries. In the Chesapeake Bay, for example, spatial and temporal dynamics of bacterial populations in bay waters relate best to water temperature, except in summer when temperatures are >20°C and bacteria are either substrate limited or affected by anoxia in the mid-bay region.[428] A longitudinal gradient is frequently observed in temperate estuaries and bays in response to allochthonous inputs of organic matter whereby bacterial production rates decline from the head to the mouth of estuaries and seawards within bays, embayments, and tidal lagoons.[163]

Physical processes have until recently been ignored when considering trophic energetics within the microbial loop in coastal waters. Tidal fluctuations in plankton abundance and microbial activity have been noted in many macrotidal estuaries in Europe and in North America. Eldridge and Sieracki[429] have modeled both the biological and hydrological regulation of pelagic microbial processes in the York River estuary. This system, a tributary within the Chesapeake Bay, undergoes tidally induced (spring-neap) stratification and destratification cyles. Good predictability of these cycles and a wealth of microbial data allowed model estimates of physical and trophic regulation of pelagic carbon cycling. The carbon model (Figure 6.4) indicates oscillation of trophic interactions and carbon fluxes cued to the stratification-destratification cycles in the estuary. Abundance of cyanobacteria and heterotrophic bacteria oscillate with the tidally induced events, but the abundance of their principal grazers, heterotrophic protists, do not.

The model predicts a close coupling, however, between cyanobacteria and grazing during destratified periods, when cyanobacterial numbers are low. When the estuary is stratified, cyanobacterial standing stocks are high because cyanobacterial growth rates outpace grazing, with little effect on cyanobacterial numbers. Heterotrophic bacterial growth and consumption are closely coupled, but the oscillations in destratification/stratification events leads to a decoupling of consumption and production.

The changes induced by these hydrodynamic cycles can result in changes in dissolved organic carbon (DOC) excretion, growth efficiency, respiration, and growth rates within the microbial loop (Table

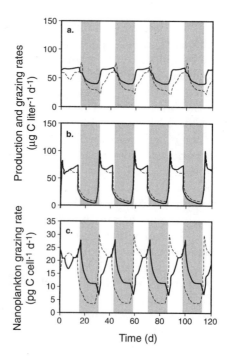

FIGURE 6.4

Predator-prey cycles simulated based on empirical data from the York River estuary in the Chesapeake Bay. (a) Bacterial production (——) and heterotrophic nanoplankton grazing (- - - -) on bacteria; (b) cyanobacterial production (——) and nanoplankton grazing (- - - -) on cyanobacteria; and (c) cell-specific grazing rates of heterotrophic nanoplankton on bacteria (——) and cyanobacteria (- - - -). Shaded areas denote destratified periods. (From Eldridge, P. M. and Sieracki, M. E., *Limnol. Oceanogr.*, 38, 1666, 1993. With permission.)

6.3). There is a noticeable shift in total DOC flux in the estuary from heterotrophic nanoplankton to phytoplankton as the estuary becomes destratified; heterotrophic protists show greater growth efficiency as reflected in lower respiration and growth rates. The importance of physical forcing events on pelagic community dynamics is clear, but remains to be demonstrated in most estuaries.

It is generally agreed that the community dynamics of pelagic heterotrophic bacteria in estuaries is balanced between production and grazing — a condition of steady state in most instances — but recent findings suggest that this is not always the case. Non steady-state behavior of pelagic communities can develop when contaminants are introduced, leading to high rates of cell mortality.[163,416]

Viruses are now considered one of the prime factors leading to bacterial mortality in coastal waters, equivalent to grazing by protists.[430] Recent evidence suggests that viruses can control bacterial

TABLE 6.3

Effect of Stratification and Destratification Events on the
Energetics of the Microbial Loop in Waters of the York River
Estuary, Chesapeake Bay, Virginia

	Stratification	Destratification
Excretion of dissolved organic carbon (DOC) (% of total flux)		
Phytoplankton	42	69
Cyanobacteria	6	3
Heterotrophic nanoflagellates (recycled)	52	28
Growth efficiency (%)		
Bacteria	30	30
Heterotrophic nanoflagellates	42	60
Respiration (μg C ml^{-1})		
Bacteria	21	15
Heterotrophic nanoflagellates	70	20
Growth rate (d^{-1})		
Cyanobacteria	1.5–3.0	1.2
Bacteria	0.2	0.3
Heterotrophic nanoflagellates	4.0	1.8–2.0

Source: From Eldridge, P. M. and Sieracki, M. E., *Limnol. Oceangr.*, 38, 1666, 1993. With permission.

abundance in some coastal waters. Fuhrman and Noble[430] found in coastal waters off southern California that 24 to 66% of bacterial losses can be attributed to viral infections. Model estimates indicate that losses to viruses leads to ~27% increases in bacterial production, but roughly a 37% loss in bacterial carbon exported to protozoan grazers and a 7% decline in macrozooplankton production. Such a demonstrable shift away from higher trophic levels leads to a significant diversion of energy and carbon flows. Whether or not viruses in coastal waters are a natural phenomenon or originate from human influences is still unclear.

Regardless of viral influences, flagellated, ciliated, and amoeboid protozoa have a significant influence in transferring heterotrophic bacterial biomass to higher trophic levels.[431] Being highly abundant in most coastal waters (10^8 to 10^{10} flagellates and 10^3 to 10^8 ciliates per m^3), protozoa feed on bacteria, other protists, and phytoplankton at rates partly determined by prey abundance and cell size. The proportion of bacterial and phytoplankton biomass that is consumed by microzooplankton varies greatly but often exceeds 50% and, in some instances, accounts for all prey biomass. Protozoans may remove a greater portion of phytoplankton biomass than larger zooplankton, exerting considerable grazing impact on phytoplankton activity in some bays, estuaries, and lagoons.[431]

Unfortunately, there are few such data from subtropical and tropical coastal waters. The study by Dagg[432] in warm, temperate Fourleague Bay, a highly productive tributary of the Atchafalaya River in Louisiana, indicates that >95% of grazing on phytoplankton was by the microzooplankton community. The role of the microbial loop may therefore be as intense, or more so, in warmer coastal waters than in cooler temperate waters. This is still open to question until more data are collected.

The degree of predatory control by metazoan zooplankton on protozoan communities has received relatively little attention. Sanders and Wickham[431] suggest that metazoans exert some control over abundance of protozoa in coastal waters. Copepods, rotifers, cladocerans, fish larvae, chaetognaths, euphausid and mysid shrimps, and other pelagic metazoans all ingest protozoans to varying degrees, as suggested by laboratory experiments and gut content analyses. Reciprocal relationships observed in the field between protozoan and metazoan population densities also suggest some predatory control, but no hard estimates exist for the proportion of protozooplankton abundance lost to predation by metazoans in coastal waters.[431]

The significance of meso- and macrozooplankton to energy flow in estuaries, bays, and tidal lagoons was established long before recognition of the importance of detritus and microbes in coastal food webs, owing to their importance as food for humans or as prey items for commercially important fish.[414] Endemic estuarine species feed on phytoplankton, protozooplankton, smaller metazoans, and larvae, but they (especially larval and juvenile stages) may feed extensively on detritus particles and associated microbes in the water column and at the sediment surface.[424,431] These plankton communities can maintain high densities throughout the year in many temperate estuaries given their dietary habits as generalists and tolerance to changes in physico-chemical conditions.

Despite the fact that the trophic roles of meso- and macrozoo-plankton are well-known (mainly in temperate estuaries and bays) how their life cycles and distribution are affected by estuarine hydro-dynamics has received much less attention until recently. Factors promoting retention of zooplankton in estuaries are still not clear. Recent evidence indicates that some zooplankton, such as the copepods *Acartia* and *Eurytempora*, actively migrate upwards in the water column on flood tides to be displaced on ebb tide in order to minimize energy loss and maximize the use of tidal energy.[433] Whether or not this behavior is in response to food needs, predator avoidance, or reproductive strategy is unclear.

Both estuarine endemics and marine invaders move within estuaries in response to changes in environmental conditions, with their relative abundance determined by tidal regime, the extent of freshwater discharge, and the intrusion of more saline waters.[433] Zooplankton may be imported or exported from a given coastal water body depending upon tidal regime, mobility, and, in estuaries, the location of the turbidity maximum. Other factors such as oxygen and nutrient concentrations play a strong regulatory role, particularly in estuaries near large population centers such as the Chesapeake Bay and Hudson River estuary. Zooplankton abundance in a given estuary or embayment is therefore a function of their responses to changes in water residence times and movements, climate, and their physiological limitations. No studies have yet attempted to consider all of these regulatory factors simultaneously, even in the long-studied estuaries of North America and Europe. Biological processes in tidal estuaries, bays, and lagoons are not divorced from changes in environmental conditions but are, in fact, ultimately controlled by them.

6.5.2   Benthic Processes

6.5.2.1 Sources and the Composition
of Sediment Organic Matter

Organic matter in the sediments of tidal estuaries, bays, and lagoons is a rich mixture of particulate and dissolved (including colloidal) material derived from various autochthonous and allochthonous sources.[434] Dead and decaying phytoplankton, epiphytes, benthic microalgae, macrophytes, and metazoans constitute the richest sources of proteins, carbohydrates, lipids, fatty acids, amino acids, and other organic compounds that sustain benthic life. Detritus derived from land or advected from offshore, on the other hand, frequently comprises a large portion of the standing amounts of sediment organic matter in the coastal zone. Living organisms constitute only a small percentage of the total sediment organic pool.[434]

The process of identifying and distinguishing the sources of organic matter in sediments (and in the water column) is complex and still in its infancy. It is also complicated by the natural temporal and spatial changes occurring in the quantity and quality of organic detritus present in coastal habitats, and by the fact that labile organic compounds are mixed, sorbed, and diluted by a large pool of refractory organic matter (residues of earlier diagenetic processes) — fulvic and humic acids — and by industrial and sewage inputs and inorganic particles derived from land. It is, therefore, not surprising that one quarter to one half of organic matter in coastal and shelf sediments remains uncharacterized, at best operationally defined based on solubility characteristics.[434]

Identification of the sources and nature of organic matter in sediments has been attempted by several methods, including stable isotopes, microscopic examination, elemental analyses, and the molecular biomarker approach. Microscopic methods are time consuming and not quantitative; elemental analyses are quantitative but offer no clue as to the quality of the organic material. Although not without problems, the stable isotope and molecular biomarker approaches offer the best chance for success in separating organic matter derived from multiple sources.[435]

Stable isotope ratios ($\delta^{13}C$, $\delta^{15}N$, $\delta^{34}S$) of organic matter in coastal waters and sediments can be used to distinguish among terrestrial, wetland, and estuarine or marine sources of organic matter (as discussed in Chapters 3 and 4 for macrophyte-dominated systems). A gradual shift in stable isotope ratios in suspended particulate matter has often been observed along the longitudinal axis of estuaries, but such shifts can be the net result of the mixing of distinct organic sources or of a shift in isotopic composition of organic matter sources downstream.[435] Intrinsic variations can be large, so interpretations must be cautious. Nevertheless, some clear distinctions have been made in some estuaries and bays, giving us some insight into the sources of organic matter and mechanisms of transport in the coastal zone.

Measurements of the isotopic composition of particulate organic matter in estuarine waters show stable isotope ratios outside the range of well-defined terrestrial and marine end-members.[435-437] The stable isotope study by Cifuentes et al.[435] observed clear seasonal changes in the nature of suspended particulate matter in the Delaware estuary and identified two major pools of organic matter — autochthonous phytoplankton and a mixture of pelagic and terrestrial detritus. Most of the suspended particulate matter was composed of planktonic debris, but biogeochemical processes occurring throughout the estuary significantly altered isotopic signatures, much more so than physical mixing. $\delta^{15}N$ values in particular appear to vary in relation to the extent of isotopic fractionation of ammonium that is assimilated by phytoplankton. Similarly, high $\delta^{13}C$ values relate to high rates of in situ primary productivity. Despite these variations, isotopic signatures for organic material in Delaware Bay sediments show a strong signal equivalent to the suspended matter. Cifuentes et al.[435] estimate that 15 to 30% of planktonic primary production was deposited to the benthos in spring. A study[436] in Dabob Bay in Washington state using both stable isotopes and molecular biomarkers showed that water-column particulates were derived from a mixture of planktonic debris and vascular plant detritus, the latter accounting for 10 to 35% of pelagic total organic carbon (TOC). Organic matter derived from land in Dabob

Bay comprises an average of 66% of sediment organic material and 33% of suspended organic matter. Roughly one half of this land-derived material is composed of gymnosperm wood with lesser amounts of gymnosperm tissues and angiosperm wood. Molecular analyses indicated that the bulk of the land-derived, vascular material had undergone extensive degradation prior to transport into the bay. Organic matter in Dabob Bay sediments is compositionally most similar to the gymnosperm wood material introduced into the bay during winter runoff. There was little residual evidence of planktonic and other vascular plant inputs that occur during the rest of the year.

A similar study of sediment in several bays and fjords in northern Newfoundland found organic carbon and nitrogen isotope values intermediate between phytoplankton and macrophytes, suggesting a mixture of at least two sources. Some land-derived material was found within at least one of the bays.[437] The nature of organic matter in estuarine, lagoon, and bay sediments is derived from the predominant sources at any given time, normally resulting in a complex mixture of compounds originating from *in situ* primary and secondary production, land, and offshore. These temporal and spatial mosaics of organic input have a significant impact on benthic faunal and biogeochemical processes which, in turn, affect pelagic production and nutrient cycling processes in the coastal zone.

6.5.2.2 Detrital Food Chains

The structure and energetics of coastal benthic food chains are greatly influenced by changes in climate, sediment type, salinity, and other environmental conditions, but also by changes in food supply.

Both on a local scale and at the systems level, suspension-feeding macrobenthos are capable of depressing phytoplankton biomass in many estuaries, bays, and tidal lagoons; filter-feeding rates often exceed local productivity of phytoplankton. The supply of phytoplankton biomass to an area is therefore important and greatly dependent on the hydrodynamics of coastal waters, particularly water residence times.[438] The biomass of benthic suspension feeders is greatest in coastal habitats where residence times are shortest, implying that clearance rates of food are a function not of particle concentration, but of the rate of water flow.

Benthic filter feeders consume a wide variety of particles other than phytoplankton cells. Recent evidence indicates that organic aggregates precipitated from macrophyte-derived dissolved organic matter (DOM) can serve as food for estuarine bivalves.[438] Dietary plasticity in which free bacterial and phytoplankton cells, detrital particles and attached microbes, and organic aggregates are consumed, is well known for many benthic suspension feeders. Considering the high rates of grazing and the variety of particles consumed, benthic suspension feeders can play a large role in nutrient recycling and in sedimentation and resuspension of particulate matter in the coastal zone, especially in upwelling regions.

Physical forces alone do not determine the distribution of benthic suspension feeders over time and space. Constraints on the local scale suggest why they are patchily distributed. The extreme limits of biomass on both sides of the environmental gradient are set by physical forces such as turbulence, vertical mixing, and horizontal advection, which optimize food intake, but the limits of median densities of suspension feeders are set by intra- and interspecific interactions, predation, and reproductive strategies. The recent studies of Herman and his colleagues[439,440] show that, in European tidal waters, suspension-feeding benthic communities can stabilize eutrophic estuarine systems by ameliorating the effects of plankton blooms and shunting excess material to biodeposits. They also serve to stabilize comparatively open systems with short residence times by retaining and thus conserving nutrients within the system, minimizing losses of organic matter. Their considerable influence on estuarine dynamics can best be discerned by the observed shift in some estuaries from phytoplankton to macroalgal-based production following mass mortality of benthic filter-feeding populations.

Deposit-feeding benthic communities are fueled by the complex mixtures of organic materials in sediments.[143] In temperate coastal zones, clear temporal changes to benthic detrital food chains occur in food resources. The seasonal variation in the fragmentation and decay of salt marsh and seagrass detritus, winter blooms of seaweeds, sedimentation of plankton blooms in spring and autumn, and growth of benthic microalgae all make up a continually changing diet of food for benthic heterotrophs, in addition to changes in microbial foods. Deposit-feeding benthic communities live in an ever-changing milieu, responding to seasonal inputs of various forms of detritus.

It is now fairly well established that infauna, including microbes, respond positively to the sedimentation of the spring phytoplankton bloom in temperate waters, as indicated by early spring bursts of larval recruitment and increased rates of benthic oxygen consumption.[143] Such events probably do not occur in tropical estuaries — at least no data exist documenting such an occurrence; tropical benthic deposit-feeders may be more affected by human- and climate-induced disturbances than by food availability.[12]

Enormous research effort has been expended since the 1960s in determining the nature of benthic deposit feeding. Arguments have raged over the relative importance of microbes vs. detritus as food sources for benthic detritivores, mired in disagreement over the role of cophrophagy and optimal foraging. A general consensus seems to have emerged that is a more realistic concept of life in sediments: diets vary greatly among detritivores, as some species feed mainly on bacteria and microalgae (either free or attached to inorganic and organic particles, including fecal pellets), while other species feed mostly on particulate detritus derived from mixed sources. Most, however, feed on a variety of living and dead organic matter in order to obtain the necessary proteins, lipids, vitamins, and other essential compounds necessary to sustain life.

Bacteria have long been considered a main food resource, but recent books[441] and reviews[5,7] demonstrate that bacteria alone are insufficient to meet the carbon requirements for most benthic detritivores, including most meiobenthic species. In fact, benthic microalgae have been recognized as a more nutritious food than bacteria owing, in part, to their more complex lipid and fatty acid contents. From a nutritional perspective, bacteria may be ingested largely as a requirement for nitrogen. As illustrated in Figure 6.5, no one factor, but rather an "interactive hierarchy" of mechanisms, regulates detrital food chains: [442]

- Quality and quantity of food resources
- Physiological constraints (bioenergetics) and tolerances
- Behavioral and life history strategies

All of these mechanisms operate within the confines of environmental cues.

6.5.2.3 Mineralization of Organic Matter

Meiofauna and macrobenthos may do more to stimulate microbial growth and detritus mineralization rates by tube-building, bioturbating, defecating, respiring, and secreting mucus than by grazing.[5,7] Few studies have demonstrated that benthic metazoans control microbial standing stocks in nature. In most sediments, rates of bacterial growth and productivity outpace rates of consumption by metazoans, suggesting microbial food webs in sediments are self limiting (a trophic dead end) or controlled by extrinsic factors.[5] The sediment matrix is an exceedingly complex environment of sharp, and sometimes subtle, chemical, geochemical, and biological gradients. For the most part, benthic metazoans modify their surroundings to avoid coming into contact with anoxic conditions.

Recognition of the importance of sediment biogeochemistry is one reason why some common insights have been reached regarding the energetics of benthic deposit feeders. Even to the present day, ecologists

Regulatory Mechanisms of Detritus Food Chains

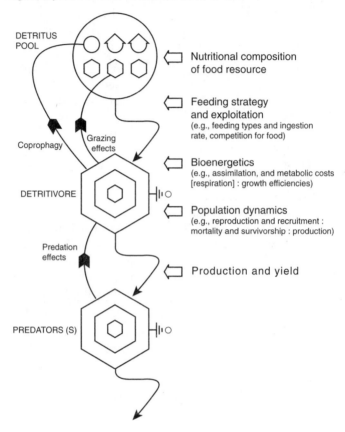

FIGURE 6.5
Hierarchy of mechanisms regulating benthic detrital food chains. (From Tenore, K. R., in *Ecology of Marine Deposit Feeders*, Lopez, G., Taghon, G., and Levinton, J., Eds., Springer-Verlag, New York, 1989, chap. 14. With permission.)

use terms such as "redox potential discontinuity (RPD) layer" when they mean to describe the oxic-anoxic interface. The RPD layer is a visual boundary that actually distinguishes by color the brown, oxidized surface sediment layers from the deeper, gray or black, anoxic sediments. Molecular oxygen penetrates only a few millimeters into the sediment surface; some of the brown sediments are actually postoxic, devoid of free oxygen but containing some bound oxygen (e.g., NO_3, PO_4). The gray or black layer observed is actually the sulfate-reducing zone where the metabolic byproducts of anoxic decomposition form iron mono- and polysulfides, the compounds that darken sediments gray to black within this zone.

The diagenetic zones within sediments are much more complicated than benthic ecologists have recognized. Advances in microbial and biogeochemical methods (and knowledge) have accelerated the pace of our understanding of the dominant role benthos play in decomposing organic matter and in recycling nutrients back to the photic zone.

Bacteria are responsible for the bulk of organic matter decomposition in coastal sediments.[5,7] Significant progress has been made in recent years to quantify the rates and pathways of organic matter degradation. The rate of early diagenesis is influenced by several factors, including:[4,6]

- Rate of organic input
- Temperature
- Intensity of bioturbation
- Quality of the organic material
- Sediment type
- Mass sediment accumulation rate

Early diagenesis in coastal sediments is most often dominated by oxygen consumption and sulfate reduction, but few studies have measured more than one or two specific processes simultaneously (see Tables 23 and 24 in Heip et al.[421]). It was commonly believed until recently that organic matter oxidation proceeds most efficiently via aerobic respiration, proceeding more slowly as organic matter is mixed and degraded within deeper, anoxic sediment layers. Recent studies indicate, however, that

- Anaerobic degradation can often be as rapid (or more so) as aerobic decomposition.
- Other anaerobic processes, such as iron and manganese reduction,[443] often are as important as sulfate reduction in postoxic and anoxic sediments.

Several field[444] and laboratory[445,446] studies have shown that aerobic and anaerobic diagenesis can proceed at equivalent rates in some coastal sediments. In Long Island Sound, Mackin and Swider[444] measured ΣCO_2 production and the distribution of various porewater solutes and their exchange across the sediment-water interface to model the pathways of early diagenesis. Their calculations showed that oxygen consumption accounted for 3 to 14% of total organic matter oxidation, whereas the contribution of sulfate reduction to total decomposition ranged from 65 to 85%. They suggested that aerobic respiration contributes more to early diagenesis under conditions of low rates of organic input, intense

TABLE 6.4

Budget for Carbon and Nitrogen in Microcosms With and Without
the Polychaete *Nereis virens*

	Nereis virens	Aerobic Control	Anaerobic Control
Carbon			
a. ΔPOC in sediment	−44.9	−17.3	−26.3
b. ΔPOC in *Nereis*	−11.1	—	—
c. $CaCO_3$ dissolution	−2.4	−1.7	−1.2
d. ΔTCO_2 in porewater	1.6	2.6	3.7
e. Loss from sediment (a + b + c + d)	−56.8	−16.4	−23.8
f. Observed TCO_2 flux	−186.5	−52.6	−64.0
g. Unaccounted for carbon (f − e)	−129.7	−36.2	−40.2
Nitrogen			
a. ΔPON in sediment	−7.56	−4.68	−5.45
b. ΔPON in *Nereis*	−2.61	—	—
c. ΔDIN in porewater	−0.01	0.78	1.13
d. Loss from sediment (a + b + c)	−10.18	−3.90	−4.32
e. Observed DIN flux	−3.07	−0.89	−3.78
f. Denitrification (d − e)	−7.11	−3.01	—
g. Nitrate reduction	−11.66	−4.93	—
h. Observed nitrate flux	−0.77	−0.51	—
I. Nitrification (g + h)	12.43	5.44	—

Note: Values are mean rates (mmol m^{-2} d^{-1}) over the 94-d incubation period.
POC = particulate organic carbon; PON = particulate organic nitrogen;
DIN = dissolved inorganic nitrogen.

Source: From Kristensen, E. and Blackburn, T. H., *J. Mar. Res.*, 45, 231, 1982.
With permission.

bioturbation, and/or where organic matter concentration declines greatly
with increasing sediment depth. The differences between relative rates
of oxygen consumption and sulfate reduction are related to sedimentation rates.[447] At comparatively low rates of sedimentation, rates of
oxygen consumption outpace rates of sulfate reduction, but the overlap
between the two data sets indicates that they proceed at roughly equal
rates at sedimentation rates exceeding 0.1 g cm^{-2} yr^{-1}.

This phenomena was explored in greater detail in laboratory experiments manipulating benthic faunal densities and the oxygen status
of the overlying water.[445,446] In one experiment, Blackburn and his
colleagues[445] constructed carbon and nitrogen budgets for microcosm
sediments with and without the polychaete, *Nereis virens* (Table 6.4).
Their results showed that:

- The presence of the polychaete enhanced the loss of particulate
  organic carbon (POC) and particulate organic nitrogen (PON) and
  fluxes of total CO_2 and dissolved nitrogen.

- Rates of carbon and nitrogen flux were significantly greater in anaerobic sediments without the worms than in the aerobic microcosms with and without the polychaete.

The differences between control and polychaete treatments were attributed to worm feeding, bioturbation, and respiration and the stimulation of microbial decomposition (including denitrification and nitrate reduction) by these activities. These experiments[448-450] show the large influence of particle reworking, feeding, burrowing, tube construction, and ventilation by benthic infauna on benthic mineralization rates and specific carbon and nitrogen processes. Micro-environments created by infauna facilitate a tight coupling and transfer of dissolved nitrogen between oxic and anoxic processes.

Oscillations in redox status are the rule rather than the exception in coastal sediments. Bioturbation processes and physical disturbances either stimulate rates or alter the pathways of organic matter decomposition in shallow coastal deposits, or both. Aller[450] demonstrated the stimulatory effects of oscillating sediment redox status by simulating changes in irrigation rates of infaunal tubes in the laboratory. Some evidence that physical disturbance also plays a role in affecting early diagenesis in coastal sediments comes from the study by Alongi[451] in the coastal zone of the central Great Barrier Reef lagoon. In the dry season, sulfate reduction dominates organic matter degradation, but in the wet season, storm-induced resuspension and monsoonal rains oxidize subsurface sediments, altering porewater chemistry to the extent that aerobic respiration dominates early diagenesis. Monsoons, therefore, play a crucial role not only in proximity to rivers, but also in very shallow shelf areas in the tropics. Unfortunately, little if any other such data exist from tropical, marine coastal areas.

Complete carbon budgets of benthic mineralization processes are available only for a few temperate coastal locations, but these studies are helpful in giving a wider perspective of the relative rates of different mineralization processes, preservation, and fate of organic material. As we observed for mudflat sediments and those of macrophyte-dominated systems (Chapters 2 and 4), present models imply that anoxic degradation accounts for the bulk of organic matter decomposition.

A carbon budget for sediments in Skan Bay in Alaska (Figure 6.6) shows that of the 22.2 mol C m^{-2} yr^{-1} of POC that is deposited in bay sediments, 82 to 100% is remineralized, and 0 to 18% is buried. Sulfate reduction is the major metabolic process, accounting for 79% of total mineralization; nitrate reduction and methanogenesis each accounted for only ~3%, with aerobic respiration making up the balance (15%) of total organic carbon oxidation.[453]

In semi-enclosed lagoons and embayments, such as Cape Lookout Bight located on the outer banks of North Carolina, rates of organic

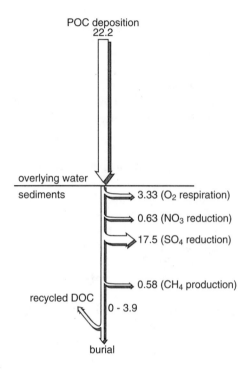

FIGURE 6.6

Annual carbon budget (mol C m^{-2} yr^{-1}) for sediments of Skan Bay, Alaska. All values represent means, except for the range of burial rates. (Data from Alperin, M. J., Reeburgh, W. S., and Devol, A. H., in *Organic Matter: Productivity, Accumulation, and Preservation in Recent and Ancient Sediments*, Whelan, J. and Farrington, J. W., Eds., Columbia University Press, New York, 1992, chap. 6.)

matter accumulation are usually rapid.[453] Such high rates of organic input induce a high state of anoxia in which a sufficient amount of labile organic matter persists below the zone of sulfate reduction to fuel methane production and promote preservation in deeper sediment layers (Figure 6.7). Of the average annual organic input of 165 mol C m^{-2} yr^{-1} to Cape Lookout Bight sediments , only ~29% is remineralized to ΣCO_2 (36.4 mol C m^{-2} yr^{-1}), DOC (2.55 mol C m^{-2} yr^{-1}), and CH$_4$ (8.64 mol C m^{-2} yr^{-1}); the bulk (~71%) is buried as particulate organic carbon. During decomposition, changes in POC and total nitrogen concentrations indicate an atomic C:N ratio of 6.6, approximating that of fresh, labile organic matter derived from bacteria, diatoms, and dinoflagellates. Martens et al.[453] were also able to trace the decomposition products of most (64%) of the degraded organic carbon; most of this material broke up into dissolved free amino acids (31%), lipids (17%), and sugars (16%), similar to their composition in fresh plankton.

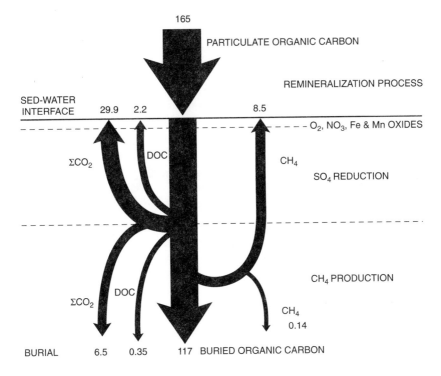

FIGURE 6.7
Annual carbon budget (mol C m^{-2} yr^{-1}) for sediments of Cape Lookout Bight, North Carolina. All values are mean rates. (From Martens, C. S., Haddad, R. I., and Chanton, J. P., in *Organic Matter Productivity, Accumulation, and Preservation in Recent and Ancient Sediments,* Whelan, J. K. and Farrington, J. W., Eds., Columbia University Press, New York, 1992, chap. 2. With permission.)

The carbon that is buried is composed mostly of refractory materials of algal and, to a lesser extent, vascular plant origin.

The fluxes of carbon and nitrogen are inexorably interlinked during the degradation of organic matter.[6,447] Many measurements have been made of individual nitrogen processes, but exceedingly few studies offer a complete picture of nitrogen flux in coastal zone sediments; none are extant for tropical subtidal sediments. Table 6.5 lists some budgets available for several coastal areas. Ammonification is usually the largest flux of nitrogen, followed by rates of NO_3^- and N_2 as a result of nitrification and denitrification, respectively.[454] A recent study by Kemp et al.[455] of sediment nitrogen cycling in the mesohaline region of Chesapeake Bay found large seasonal differences in transformation processes and the rates of input, release, and storage of nitrogen (Table 6.6). Rates of nitrification and denitrification peak in spring and autumn with virtually unmeasurable rates during summer. Both processes are

TABLE 6.5

Rates of Benthic Nitrogen Recycling in Some Representative
Coastal Environments

| Location | Benthic N Flux | | Denitrification | % of Plankton N Requirements |
	NH_4^+	NO_x		
Belgian coastal zone	25	32	17 (23%)	78
North Sea Bight	13	17	5 (15%)	38
Great Belt (Denmark)	10	7	49 (16%)	36
Kattegat (Denmark)	13	5	69 (22%)	42
E. Kattegat (Denmark)	20	5	28 (8%)	53
Limfjord (Denmark)	25	13	188 (26%)	55
La Jolla Bight (U.S.)	12	2	—	7
Narragansett Bay (U.S.)	34	4	33 (16%)	—
South River (U.S.)	38	0.5	—	29
Neurse River (U.S.)	76	1	—	26
Georgia Bight (U.S.)	55	3	—	16
Buzzards Bay (U.S.)	18	1	—	70
Cap Blanc (Africa)	78	58	—	39
Missionary Bay (Australia)	6	1	—	6
Great Barrier Reef shelf (Australia)	12	1	—	13

Note: Values are expressed as mg N m^{-2} d^{-1}. Values in parentheses are percentage
of total benthic nitrogen mineralization.

Source: Adapted from Billen and Lancelot,[454] with additional data from references
provided in Alongi.[12]

coupled, leading to a situation in which inhibition of denitrification
enhances increased rates of ammonium flux from the sediments to
support further primary production.

Denitrification has major ecological and geochemical consequences
because the process represents a sink or loss of nitrogen from the
marine nitrogen cycle. Seitzinger[456] estimates that loss of N_2 accounts
for roughly 15 to 70% of nitrogen flux from estuarine and coastal
sediments and, on average, decreases by 40% the amount of nitrogen
that is transported to offshore waters; loss rates increase proportionally with increases in external nitrogen loading. The efficiency of nitrogen removed via denitrification varies greatly among estuaries. For
instance, in a study of several estuarine tributaries of Galveston, Cor-
pus Christi, and San Antonio Bays on the Texas coast, Zimmerman and
Benner[457] found that the percentage of nitrogen lost via denitrification
ranges among estuaries from 14 to 136% (Table 6.7).

Bacterial mineralization of organic nitrogen and subsequent release of solutes supplies from 7 to 78% of nitrogen required by pelagic
primary producers in coastal seas.[454] As discussed in the next section,

TABLE 6.6

Seasonal Changes of Major Nitrogen Transformations
and Fluxes from Sediments in the Mesohaline Region
of Chesapeake Bay

	Spring	Summer	Autumn
Internal Nitrogen Processes			
Net ammonification	2.4	9.0	1.2
Net nitrification	0.9	0.0	1.7
Denitrification	0.3–2.0	0.0-0.1	0.6–1.2
Inputs, Outputs, Δ Storage			
Particulate organic nitrogen (PON) deposition	11.1	7.7	6.7
Burial	–1.8	–1.8	–1.8
Δ PON pools	–2.9	1.8	–1.4
Δ NH_4^+ pools	–0.4	–0.8	0.3
Δ Macrofaunal nitrogen	–0.5	0.3	0.1
NH_4^+ efflux	–1.1	–8.2	0.2
NO_3^- efflux	1.1	0.0	–0.5
N_2 efflux	–2.0	0.0	–1.2
Net difference	3.5	-1.0	2.4

Note: Mean values are mmol N m^{-2} d^{-1}. Losses from sediments are
depicted as negative values.

Source: From Kemp, W. M. et al., *Limnol. Oceangr.*, 35, 1545, 1990.
With permission.

benthic-pelagic coupling is critical for the recycling of nutrients and
energy transfer in the coastal zone.

6.5.3 Whole-System Budgets and Fisheries Yield: Some Perspectives

Processes occurring on and within sediments are coupled to processes
occurring in the shallow, overlying water column. Physical processes
ultimately affect the nature of benthic and pelagic food webs, but it is
the biological and biogeochemical links between benthic and pelagic
communities that lead to properties of ecosystem structure that sustain
high fishery yields in the coastal zone.

6.5.3.1 Benthic-Pelagic Coupling

The strongest functional link between benthic and pelagic food webs
is the deposition of organic matter to the benthos and its subsequent

TABLE 6.7

Efficiency of Nitrogen Removal by Denitrification from Some Estuaries
in the Gulf of Mexico

	Mass Nitrogen Loading		Nitrogen Loss by Denitrification	
Location	River Load	Total Load	Amount (g N m^{-2} yr^{-1})	% of Total N Removal
Nueces estuary	1.5	4.7	6.4	136
Guadalupe estuary	16.9	21.7	4.9	23
Trinity-San Jacinto	17.0	32.9	4.5	14
Ochlockonee Bay	16.6	17.3	9.0	52

Source: From Zimmerman, A. R. and Benner, R., *Mar. Ecol. Prog. Ser.*, 114, 275, 1994.
With permission.

return in dissolved form to the water column to sustain pelagic primary production. The magnitude of both processes is well known,[458] but factors regulating variation in the supply and biochemical quality of this material are less understood. Measurement of detritus flux to the benthos by using sediment traps is equivocal, owing to differences in flux rates depending on trap design. Further, traps placed in shallow waters often lead to biased results, as small-scale turbulence can cause low capture efficiency of settling particles. Deposition of resuspended sediments can also obscure true rates of newly settled particulates. For these reasons it is often very difficult to measure directly the rates of organic input to shallow coastal sediments.

Rates of organic input to the seabed can be indirectly estimated by geochemical modeling of naturally occurring isotopes of conservative elements (e.g., ^{210}Pb) or nutrients (e.g., TOC). Many such model estimates exist, but few attempts have been made either to link benthic decomposition rates with spatial and temporal changes in input rates of suspended particulate matter or to trace the sources and decomposition of this material as it settles out of the water column. Lucotte et al.[459] made such an attempt by tracing the origin and flux of suspended particulate material in the lower St. Lawrence estuary. They found that, from spring to autumn, isotopically lighter particles in the water column are replaced by more enriched particles coinciding with the early summer phytoplankton bloom and related bursts of zooplankton population growth. Less than 30% of the annual phytoplankton production settles to the estuarine sediments, mostly as copepod fecal pellets, with the remainder being remineralized in the water column. Organic carbon in surface sediments is composed of a nearly equal mixture of terrigenous and marine material, reflecting the fact that approximately 75% of the terrigenous POC flowing from the upper estuary deposits within the lower estuary. Most of this material is refractory, composed of lignocellulosic material and pollen grains. A first-order budget (10^3 t C yr^{-1}) estimated POC input to the lower St. Lawrence estuary from primary production (1000) and, to a lesser extent, from sedimentation of autochthonous (280) and allochthonous (220) particles. Benthic activity was not measured but presumably was linked to labile POC derived from the plankton blooms.

Concurrent measurements of organic carbon sedimentation and mineralization in Chesapeake Bay[460,461] link rates of mineralization to rates of input, although this link is complicated by time lags, variations in pelagic mineralization and water chemistry, and seasonal effects of anoxia. In the mesohaline region of the bay, POC deposition to the sediments (15 to 31 mol C m^{-2} yr^{-1}) equates to 36 to 74% of average primary production in the water column. Mineralization by aerobic and sulfate reduction pathways account for 14 to 32% and 47 to 72% of the annual POC deposition; the remainder (14 to 21%) is buried within deeper deposits. Pelagic productivity and subsequent benthic mineralization peaks during the warm months (May to September) in the mid-bay in contrast to other temperate coastal systems where maximal activity occurs during and immediately after the spring bloom. Modeling of the sedimentary POC pools and their turnover in the bay indicate that only 3 to 4% of the total POC pool is readily labile — corresponding to the most readily decomposible fraction of phyto-detritus — but nearly one third of total benthic carbon mineralization is fueled by the less metabolizable material.[461]

Anoxia in mid-Chesapeake Bay is driven mainly by planktonic respiration and by accumulation and subsequent oxidation of sulfides (produced as a byproduct of sulfate reduction) released from the porewater during summer stratification; pelagic respiration accounts for nearly two thirds of total O_2 depletion.[462] This biological depletion of oxygen is tightly coupled to, and balanced with, physical exchange processes that ultimately regulate oxygen availability.

Combining data from various coastal systems, Kemp et al.[462] noted that benthic respiration and the proportion of total system respiration accounted for by benthic respiration decrease with increasing water depth. This suggests that as the water column deepens, the sinking of particulate material to the benthos takes longer, allowing more time for planktonic food chains to consume the most labile fractions. Benthic mineralization therefore dominates total system respiration in coastal zones ≤ 5 m in depth. This relationship is robust considering the variety of coastal environments included.

The coupling of benthic and pelagic processes is best seen in the strong positive relationship between benthic respiration and pelagic primary productivity (Figure 6.8). The slope of the regression indicates that, on average, ~24% of pelagic primary production is mineralized

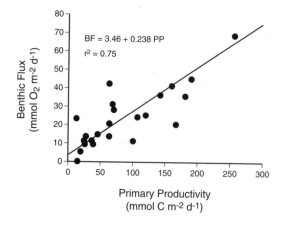

FIGURE 6.8

The positive relationship between benthic oxygen consumption and pelagic primary production from a variety of coastal zones. (From Dollar, S. J. et al., *Mar. Ecol. Prog. Ser.*, 79, 115, 1991. With permission.)

by benthic aerobic processes alone, at least in the coastal systems included in the analysis. This linkage is even stronger in some systems. For example, in Tomales Bay, Dollar et al.[463] found that the fallout of particulate material during summer leads to a strong postsettlement peak in benthic respiration in late summer-early autumn, with most pelagic primary production fueling benthic mineralization rates. In tropical coastal areas, benthic respiration rates are fairly continuous, suggesting a narrow seasonal range in water temperatures, but also a continuous supply of food from the water column.[12] Unfortunately, few data exist linking pelagic primary productivity and benthic mineralization rates in tropical coastal environments.

6.5.3.2 Carbon and Nitrogen Budgets: Trophic Inferences

Complete nutrient budgets exist for a number of temperate estuaries and bays. Construction of most of these budgets involved some educated guesses, assumptions, and extrapolations from similar ecosystems. Nevertheless, they are instructive in providing some clues as to the major and minor pathways of nutrient flow within coastal zone ecosystems and the relative differences in the behavior of carbon and nitrogen. Baird and Ulanowitcz[464] compared the ecosystem properties and the pathways and magnitude of carbon flow of four tidal estuaries: the Ythan in Scotland, the Ems-Dollard in The Netherlands, and the Swartkops and Kromme estuaries in South Africa. The basic structure of the four estuaries is similar, but rates of net primary production range from 203 mg C m^{-2} d^{-1} in the Ems-Dollard system to a narrow spread of 1729 to 2312 mg C m^{-2} d^{-1} among the other three systems. Primary production in the Kromme, Ythan, and Swartkops estuaries is dominated by benthic autotrophs (ratio of pelagic to benthic production is 1:>5), but in the Ems estuary phytoplankton account for ~50% of net primary production.

Differences in the standing stocks and primary productivity among the estuaries may be due to the strong seasonal changes experienced by the Ythan and Ems-Dollard systems. Direct utilization of fixed carbon is very high (98%) in the Ems, but lower in the Swartkops (38%), and much lower in the Ythan (10.5%) and Kromme (9.2%) estuaries. Differences in net primary production efficiencies and in the ratio of detritivory to herbivory and the numbers of internal transfers of organic carbon (14.6 to 26) can be attributed to differences in trophic structure and composition. For instance, in the Ythan and Kromme estuaries, carnivorous fish obtain more of their nourishment indirectly via suspended and sedimentary POC and benthic autotrophs than they do in the other two systems. Low efficiencies of net primary production coupled with a high detritivory-to-herbivory ratio indicate greater use of detritus and recycled materials in the Ythan and Kromme systems. The converse suggests that a greater array of resources are available within the other two estuaries. This is despite the fact that all four estuaries recycle resuspended particulate material through the sedimentary POC pool to other trophic groups. All four estuaries show low rates of trophic efficiency (4.0 to 12.5%), with the lowest efficiency and shortest food chain lengths found in the Swartkops — the most polluted estuary.

Carbon and nitrogen budgets for Chesapeake Bay and Tomales Bay show carbon flow features similar to those of the four tidal estuaries above but indicate more efficient recycling and retention of nitrogen than carbon.[464] In the Chesapeake Bay, seasonal differences in carbon flux are huge, with significantly higher rates of biological activity in spring and summer than in autumn and winter. In summer, grazing on primary producers is reduced by predation, mainly by ctenophores and sea nettles. Ungrazed phytoplankton is consumed within the microbial loop, with the remainder depositing to the bay floor where it is largely consumed. Network analyses[465] indicate that many of the keystone predators, such as bluefish and striped bass, rely on indirect pathways for survival. These analyses indicate that 70% of detritus is derived from internal recycling of organic matter, as detritivory is nearly ten times greater than herbivory within the food webs of the bay. Average trophic efficiency is similar to other systems, nearly 10%. Supporting the results from smaller-scale empirical studies, the models indicate that most benthic-pelagic coupling involves sedimentation of phytodetritus, but they also indicate that benthic

TABLE 6.8

Seasonal Nitrogen Demand and Supply Budget for the Entire
Chesapeake Bay System

Demand/Supply	Season			
	Spring	Summer	Autumn	Winter
Demand				
Phytoplankton	130.7 (57%)	232.5 (56%)	175.3 (80%)	71.0 (79%)
Benthic algae	31.3 (14%)	46.5 (11%)	7.6 (4%)	2.4 (2%)
Bacterioplankton	66.6 (29%)	139.3 (33%)	35.6 (16%)	17.0 (19%)
Total	228.6	418.3	218.5	90.4
Supply				
Regenerated nitrogen	78.4 (37%)	210.5 (80%)	79.8 (52%)	35.6 (18%)
"New" nitrogen	211.0 (63%)	53.2 (20%)	73.6 (48%)	166.2 (82%)
Total	289.4	263.7	153.4	201.8
F ratio[a]	2.7	0.3	0.9	4.7

[a] F ratio is "new" nitrogen:regenerated nitrogen.

Note: Units are mg N m^{-2} d^{-1}.

Source: From Baird, D. et al., *Estuarine Coastal Shelf Sci.*, 41, 137, 1995. With permission.

filter feeders and fish are important links between benthic and pelagic carbon flow. Further, rather than serving as a trophic link, the microbial loop acts to dissipate energy and carbon from the system, mostly as respired CO_2.

Seasonal patterns of nitrogen also reflect the greater rates of biological activity in the warmer months (Table 6.8), with both "new" and regenerated nitrogen meeting the needs of primary producers. Regenerated nitrogen is supplied mostly from release of dissolved nitrogen from the sediments, with a lesser contribution from pelagic microheterotrophs. Nitrogen is more efficiently retained within the system compared to carbon. The rate of nitrogen recycling is ~80% of total flux during summer, but ranges from only 18 to 52% during the remainder of the year.[466] The pathways of nitrogen conservation are complicated and numerous, involving both pelagic and benthic, microbial food webs. Microbes act as a sink for carbon but help to retain nitrogen.

A nitrogen budget[467] for Carmarthen Bay, a shallow (≤17 m) macrotidal and vertically mixed embayment located in South Wales, is a good example of the influence of hydrography on nutrient cycling in coastal waters. Advective inputs of nitrate from the Bristol Channel dominate the nitrogen cycle in this system. The bay is more open to the sea than the Chesapeake Bay and is continually exposed to wave energy from prevailing southwesterly swells.[467] Most (83%) nitrogen comes from the channel with local rivers contributing the remaining 17%; nitrate is the preferred form of DIN for phytoplankton, underscoring the link to advective inputs from the Bristol Channel. In contrast, ammonium inputs are derived mostly from pelagic microbial remineralization (50%) and from benthic regeneration (35%). Zooplankton excretion of ammonium provides up to 12% of phytoplankton nitrogen requirements, exceeding river inputs. Most (82.5%) of this regenerated ammonium is reassimilated by the phytoplankton, with the remainder imported via advective exchange with the channel. The efficient recycling of nitrogen is evident by comparing inputs and outputs: 0.43 mmol N m^{-3} d^{-1} enters the bay (mostly as nitrate), but only 0.03 mmol N m^{-3} d^{-1} leaves the bay, mainly in the form of ammonium.

Other coastal bays, such as Narragansett Bay in Rhode Island and Tomales Bay in California, are just as strongly influenced by nutrient inputs from terrigenous and oceanic sources. Smith et al.[468,469] and Nixon et al.[470] constructed detailed budgets for the behavior of carbon, nitrogen, and phosphorus in Narragansett Bay and Tomales Bay, respectively. The inner portion of Tomales Bay receives little freshwater input, with salinity slightly above that of ocean water. Its circulation is driven by winds and, to a lesser extent, evaporation; average tidal range is 1.1 m. A nitrogen model, representing the dry season situation, indicates that the bay imports dissolved fixed nitrogen (DON + NH_4^+). The bay is net heterotrophic, with <2% of organic carbon production being lost by denitrification. This excess heterotrophy liberates dissolved inorganic nitrogen which fuels denitrification, thus creating a system in which carbon metabolism is causing a deficit in nitrogen for primary productivity — that is, carbon-controlled nitrogen cycling. Tomales Bay exports dissolved phosphorus (as PO_4) and carbon (as CO_2) as a direct result of the net oxidation of organic matter, which is either of marine or terrestrial origin. Denitrification plays a minor role in the carbon cycle but is the major loss of nitrogen from the bay. Modeling exercises[469] imply that primary production in Tomales Bay is regulated by availability of phosphorus, as internal recycling becomes important and phosphorus controls availability of nitrogen via nitrogen fixation. The supply of nitrogen equals the supply of phosphorus as denitrification and nitrogen fixation act to counterbalance one another.

The mass balance estimates for Narragansett Bay,[470] a well-mixed, high-salinity coastal estuary, indicate that primary production is the major source of carbon, but land runoff, sewage, and fertilizer are the major sources of nitrogen. Offshore bottom water flowing into the bay is the largest source of inorganic phosphorus. Despite different routes of input, all three elements are retained within the bay at levels ≤20% of total annual inputs. The major loss of carbon is respired CO_2; denitrification removes roughly 10 to 25% of the total nitrogen input, but a large fraction of nitrogen and phosphorus (10 to 20% of the organic matter) is exported offshore. Unlike Tomales Bay, Narragansett Bay appears to be net autotrophic, with a positive carbon balance of 700 to 1170×10^6 mol yr^{-1}, equivalent to 7 to 12% of net carbon fixation.

Differences in net nutrient balance between Tomales Bay and Narragansett Bay can be attributed to several factors, such as differences in water residence times, but Nixon et al. [470] suggest that anthropogenic enrichment, particularly of nitrogen, may have shifted the ecosystem of Narragansett Bay from one of overall balance to net autotrophy. These mass balance estimates explain much of the observed behavior of nitrogen and phosphorus relative to organic carbon fixation and oxidation, but it remains to be seen whether or not these nutrients behave similarly in tropical and other temperate coastal systems with shorter or longer residence times.

6.5.3.3 Factors Regulating Coastal Fisheries

Coastal regions comprising large rivers, wetlands, and estuaries tend to support high yields of economically important fish, crustaceans, and molluscs. Why this is so has eluded scientists, managers, and fishermen alike. Many viable hypotheses have been offered to explain this phenomenon, including:[414]

- High rates of coastal primary productivity, especially macrophytes, which suggest high availability of pelagic and/or benthic prey
- High level of habitat diversity, structural complexity, and turbidity, offering increased partitioning of resources and shelter from predators
- Physiological/behavioral attraction to river discharge and precipitation (lower salinity preferences for part/whole of life cycle)

The relative significance of each hypothesis probably varies with the uniqueness of each coastal habitat and associated harvestable communities.

The remarkable similarity in fishery yields to humans among various marine ecosystems seems to be related to rates of primary productivity (Figure 6.9), but other factors play key roles in sustaining coastal fisheries. For instance, St. Georges Bay situated in the southern Gulf of St. Lawrence, is a successful nursery ground for pelagic spawners (lobster, hake, and mackerel) because of warm temperatures every summer and a rich assortment of pelagic and

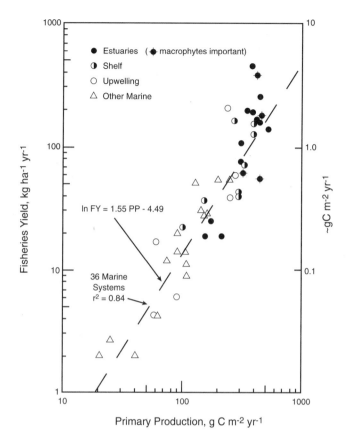

FIGURE 6.9
The relationship between primary production and fisheries yield for different aquatic environments. (Adapted from Day, Jr., J. W. et al., *Estuarine Ecology*, John Wiley & Sons, New York, 1989, 558.)

epibenthic foods.[471] Considering the large temporal and spatial variations characteristic of tidal estuaries, bays, and lagoons, it would be ironic indeed if the long-term predictability of such changes is one of the major factors driving the high fisheries yields in the world's coastal zones.

6.6   COASTAL LAGOONS (TYPE VII)

The primacy of wave energy separates coastal lagoons from tidally dominated lagoons and other coastal systems under Dronker's classification[412] (Table 6.1). An excellent book has recently been published on coastal lagoon processes,[472] so much of the discussion which follows is abstracted from various chapters contributed to the volume. A practicable definition of a coastal lagoon is[472]

> A shallow water body separated from the ocean by a barrier, connected at least intermittently to the ocean by one or more restricted inlets, and usually oriented shore-parallel.

Worldwide, coastal lagoons occupy by area approximately 13% of the coastal zone and include large water bodies, such as the Baltic Sea and the Wadden Sea in Europe. Physical processes within lagoons are greatly influenced by:

- Catchment size
- Water depth
- Inlet size and configuration
- Orientation with respect to prevailing winds

Advective transport regulates lagoonal circulation balanced by precipitation, evaporation, groundwater seepage, and surface runoff. Seasonality is induced by wet and dry seasonal changes, higher evaporation in summer, and temporal changes in wind direction. Tides are important mainly near inlets, but residual tidal flow can be important in the long-term; however, advection in response to local wind forcing is usually dominant. For these reasons, most coastal lagoons accumulate organic and inorganic materials, including sediments, and serve as material traps. These processes greatly influence the structure and energetics of food webs, including nutrient dynamics and mass balances. Occasionally, I will violate Dronker's distinctions[412] by discussing certain aspects of tidal lagoons in order to fully illustrate the gamut of lagoonal processes.

6.6.1   Pelagic Food Chain Dynamics

Three types of coastal lagoons may be distinguished on the basis on their primary producers:[473]

- Phytoplankton
- Benthic macroalgae and macrophytes
- Algal mats

Each of these systems have some contributions from the other autotrophs, but the majority of coastal lagoons are phytoplankton based.

Attempts to link autotrophic biomass and production with one critical physical factor have been unsuccessful, as it is highly likely that multiple physiographic and hydrological factors and the synergistic/ antagonistic interactions among them regulate primary producers. The main key factors (water residence time, water depth, tides, wind, river inputs, area-to-volume ratio, to name but a few) appear to vary depending upon whether or not the lagoon entrances are restricted to ocean exchange. For instance, in restricted lagoons in the dry tropics, the overlying waters become hypersaline, and prolonged exposure of large intertidal areas gives way to the development of extensive algal mats.

Knoppers[473] attempted to correlate estimates of primary production rates in all three coastal lagoon types with residence time of water and water depth. The relationship with residence time shows no clear patterns, as there is a large scatter of the data: production rates range from 50 to 1850 g C m^{-2} yr^{-1}. On average, subtropical and tropical lagoons are more productive (grand mean of ~400 g C m^{-2} yr^{-1}) than their temperate counterparts (grand mean of ~270 g C m^{-2} yr^{-1}), although their ranges overlap. On a global basis, total net primary productivity in coastal lagoons amounts to ~10^{11} kg C yr^{-1}, or roughly 25% of total primary production contributed to the world ocean by estuaries.[473] Rates of primary productivity in both phytoplankton- and macrophyte-based lagoons do, however, relate well to water depth, with maximal phytoplankton production in shallowest waters and peak rates of macrophyte production in deeper lagoon waters.

Recent evidence suggests that primary productivity and the size structure of the planktonic community in tropical and subtropical lagoons change in response to hydrological variations between wet and dry seasons and, on shorter time-scales, by wind-generated resuspension events. For instance, in Ebrie Lagoon on the Ivory Coast of central Africa, wind-induced resuspension, particularly during the wet season, results in a release of dissolved and particulate material from the seabed which stimulates production of picoplankton and nanoplankton.[474,475] This lagoon is an elongated system that receives extensive domestic waste inputs from the city of Abidjan. The system is eutrophic in most parts, with the deeper central basin subjected to intermittent stratification and anoxia events. Presently, tidal exchange is sufficient to minimize the effects of cultural eutrophication. This is likely to change given the rapid increases in human population growth forecast for this region.[475]

The best-studied tropical lagoon is Laguna de Terminos in the southern part of the Yucatan peninsula in Mexico.[476] This system is turbid year-round, despite distinct wet and dry seasons. Mangroves and seagrasses dominate the intertidal and shallow subtidal areas, but phytoplankton production (mean = 228 g C m^{-2} yr^{-1}) dominates total primary production. As in most other tropical lagoons, primary productivity is highest in the wet season, stimulated by river runoff and by erosion of nutrient-rich material from seagrass beds and mangrove forests. Like Ebrie Lagoon, eutrophication is ameliorated by its size and sufficient tidal exchange.[476]

Groundwater inputs are an important source of nutrients in many lagoons in tropical areas with karstic, highly permeable soils, such as in Hawaii, Western Australia, and on the northern Yucatan peninsula in Mexico. In Celestun lagoon on the Yucatan peninsula, freshwater input occurs mostly as groundwater discharge as there are no rivers, but owing to the size of the lagoon, it is difficult to relate changes in primary production directly to these inputs.[477,478] Further, both macrophytes and phytoplankton co-dominate total system production and have different optimal growth conditions. In the dry season, remineralization processes in the water column and sediments support high biomass of submerged macrophytes (mostly seagrasses); during the rainy season, groundwater discharge stimulates phytoplankton blooms. A significant fraction of phytoplankton biomass is deposited to lagoon sediments and is decomposed to nourish macrophyte growth in the dry season.[478]

Pelagic heterotrophs in coastal lagoons are fueled by carbon fixed by primary producers. Members of the microbial loop have only been examined in one tropical lagoon. In Ebrie lagoon, maximum bacterioplankton standing stocks (1 to 3 × 10^{10} cells l^{-1}) and productivity (2.4 to 5.8 g C m^{-2} d^{-1}) were measured during destratification events,[479] with lower values in deeper, stratified waters. Moreover, rates of phytoplankton production were not sufficient to meet bacterial carbon needs, suggesting use of carbon inputs from macrophytes. In most temperate lagoons, bacterial productivity is cued to primary productivity, with community dynamics also controlled by protozoan grazing.[480] In boreal areas, such as in the Gulf of Bothnia,[481] the annual cycle of bacterioplankton communities is controlled by a complex interaction between water temperature effects on growth and grazing by protozooplankton. Little of the phytoplankton produced during the spring bloom in the northern Baltic funnels through the microbial loop, as most settles to the sea floor.

The community dynamics of zooplankton in coastal lagoons is commonly associated with variations in phytoplankton biomass. A good example is the annual cycle of plankton communities in Laguna Joyuda on the southwestern coast of Puerto Rico.[482] Net phytoplankton productivity in the lagoon is positively related to variations in rainfall and subsequent DIN inputs and to water temperature. This coincides with low abundance of nongelatinous zooplankton which are controlled by zooplanktivorous medusae (mostly the scyphomedusae, *Phyllorhiza punctata*).[482] Grazing by nongelatinous zooplankton, particularly by the calanoid copepod *Acartia tonsa* is, therefore, less intense during the rainy season and may be one indirect cause for the wet season blooms of phytoplankton. Nutrient regeneration by zooplankton may also play a crucial role in sustaining phytoplankton production, particularly when inputs from allochthonous sources are limited. The plankton cycle described above is similar to those observed during late summer in temperate bays and estuaries, in which plankton communities are controlled by the coupling of biological and climatological forces.

6.6.2   Benthic Processes

Little is known about benthic food chain and nutrient processes in tropical and most temperate coastal lagoons. Secondary production estimates exist for some suspension feeders and commercially important molluscs, but with the exception of life history and community structure studies, essentially nothing is known of the energetics of small infauna in coastal lagoons.[483] The few data available for benthic energetics in tropical lagoons indicate very high rates (1 to 5 g C m^{-2} d^{-1}) of organic matter decomposition, particularly in lagoons with restricted entrances, suggesting rapid accumulation of organic matter. The best-studied lagoons remain the large systems, such as the Baltic and Wadden Seas.

What little knowledge does exist of benthic dynamics and nutrient cycles in coastal lagoons comes mainly from detailed studies of bays and small lagoons within Danish waters. In Faellestrand, a shallow oligotrophic lagoon on the island of Fyn, seasonal dynamics of benthic primary production and consumption show that temperature, day length, and nutrient availability control primary production, whereas benthic respiration is controlled solely by temperature.[484] During spring and summer, primary productivity peaks, outpacing consumption, and leads to autotrophic conditions favorable for enhanced secondary production of the benthic infauna, particularly the polychaete *Nereis diversicolor* and the amphipod *Corophium volutator*. The reactive pool of organic matter fueling consumption accounts for 12% of sediment organic carbon, with an average turnover time of <1 month in the warmer months. In autumn and winter, the lagoon becomes predominantly heterotrophic as microalgal production declines sharply due to light and nutrient limitation.

TABLE 6.9

Annual Carbon and Oxygen Budget for Sediments
of Organic-Poor Faellesstrand Lagoon, Denmark

	CO_2 (mol m^{-2} yr^{-1})	O_2 (mol m^{-2} yr^{-1})
Gross primary production	14.6	12.5
Total respiration	11.7	12.2
Animal respiration (included in total respiration)	—	2.8
Net primary production	3.0	0.3
Secondary production	1.6	—

Source: From Kristensen, E., *Estuarine Coastal Shelf Sci.*, 36, 565, 1993. With permission.

On an annual basis (Table 6.9), organic matter is rapidly turned over in these Danish lagoons, with ~80% of fixed carbon lost via respiration. Subsequent laboratory experiments[485] demonstrate that the activities of *Nereis diversicolor* stimulate the diagenesis of even refractory organic matter, with most carbon lost as respiration and secreted DOC. Microbial decomposition of labile detritus is temperature dependent but lower in the presence of the worms compared to rates in control microcosms, indicating competition between the worms and microbes for labile material during warmer months.

The cycling of nitrogen and phosphorus in Aarhus Bay indicates very dynamic nutrient pools in lagoon sediments, linked to sedimentation events, benthic activity, and complex kinetics of solute-solid-phase interactions.[486,487] Blackburn and Henriksen[486] traced the fate of sedimentary nitrogen (Figure 6.10) and found that bacterial mineralization of organic nitrogen, related to seasonal changes in the C:N content of the source detritus, is the dominant process. Less nitrogen is exchanged at the sediment-water interface in summer than in autumn — despite higher rates of microbial activity — as proportionately more mineralized nitrogen is exchanged and bound to organic and inorganic materials in the sediment. On an annual basis, however, 44 to 66% of the net mineralized nitrogen is transferred to the water column, supplying 30 to 82% of the nitrogen required by phytoplankton.

The phosphorus cycle in Aarhus Bay sediments (Figure 6.11) is fueled by sedimentation from the water column at an average annual rate of 51 to 63 mmol m^{-2} yr^{-1}; phosphorus is released at a rate of 34 mmol m^{-2} yr^{-1}, leading to a net burial rate of 18 mmol m^{-2} yr^{-1}. Adsorption into iron oxyhydroxides is the most important factor controlling

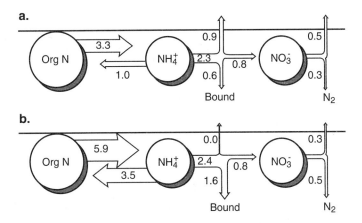

FIGURE 6.10
Sediment nitrogen cycling in Aarhus Bay, Denmark, during autumn (a) and summer (b). Rates = mmol N m^{-2} d^{-1}. (From Blackburn, T. H. and Henriksen, K., *Limnol. Oceanogr.*, 28, 477, 1983. With permission.)

phosphorus release. Seasonally, phosphorus that sediments during the spring bloom is bound to iron and retained until autumn when 50% of the phosphorus is released. This enhanced efflux from sediments appears to be related positively to sulfate reduction; increased production of sulfide results in reduction of iron oxides and net release of iron-bound phosphorus. Precipitation of FeS and pyrite may also result in release of dissolved reactive phosphate due to saturation of sorption sites for phosphorus on iron oxyhydroxides in surface sediments. Phosphorus is solubilized mostly by bacteria and some is derived from animal excretion. However, once dissolved in the interstitial water, phosphorus becomes inextricably complexed with the iron and sulfur cycles.[487] The pools of iron, manganese, and sulfur cycle rapidly in sediments and between the sediments and overlying water column, partly due to the mineralization activities of sulfate-reducing and iron- and manganese-reducing bacteria.

The significance of benthic processes driving the nitrogen cycle has been examined in other lagoons, such as the Tancada lagoon within the Ebro River delta, in northeast Spain. Using network analyses, Fores et al.[488] estimated that 60 to 80% of the nitrogen required to sustain phytoplankton production in the lagoon comes indirectly from sediment ammonium release; denitrification and phytoplankton were the major exports of nitrogen. The nitrogen cycle in the lagoon is highly dependent upon internal recycling mechanisms, despite the fact that the decomposition of macrophytes drives the cycle.

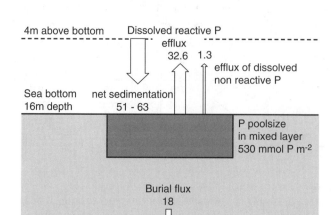

FIGURE 6.11

Annual phosphorus cycle in Aarhus Bay, Denmark. Fluxes = mmol P m^{-2} yr^{-1}. (From Jensen, H. S. et al., *Limnol. Oceanogr.*, 40, 908, 1995. With permission.)

6.6.3   Whole-Lagoon Budgets and Fisheries Implications

With respect to nutrient supply to coastal lagoons, Knoppers[473] indicates that:

- Choked lagoons receive their major inputs from rivers.
- In restricted to leaky lagoons, riverine and marine inputs are equally important.
- In arid-zone lagoons, new nutrients are supplied from marine (including intrusions of upwelled water) and atmospheric sources.

The pelagic and benthic nutrient budgets discussed in previous sections show that nutrient sources and sinks vary greatly among coastal lagoons even among the same morphological types. Clarification of the relative importance of particular sources is limited by the lack of complete budgets for lagoons, particularly those in tropical and subtropical regions.

The few attempts to estimate system-level nutrient budgets in tropical lagoons indicates great complexity and variation over time and space. How nutrients are supplied and cycled through one part of a lagoon is likely to be very different from how they are supplied and used within another part. For example, nutrient flows between different areas of Celestun lagoon on the northwest Yucatan Peninsula differ greatly and also differ between wet and dry seasons.[489] In the rainy season, large inputs of nitrate, silicon, and soluble reactive phosphorus (SRP) enter the lagoon via groundwater seepage. Nitrate and SRP are retained in the lagoon; ammonium and suspended particulate matter are delivered to the water column and circulated throughout the lagoon. In the dry season, when water residence times are long, SRP is released in the inner zone while being slowly exported from the outer zone to the sea. On an annual basis, nitrogen, phosphorus, and silicon are exported to adjacent shelf waters. In lagoonal systems along other parts of the Mexican coast, exchange rates of organic carbon, nitrogen, and phosphorus are higher than those measured from most salt marshes and mangrove forests, with most exchanges occurring during extreme wet seasons.[490]

The most detailed budgets for carbon and nitrogen are of the Baltic and Wadden Seas. Carbon flux models (Figure 6.12) of the non-tidal Veerse Meer and Grevelingen lagoons on the Dutch coast show that macrophytes, epiphytes, benthic microalgae, and phytoplankton drive high rates of microbial mineralization in the lagoon sediments to the extent that little organic matter is lost by burial.[491] The role of the microbial loop is not clear, but nearly 75% of total carbon is lost via respiration by pelagic and benthic heterotrophs in both lagoons. The similarity in biomass and carbon flux rates is remarkable given that nutrient loadings into Veerse Meer are greater than those into Grevelingen. Pelagic and benthic heterotrophs appear to consume the additional autotrophic biomass in Veerse Meer. Much of the detritus in both lagoons undergoes a cycle of suspension, sedimentation, and resuspension. Such a cycle likely facilitates continuing availability of the detritus to consumers in the water column and on the sediment surface. Veerse Meer is more susceptible to anoxia, but it appears that anaerobic bacteria are as efficient as aerobic bacteria in mineralizing organic matter.

Both lagoons are important feeding and nursery locations for birds and fish. Closure of Grevelingen lagoon in 1971 resulted in a trophic shift in higher consumers from a loss of pelagic herbivores to an increase in macrophyte feeders — species diversity declined drastically for most trophic groups. Not surprisingly, bird and fish production is greater in the Veerse Meer ($\sim$10 g C m^{-2} yr^{-1}) than in the Grevelingen ($\sim$6 g C m^{-2} yr^{-1}) lagoon.

In the Baltic,[492] primary production is dominated by phytoplankton (39×10^8 kg C yr^{-1}) with lesser inputs from benthic microalgae and macrophytes (5.4×10^8 kg C yr^{-1}). Nearly 68% of fixed carbon is

A. VEERSE MEER LAGOON

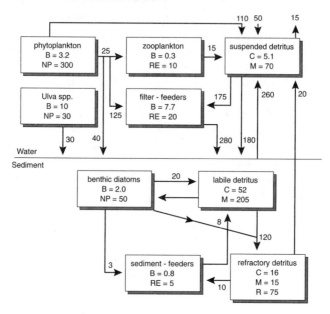

B. GREVELINGEN LAGOON

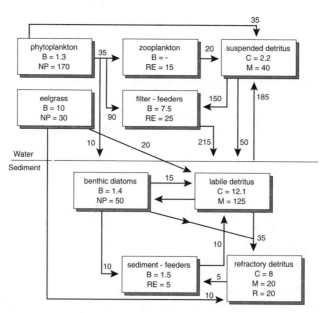

FIGURE 6.12
Annual carbon cycle in the (A) Veerse Meer and (B) Grevelingen Lagoons, The Netherlands. Biomass = g C m^{-2}; fluxes = g C m^{-2} yr^{-1}. B = biomass; NP = net primary production; RE = respiration and excretion; M = heterotrophic mineralization; C = concentration ; R = refractory. (Adapted from Nienhuis, P. H., *Vie Milieu*, 42, 59, 1992.)

TABLE 6.10

Fishery Yields in Coastal Lagoons in Relation to Yields from Other Aquatic Ecosystems

Systems	Yield (t km^{-2} yr^{-1}) Median	Mean
Coastal lagoons	5.1	11.3
Continental shelves	4.8	5.9
African/Asian reservoirs	4.2	7.5
Coral reefs	4.1	4.9
River floodplains	3.2	4.0
Reservoirs (U.S.)	1.3	2.4
Natural lakes	0.5	2.8

Source: Adapted from Kapetsky, J. M., in *Management of Coastal Lagoon Fisheries,* Kapetsky, J. M. and Lasserre, G., Eds., FAO Stud. Rev. GFCM No. 61, Vol. 1, Food and Agricultural Organization of the United Nations, Rome, 1984, 97.

processed equally through the pelagic food web and benthic suspension feeders. The rest settles to the lagoon floor, where it is estimated that macrobenthic and meiofaunal communities utilize 20 and 4 g C m^{-2} yr^{-1} of this material, respectively. Fish consume roughly 5 g C m^{-2} yr^{-1}, with an annual yield to humans of ~0.2 g C m^{-2} yr^{-1}. Only ~2 g C m^{-2} yr^{-1} is buried in sediments. Mineralization by sediment bacteria were not included in these estimates,[492] but if we take the difference between the total available pool (48 g C m^{-2} yr^{-1}) and that which has been accounted for (31.2 g C m^{-2} yr^{-1}), at minimum, 16.8 g C m^{-2} yr^{-1} or 35% of the available carbon pool is immediately available to bacteria. This value is likely to be much greater considering that a considerable fraction of secondary production is ultimately remineralized in sediments. Consider further that, as deduced from other coastal ecosystems, at least 50% must ultimately be lost as respiration. Sadly, this proportion is likely to increase in the future as the Baltic is one of the most polluted coastal ecosystems on Earth.

Coastal lagoons, partly because of the high rates of primary production and other abundant resources, are rich fisheries grounds and are increasingly favored sites for aquaculture. Kapetsky[493] estimates that coastal lagoons are, on average, more productive in terms of fishery yields per unit area than any other aquatic ecosystem (Table 6.10). However, not all lagoons yield high annual catches, as catch frequencies are highly skewed with most lagoons yielding ≤5 and ≤1 metric tons km^{-2} yr^{-1} of finfish and shrimp, respectively.[493] Unproductive

lagoons include those with extreme temperature and salinity variations, seasonal anoxia, turbidity, and high waste discharge.

The factors cited earlier to explain high fishery yields in coastal
plain estuaries, tidal lagoons, and bays (Section 6.4.3.3) are equally
applicable for coastal lagoons. Similarly, not one of these factors has
been conclusively demonstrated to be responsible for the high rates of
recruitment, growth, and production of harvestable species of finfish,
crustaceans, or molluscs. Energy and nutrient budgets offer few direct
clues. A better understanding of what drives coastal fisheries can come
not only from better knowledge of life history strategies, physiological
tolerances, and bioenergetics of individual species, but also by inte-
grating species responses within the dynamics of the entire ecosystem.
Methods such as network analysis are proving to be effective, plural-
istic approaches to fishery problems.

6.7 RIVER-DOMINATED SYSTEMS (TYPES I, II, AND III)

Rivers are the chief conduits of water, dissolved and particulate organic matter, salt, and other inorganic materials from the continents to
the sea. It is somewhat of an understatement to say that they greatly
impact water circulation, element cycles, and food webs along the
world's continental margins. It is somewhat surprising, then, how
ignorant we are of their myriad contributions to global ocean fluxes,
particularly in tropical seas.

6.7.1   Global Estimates of River Loads

It has been estimated that the world's rivers drain nearly 60% of the
entire land area (1.5×10^8 km^2) on Earth, annually discharging roughly
26×10^9 metric tons of sediment and nearly 38×10^3 km^3 of freshwater
to the world ocean.[494] Small rivers (drainage basins <10,000 km^2) account for a small proportion (~31%) of the total drainage basin area,
but discharge approximately half of the total sediment load, reflecting
higher yields than large (drainage basins >10,000 km^2) rivers.[495] Milliman
and Syvitski[495] contend that most large rivers (e.g., Amazon, Mississippi, Yangtze) act as point sources for sediment and water discharge
and form large deltas, as they generally flow into coastal seas along
passive continental margins. In general, smaller rivers drain narrow,
collision and active margins and mountainous islands, resulting in the
transport of sediment burden equivalent to those carried by larger
rivers. A larger proportion of the sediment loads of small mountainous
rivers transported to sea is therefore more likely to bypass narrow

continental margins and escape to the deep ocean basins.[495] The small rivers of Indonesia and Malaysia that discharge into the shallow Sunda Sea are an exception.

These estimates are crude, considering large variations in climate and that little data exist for most small and large tropical rivers. Moreover, it is likely that these estimates are no longer valid and are probably overestimates, as dams and subsequent diversions for industrial and agricultural purposes have resulted in a decline in outflow to the sea as illustrated by changes in the flow of the Nile River. This trend may be complicated in the future by expected increases in erosion due to deforestation. These changes will have a severe impact on coastal ecosystem processes.

Meybeck[496] has estimated the global riverine inputs of carbon, nitrogen, phosphorus, and sulfur to the ocean (Table 6.11). Organic and inorganic carbon are carried nearly equally in dissolved and particulate form. Atmospheric input of CO_2 is a larger source (67%) of dissolved inorganic carbon than is weathering. DOC and POC are in nearly equal proportion, but only a small fraction is labile. Ittekkot[497] estimated that 35% of the POC is labile, presumably consumed within the coastal zone, with the rest being refractory and potentially available for accumulation in marine sediments. The estimates for the other elements are crude but indicate that nitrogen and phosphorus are transported mainly in particulate form, as indicated from the soluble transport index. In contrast, 91% of sulfur carried to the sea is in the form of sulfate.

The incorporation of these elements into coastal food chains and clues to their eventual fate are the subject of the remainder of this chapter. I will focus primarily on a few large rivers that have been comparatively well studied to illustrate recent advances in river-ocean processes. Some estuarine plumes and fronts that occur on temperate shelves where across-shelf patterns and exchanges are well-understood (e.g., the New York and Georgia Bights, the Irish and North Seas) will be covered in the next chapter.

6.7.2 River Deltas With and Without Coastal Barriers (Types I and II)

The lower Mississippi River system is arguably the best-studied river delta of the type in which freshwater discharge dominates all other physical forces (e.g., tides, wind). The Mississippi-Atchafalaya River discharges roughly 210×10^6 metric tons of sediment and 570 km^3 of freshwater into the northern Gulf of Mexico annually,[495] being the largest river in North America. Nearly two thirds of this material

TABLE 6.11

Estimates of Global Budgets (in 10^{12} g yr^{-1}) of River-Borne Carbon, Phosphorus, Nitrogen, and Sulfur to the Oceans

Element	Dissolved Species			Particulate Matter			STI (%)		
	DIC[a]	DIC[b]	DOC	PIC	POC[c]	POC[d]	IC	OC	TC
Carbon	126	255	198	170	78	94	69	53.5	63
	DIN		DON			PN			TN
Nitrogen	4.5		10.0			21			41
	PO_4^3		DOP	PIP	POP		IP	OP	TP
Phosphorus	0.4		0.6	12	8		3	7	5
	SO_4^2					PS			TS
Sulfur	108					10			91

[a] From weathering.

[b] From atmospheric CO_2.

[c] From rock.

[d] From soil.

Note: Particulate nitrogen is derived from particulate organic carbon (POC) budget using a C:N ratio of 8.5. Particulate S budget is estimated assuming that PS content in suspended matter is ~600 × 10^{-6} g g^{-1}. The soluble transport index (STI) is the proportion of each element carried in dissolved phase compared to total river load. Values do not include estimated anthropogenic inputs. DIC = dissolved inorganic carbon; DOC = dissolved organic carbon; PIC = particulate inorganic carbon; POC = particulate organic carbon; OC = organic carbon; TC = total carbon; DIN = dissolved inorganic nitrogen; DON = dissolved organic nitrogen; PN = particulate nitrogen; TN = total nitrogen; DOP = dissolved organic phosphorus; PIP = particulate inorganic phosphorus; POP = particulate organic phosphorus; IP = inorganic phosphorus; OP = organic phosphorus; TP = total phosphorus; PS = particulate sulfur; TS = total sulfur.

Source: Adapted from Meybeck, M., in *Interactions of C, N, P, and S Biogeochemical Cycles and Global Change,* Wollast, R., MacKenzie, F. T., and Chou, L., Eds., Springer-Verlag, Berlin, 1993, 163.

debouches onto the shallow shelf west of the river delta and continues to move in a westerly direction parallel to the coast due to prevailing winds, favorable currents, and topography.

The plumes from the Mississippi River are turbid and rich in nutrients; nitrate concentrations ≥100 μM are typical. In the most turbid plume water in close proximity to the delta, primary productivity is severely light limited. Farther away from the delta, a zone of intermediate (15 to 30) salinity exists in which the high suspended loads have settled sufficiently to result in sufficient light penetration; the high dissolved nutrient concentrations persist, facilitating exceedingly high rates of phytoplankton production (up to 5 g C m^{-2} d^{-1}) and biomass.[498] Phytoplankton activity and standing stocks decline farther away from the river delta as nutrients are depleted and losses due to grazing and sedimentation to the shelf floor take their toll. These plumes and attendant processes operate on a scale of days and over tens of kilometers. Annual patterns vary greatly from year to year with changes in the strength of river flow.

A mass balance model suggests that light is the most important factor limiting phytoplankton within the Mississippi River plume and further indicates that nutrient regeneration within the water column contributes significantly to the maintenance of phytoplankton production.[499] The highest rates of ammonium regeneration, bacterial production, and oxygen consumption coincide with the maximum rates of primary production within the plumes of intermediate salinity. Chin-Leo and Benner[500] found that near the Mississippi delta, bacteria are carbon limited, but they show phosphorus and nitrogen limitation within the intermediate salinity zone; bacterial productivity is higher in summer (mean = 443 mg C m^{-2} d^{-1}) than in winter (mean = 226 mg C m^{-2} d^{-1}). Seasonal changes in phytoplankton and bacterial productivity show that in summer 10 to 58% of phytoplankton production is capable of supporting the bacteria but is insufficient in winter, suggesting that other sources of organic carbon — most likely river-borne material — are utilized by the pelagic bacteria. Bacterial mineralization of particulate organic nitrogen originating from phytoplankton results in rapid ammonium regeneration within the intermediate salinity waters of the plume. Recent observations indicate that intense nitrification occurring within this salinity regime accounts for 20 to 64% of pelagic oxygen consumption.[501] These intense rates of nitrification appear to be fueled by the high rates of ammonium regenerated in excess of phytoplankton requirements and can partly explain the deviation of nitrate concentrations from a conservative mixing relationship. Larger zooplankton and other planktonic heterotrophs also contribute to nitrogen regeneration by supplying excreted nitrogen derived from phytoplankton and other prey. Copepods and microzooplankton graze phytoplankton cells at rates equivalent to 4 to 90% of daily phytoplankton production,[502,503] suggesting that they can have a significant impact on the recycling of carbon and nitrogen in the river plume.

High rates of community respiration in waters overlying the inner shelf,[504] intensified by these rapid transformations of nitrogen, contribute to the formation of hypoxia in these waters, particularly in summer, when water-column stratification intensifies. The distribution of hypoxic bottom-waters can be extensive (Figure 6.13), varying from

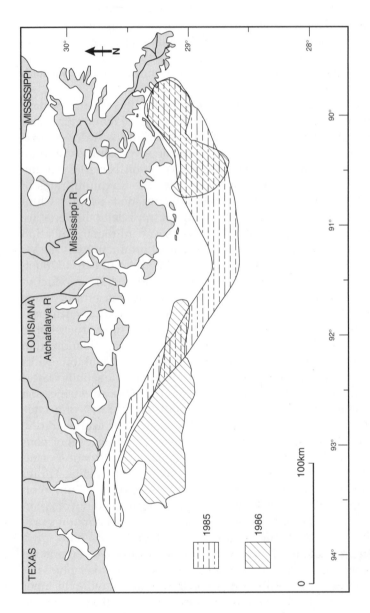

FIGURE 6.13

The distribution of hypoxic bottom waters on the Louisiana continental shelf in relation to the Mississippi delta, July 1985 and July 1986. (Adapted from Rabalais, N. N. et al., in *Modern and Ancient Continental Shelf Anoxia*, Tyson, R. V. and Pearson, T. H., Eds., Geological Society Special Publ. No. 58, London, 1991, 35.)

year-to-year in response to regional wind fields and currents and river discharge.[505] The hypoxia results in high levels of mortality of benthic metazoan life; recolonization and rapid recovery of benthic communities begins immediately after cessation of the hypoxia event. Anoxic conditions on the shelf floor are actually caused by the depleted oxygen conditions in near-bottom waters, coupled with large inputs of phytodetritus (mainly of diatom origin) fueling rapid rates of consumption of the remaining oxygen at the sediment-water boundary. Turner and Rabalais[506] provide some evidence that the increasing severity of these events is linked to eutrophication, as nitrogen loading in the Mississippi has doubled from the turn of the century to 1980.

The sedimentary record from the inner Louisiana shelf supports the notion that coastal eutrophication has been increasing because of nutrient-enhanced discharge from the Mississippi. Stable isotopes and other organic tracers in sediment cores taken from within the area of chronic hypoxia and from near the river mouth, show changes in accordance with increased nutrient inputs beginning in the mid-1950s, accelerating in the mid-1960s, and leveling off in the mid-1980s.[507] Variations down-core for total organic carbon concentrations and stable isotopes suggest that roughly 20 to 50% of TOC settling onto the shelf floor is buried. The remainder is presumably either remineralized or exported to the continental slope. The $\delta^{13}C$ data indicates that <40% of the sedimentary TOC that is buried is of terrestrial origin. By far, most of the terrestrially derived carbon deposits close (within 60 to 70 km) to the Mississippi River mouth, with organic carbon originating from phytodetritus increasing in proportion farther west from the delta.[506]

This material is rapidly decomposed in the sediments. Lin and Morse[508] measured maximum rates of sulfate reduction (up to 13.8 mmol m^{-2} d^{-1}) in close proximity to the delta with decreasing rates (minimum of 0.6 mmol m^{-2} d^{-1}) distally. Rates of sulfate reduction correlated positively with sedimentation rates and organic carbon concentrations. Sulfate reduction rates are slightly higher than expected given the ambient TOC concentrations, probably reflecting the lability of phytodetritus that sinks to the shelf bottom. Intensive bioturbation also plays an important role in stimulating anaerobic decomposition in these deposits. As a consequence of the high sulfate reduction rates, the Mississippi delta and adjacent Louisiana shelf sediments are a major sink for reduced sulfur that is accumulating as pyrite at rates of 0.9 to 4.8×10^{11} g yr^{-1}. This input represents about 96% of the pyrite burial along the entire continental margin within the Gulf of Mexico.[508]

The pattern of enhanced nutrient concentrations and pelagic and benthic community activities, in response to river plumes, is a recurring feature in coastal systems receiving discharges from rivers with coastal barriers, such as the Nile, Rhone, and Po Rivers; however, each situation is unique, given the different physical settings and the size of each river. The Rhone River, for instance, is considerably smaller than the Mississippi, with a drainage basin area and annual sediment load of only 3 and 15%, respectively, of the Mississippi. Unlike most other outflows, the Rhone River plume, due to the absence of tides and the juxtaposition of different Mediterranean mass masses, is defined by sharp boundaries and is stratified into a three-tiered structure: a surface freshwater layer, one meter thick, rich in suspended particulate matter (>20 mg l^{-1}) and dissolved nutrients, and a subsurface, seawater layer depleted in nutrients, separated by a frontal boundary that is discontinuous.[509] The plume is normally 1 to 6 km in length. Further, the Rhone River empties onto a relatively narrow shelf margin.

The three layers are not only physically and chemically distinct but appear to be trophically distinct, as well. Phytoplankton biomass and production is maximal within the frontal layer where phytoplankton activity and nutrient cycles intensify in summer. Spatial patterns of bacterial and microplankton abundance mirror those of the phytoplankton, but temporal patterns of each trophic group are different. In spring, diatoms and autotrophic flagellates dominate the surface plume and discontinuity layers. Bacterial densities and productivity peak in summer as autotrophs decline until autumn. In winter, when light intensity is low, cyanobacterial densities peak as picoplankton dominate autotrophic biomass. Nanoplankton are dominant in the plume layer, whereas microplankton dominate the frontal boundary. Bacterial production is highest in the frontal layer but does not mirror the patterns of primary production.[510] This suggests that bacterial nutrient requirements in the plume layers are supplemented by riverine inputs. Like the Mississippi plume, nitrifying and denitrifying bacteria rapidly utilize river-derived ammonium within plume waters, leading to significant changes in concentrations and recycling of dissolved nitrogen.[510]

Uncoupling of bacterial and phytoplankton productivities has been observed in other river plumes.[511] In the Hudson River plume in spring, Bianchi et al.[511] estimate that bacterial carbon requirements exceed the carbon available from phytoplankton by 100%, but in the discontinuity layer, phytoplankton carbon was sufficient to supply half of the bacterial carbon demand. Protozooplankton biomass patterns suggest that bacteria in the surface plume layer are not closely controlled by predation but are dependent on temperature and nutrient concentrations.[511] The benthic response to the Hudson River plume is poorly understood and complicated by pollution.

The situation for the Po River delta is also unique in that it is smaller than the other rivers (drainage basin area of 54,000 km^2, sediment load of 13×10^6 t yr^{-1}) and empties into the northern end of the semi-enclosed and narrow Adriatic Sea. The Po River drains one of the most agriculturally productive regions in southern Europe. A cyclonic circulation pattern in the Adriatic means that discharge from the Po and smaller Italian rivers flows south along the Italian coastline for most of the year. In summer, the circulation pattern changes and intense stratification results in the Po discharge being advected eastwards across the narrow Adriatic.[512]

The eastward advection of Po waters results in a shift in biomass and relative size of the phytoplankton community. All size classes increase in biomass as a result of the elevated levels of nutrients from the river, but the size spectrum shifts to a relative increase in the importance of microplankton at the expense of pico- and nanoplankton. More intensive investigations[513] show that nutrient and phytoplankton concentrations correlate positively and that primary productivity is higher (mean = 330 mg C m^{-2} d^{-1}) near the coast than in the central passage (mean = 150 mg C m^{-2} d^{-1}), supporting the notion of a narrow belt of greater productivity along the Italian coast south of the Po.[513] Diatoms and phytoflagellates dominate both the inshore and offshore phytoplankton assemblages, but photosynthetic efficiency is significantly higher in the coastal communities than in those farther from the coast. Photosynthetic efficiency is linear with respect to phosphorus but hyperbolic with respect to nitrogen, indicating that phosphate is the limiting nutrient, in agreement with other circumstantial evidence.[514]

In contrast to the higher primary production in surface waters, summer stratification and high nutrient loads combine to create hypoxic conditions in bottom waters of the northern Adriatic, causing mass mortality of benthic organisms and declining fishery yields. These events are becoming more frequent and widespread (up to 50 km^2). Hypoxia develops as a direct result of the depletion of bottom water oxygen as the organic matter (accumulated from algal blooms brought on by the eutrophic conditions) on the sea floor decays. Stratification of the water column prevents oxygen replenishment near and on the seabed. Some evidence indicates that the benthic communities have been in a state of decline (both in densities and numbers of species) since the first large-scale hypoxia was detected in 1969.[515] These events have a similar impact on planktonic communities, particularly those living near the sea floor.

Biogeochemical studies show that the highest rates of remineralization in sediments occur at sites near the Po Delta, which are characterized by high (up to 1 g cm^{-2} yr^{-1}) accumulation rates, low bioturbation, and high rates of burial, roughly half of the organic input.[516] At stations south of the delta that receive a high influx of planktonic debris, sediments accumulate organic matter at significantly lower (0.1 to 0.2

g cm^{-2} yr^{-1}) rates than near the delta. The deposits are characterized by high rates of bioturbation and remineralization, with correspondingly low rates of burial. Farther offshore in the northern Adriatic, recyling and burial rates of nutrients are similar to those south of the Po Delta. Budget estimates indicate that 30 to 50% of the nitrogen is lost by denitrification. Benthic anaerobic activity is more intense during the summer stratification period.[516]

The Adriatic Sea is an important fishing ground within the Mediterranean, especially for hake and bivalves such as striped venus, *Chamelea gallina*. During the 1970s and 1980s, this region suffered a large decline in total catch of both demersal and pelagic fish stocks, attributed mainly to over-fishing and mortality as a result of eutrophication and hypoxia.[516] A continuing decline in the pelagic fishery up to the present day has been ascribed to changes in climate, drought conditions, and higher-than-average temperatures. Discharge from the Po and small rivers has decreased by 12% in the past few years, resulting in less nutrient input into the Adriatic and lower primary productivity and a shift in pelagic community composition to microflagellates.

The fisheries crisis in the Adriatic is complex, as it is difficult to distinguish the effects of human activities from natural variability. Such appears to be the case for many river-dominated coastal regions draining land inhabited by dense human populations. The Nile is perhaps one of the best-documented examples. The northern Nile delta has changed dramatically within the twentieth century due to erosion, salinization, and pollution.[517] The future looks bleak for the Nile. Conservation measures have proved to be inadequate against the human population which continues to swell in countries bordering the drainage basin of the river.

Biogeochemical approaches may help to understand exactly how eutrophication results in changes in pelagic and benthic food web structure. Recently, Justic and his colleagues[518] analyzed the changes in nutrient stoichiometric balances in the northern Adriatic Sea and the northern Gulf of Mexico. They observed that significant alterations in nutrient structure have occurred in the past 30 years — the Si:N ratios have decreased, implying a potential increase in silicon limitation for primary producers. Further, the N:P and Si:P ratios have also undergone drastic changes to the degree that phytoplankton growth is unlikely to be limited by nitrogen and phosphorus. They hypothesize that under unpolluted conditions, rivers normally supply silicate at a rate that is nonlimiting for growth of primary producers, that is, in excess of their supply of dissolved nitrogen and phosphorus. In contrast, rivers that are impacted appear to be supplying nutrients in stoichiometric balance or relatively deficient in silicon. This implies a shift in nutrient limitation such that toxic algal blooms will become more frequent as coastal eutrophication intensifies. Their hypothesis is intriguing and suggests that restoring the nutrient balance is one of the key processes to be used in controlling eutrophication in river-dominated coastal waters.

6.7.3   Tidal Rivers and Their Dominance in the Tropical Ocean (Type III)

Tidal rivers large and small occupy a significant fraction of the world's coastlines, but it is the tidal rivers located in the wet tropical regions that have the most impact on the economy of the global ocean. Slightly more than half of the world's river water — laden with more than 60 and 65%, respectively, of the world's ocean-bound particulate matter and organic carbon load — enters the tropical coastal ocean annually.[496] Most enters from the Indo-West Pacific margin where the rainfall is highest; other major river/ocean boundary regions are located in northeastern South America and west-central Africa. Large, low-latitude rivers account for a fair share of the world's freshwater (32%) and sediment (17%) discharge (Table 6.12) — the Amazon alone accounts for most of the flux — but the smaller mountainous rivers, ignored until recently, account for a greater proportion (43%) of the present sediment discharge estimates (Table 6.12).

A good example of the importance of small rivers to the coastal ocean is the island of New Guinea. Recently, Milliman[519] estimated that 1.7×10^9 metric tons of sediment enter the adjacent coastal zone from New Guinea rivers each year, but the ten largest rivers — including the Fly, Sepik, Purari, Digul, Bamu, Mamberamo, and Kikori — contribute only 35% to this figure; discharge from roughly 240 smaller rivers makes up the balance. This total discharge is enormous and much greater than previously thought. It also implies that total discharge rates for other tropical coastal regions have been similarly underestimated.

Regardless of river size, a great variety of physical processes, ultimately driven by climate, make these tropical coastal margins unique compared to coastal settings in the higher latitudes.[12,520] The climate of the equatorial region is characterized by sustainably high rates of rainfall, sunlight, and temperature (Table 6.13). By virtue of these characteristics and global position, Coriolis forces are small and winds are dominated by the easterly trades. These physical forces, coupled with the enormous loads of freshwater and sediment draining from the land, produce extensive buoyant plumes (in some instances, extending beyond the shelf edge), rapid rates of sediment accumulation, and high rates of nutrient flux and primary productivity, to name but a few of the unique characteristics of river-dominated tropical coastal systems.

TABLE 6.12

Estimates of Water (km³ yr⁻¹) and Sediment (10⁶ metric ton yr⁻¹) Discharge Rates from Large Tropical and Some Subtropical Rivers

River	Country	Area	Water Discharge	Sediment Discharge
Amazon	Brazil	6.1	5700	1200
Zaire	Zaire	3.8	1292	43
Nile	Egypt	3.0	30	120
Zambesi	Mozambique	1.4	546	48
Niger	Nigeria	1.2	192	40
Tigris-Euphrates	Iraq	1.05	47	55
Orinoco	Venezuela	0.99	1080	150
Ganges	Bangladesh	0.98	1280	520
Indus	Pakistan	0.97	237	250
Mekong	Vietnam	0.79	466	160
Brahmaputra	Bangladesh	0.61	690	540
Irrawaddy	Burma	0.43	428	260
Limpopo	Mozambique	0.41	5	33
Volta	Ghana	0.40	37	18
Godavari	India	0.31	84	170
Senegal	Senegal	0.27	13	2
Krishna	India	0.25	35	64
Magdalena	Colombia	0.24	237	220
Rufiji	Tanzania	0.18	9	17
Chao Phya	Thailand	0.16	30	11
Mahandi	India	0.14	72	60
Hungho	Vietnam	0.12	123	130
Narmada	India	0.089	45	125
Fly	New Guinea	0.076	110	115
Moulouya	Morocco	0.051	2	7
Sebou	Morocco	0.04	5	26
Tana	Kenya	0.032	5	32
Purari	New Guinea	0.031	78	105
Cheliff	Algeria	0.022	2	3
Meddjerdah	Algeria	0.021	2	13
Damodar	India	0.02	10	28
Solo	Indonesia	0.016	8	19
Porong	Indonesia	0.012	6	20
Total estimated discharge in tropics			**12,906**	**4604**
Percentage of estimated world total			**32%**	**17%**

Note: Rivers are listed by decreasing drainage basin area (×10⁶ km²).

Source: Updated from Alongi,[12] with data from Milliman and Syvitski.[495]

The immense Amazon and smaller, southern Papua New Guinea (Gulf of Papua) shelves are the best-studied, large-river-dominated systems of the wet tropics, so I will focus on the fluxes and fate of

TABLE 6.13

Salient Characteristics of the Coastal Ocean of the Wet Tropics

As a Direct Result of Latitudinal Location

Solar radiation	High and stable
Temperature	High and stable
Precipitation	High and stable
Runoff	High and stable
Winds	Easterly trades
Boriolis parameter	Small

Associated with Latitudinal Location

Buoyancy flux	Large
Rossby radius	Large
Rotational constraint on benthic boundary layer	Weak
Frontal storms	Uncommon or absent (within 5° of equator)
Terrestrial weathering	Intense
Solute input	Large
Particulate organic carbon (POC) input	Large
Photochemical processes	Intense
Primary productivity	High and stable
Coastal vegetation	Mangroves
Particulate input	Large
Carbonate sediments	Present where particulate fluxes are small

Potentially Resulting from Above Characteristics

Particle coatings	Enriched in Fe, Mn, and Al oxides
Common minerals	Kaolinite and gibbsite
Redox diagenesis	Fe- and Mn-based
Scavenging of dissolved oceanic components	Intense
Primary productivity	Displaced by high turbidity
Benthic fauna and bioturbation	Reduced
Carbon burial	High
Concentrations of suspended sediment	High
Particle grain size	Small; clay and silt
Sediment accumulation rate	High
Resolution of geological record	High

Source: From Nittrouer, C. A. et al., *Geo-Mar. Lett.,* 15, 121, 1995. With permission.

organic matter debouching from these river systems, as well as their impact on coastal food chains and fisheries. Further, I will compare and contrast the Brazilian and New Guinea shelves with the Changjiang

River system in China. The Changjiang River, also classified as a Type III coastal system, is the ninth largest river in the world by drainage basin area and annually discharges nearly 880 km³ of freshwater and 480×10^6 metric tons of sediment into the East China Sea.[521] This system places the other two tidal river systems in perspective, as it discharges onto a broad continental shelf but is located in a warm temperate zone subject to monsoonal climate and receives substantial inputs from agricultural activities. The Changjiang is also one of the few well-studied large tidal river systems with respect to food chain dynamics and nutrient cycles.[521]

6.7.3.1 Physical Characteristics

In contrast to the Amazon and Changjiang systems, the Gulf of Papua receives its sediment and freshwater burden from several rivers — the Bamu, Bebea, Turama, Omati, Aird, Fly, and Purari, the latter two rivers being the largest. These rivers, located on the northwest to north coast *in toto* discharge about 473 km³ of freshwater and 330×10^6 metric tons of sediment annually into the gulf.[522] The Gulf of Papua is half-moon shaped, broad in the center and narrow at both distal ends where the slope drops off sharply into the northern Coral Sea Basin. Runoff varies little seasonally, with the entire gulf stratified over the upper 20 m. Strong winds sharpen the halocline and even inhibit tidal mixing in shallow waters where tidal currents are strong. A counter-clockwise-rotating eddy in the gulf is predominantly driven by the Coral Sea coastal current flowing eastwards from the northwest Coral Sea. This circulation pattern and the dominant wind field results in less saline water leaving the gulf at its distal ends. Residence time of river runoff in the gulf is approximately 2 months. These circulation processes result in most of the mud and organic detritus exiting from the rivers to deposit as a narrow band along the inner shelf; most of the middle and outer shelf is dominated by hard carbonate debris (Figure 6.14B).

The Amazon shelf is similarly influenced by high energy processes,[523] but unlike the Gulf of Papua with overlapping plumes from several rivers, the Amazon River serves as the point source for discharge onto a broad (300 km) shelf. Its plume is advected to the northwest as a result of wind stress from the northeasterly trade winds and strong flow along the shelf associated with the North Brazil Current. The surface plume, 3 to 10 m thick, is well-mixed inshore but is stratified on the outer shelf. The boundary between these two salinity fronts is influenced by tidal mixing. These inshore mixing processes trap fine sediments to form fluid mud suspensions that are usually 1 to 2 m thick. Buoyancy-driven upwelling occurs on the outer shelf and is drawn landwards by the inshore processes and by Eckman transport coupled with along-shelf flow on the outer shelf. The immense physical forces on the Amazon shelf vary greatly over space and time and clearly control sediment transport, as observed for the nearshore fluid muds in which massive sediment resuspension, transportation, and deposition cycles occur quickly and episodically; relict sediments are located farther offshore (Figure 6.14A). The great discharge from the Amazon coupled to the high energy of the equatorial shelf results in dilution of shelf water out to at least the shelf edge.

The Changjiang (Figure 6.14C) — physically more similar to the Amazon shelf — is a point source of discharge varying with season and debouching onto the broad East China Sea. The Changjiang has a monsoonal climate in which nearly 60% of the annual precipitation falls from April to October. Discharge has declined in the past few decades because of damming in the upper tributaries, but immense volumes of water and sediment still discharge into the East China Sea. Hydrographic observations indicate that the structure of the river's discharge is strongly influenced by an along-shelf current (Taiwan Warm Current) from the south, strong tidal currents locally, and complex shelf topography. In summer, when river runoff is high and winds are weak, it splits to form a freshwater band along the coast south of the river mouth and a shallow plume extending across-shelf to the northeast; warmer, saline water flows into the Changjiang estuary along a relic river valley. In winter, when discharge is reduced, the main direction of water motion is southerly along the coast due to wind-driven surface waves. Like the general pattern on the Amazon and Papuan shelves, modern fine sediments are deposited inshore, with relict coarse deposits distal to the river mouth (Figure 6.14C).

6.7.3.2 Pelagic Processes

A schematic representation of primary productivity in relation to salinity (Figure 6.15) for the large tidal rivers reveal two general patterns: algal blooms occur either within the river plumes (e.g., Amazon, Fly) or offshore (e.g., Changjiang, Zaire). The difference in location of the algal blooms is related to shelf topography. At or near the mouths of the Amazon and Fly Rivers, a shoal area results in rapid sedimentation and lower turbidity, sufficient to permit rapid phytoplankton growth inshore. In contrast, sediments remain in suspension at much higher salinities off the Changjiang and Zaire Rivers, owing to a deeper water column off the mouths of both estuaries. Despite clear differences in spatial and seasonal patterns of primary production and end-member concentrations of nutrients, the range of rates of primary productivity is similar among the rivers (Table 6.14).

Like other estuaries, primary productivity near large tidal rivers is regulated by availability of light and nutrients, but it has been difficult

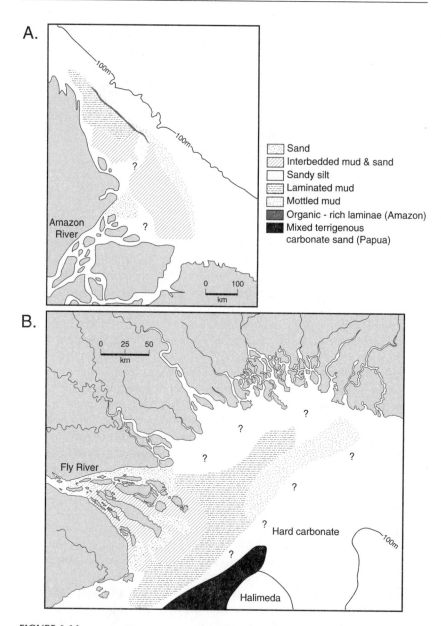

FIGURE 6.14
Distribution of sedimentary facies on the (A) Amazon, (B) Gulf of Papua, and (C) East China Sea shelves. (Data from Milliman and Qingming;[521] also Alongi and Robertson[524] and references therein).

to reconcile which factor plays the dominant role and why. Most studies indicate that in highly turbid riverine waters, low-light limits primary productivity. The situation becomes more complex, however,

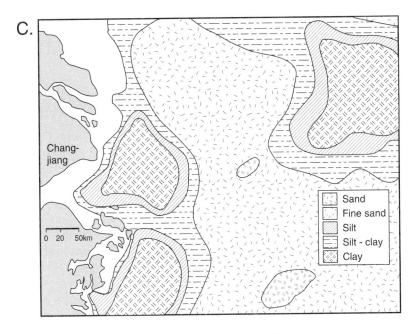

FIGURE 6.14
(continued)

in less turbid waters where nutrient concentrations often deviate from conservative mixing either due to biological uptake and/or uptake or release from particles. On the Amazon shelf, light limits production

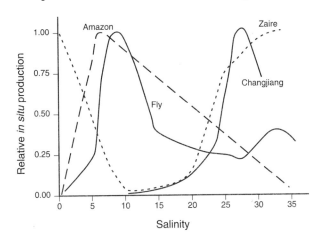

FIGURE 6.15
Patterns of pelagic primary production in relation to salinity in the plumes of the Amazon, Changjiang, Zaire, and Fly rivers. (Data from Nittrouer and DeMaster;[523] Robertson et al.;[525, 526] van Bennekon et al.;[527] and Milliman and Quingming.[521])

TABLE 6.14

Comparison of River End-Member Concentrations of Nutrients (μM) and
Suspended Solids (mg l^{-1}), and Rates (mg C m^{-3} hr^{-1}) of Primary
Productivity (Range) in Major River Plumes

River	Silicate	Phosphate	Nitrate	Suspended Solids	Primary Production	Ref.
Amazon	144	0.7	16	200	0.5–20.0	528, 529
Orinoco	110	0.3	8	100	—	530
Changjiang	110	0.8	~90	100	0.4–24.3	531
Zaire	165	0.8	7	32	0.5–25.0	527
Niger	250	0.8	5	153	—	527
Fly	80	4.0	7	52–456	0.4–28.0	525, 526

within the inshore plume, but on the outer shelf, nitrate and water
residence time apparently regulate production; phosphate and silicate
do not appear to be limiting.[528] In several of the Chinese rivers (e.g.,
Changjiang and Huanghe), phosphate appears to be limiting to algal
growth as inorganic nitrogen concentrations are high because of anthropogenic inputs (Table 6.14). In other large rivers such as the Mississippi, silicate and nitrate supply govern rates of phytoplankton
growth.[532] It is not known what nutrients limit phytoplankton growth
off most of the large tropical rivers, such as the Zaire and Fly, owing
to a lack of knowledge of biogeochemical processes.

A comparison of nutrient concentrations among the tidal rivers
(Table 6.14) implies that phytoplankton growth may be phosphate
and/or nitrate limited in the African rivers and nitrate limited for
phytoplankton in the Papuan waters. This speculation must be tested
by manipulative experiments in the field as nutrient regulation is
likely to vary with trophic composition (diatoms vs. flagellates) and
over time and space.

The biogeochemistry of nutrient elements in river/ocean mixing
zones is best understood for the Amazon,[529] where buoyancy-driven,
shelf-edge upwelling drives plankton production and nitrogen cycling. The strength of upwelling varies with intensity of river discharge. During rising river discharge, algal blooms develop, coinciding with peak flux of nitrate; long residence time allows phytoplankton
to maximize growth and production within the region of relatively low
turbidity and high nutrients. Inshore, sediment resuspension is a primary mechanism of nutrient flux to the water column, but on the shelf
proper, the chief source of nitrogen enhancing productivity is likely to
be the shelf-edge upwelling rather than the river, per se.

Silicon exhibits the most complex nutrient behavior and often
shows depletion in relation to biological activity on the Amazon shelf.

Biogenic silica and silicate are decoupled, probably as a result of particles settling out of the photic zone — by aggregation, grazing, and/or nutrient limitation — and by advection landwards by the nearshore/offshore circulation patterns as induced by the upwelling.

Rates of heterotrophic respiration and bacterial activity in tidal river waters are not well known but presumably are high, considering the relatively high nutrient concentrations and rates of primary productivity. Robertson et al.[525,526] examined heterotrophic processes within the Fly delta and adjacent Gulf of Papua and found that, on average, bacterial respiration could account for 2 to 49% of total pelagic respiration. The waters of this region are net heterotrophic and microbially dominated; rates of respiration (in the range of 0.8 to 36 g C m^{-2} d^{-1}) far exceed those of primary productivity. Zooplankton abundance is very patchy but greater on average than in most other tidal rivers; peak densities coincide with peak phytoplankton biomass, indicating herbivory. Bacterial biomass is very low, implying intense grazing within the microbial loop. Pelagic food chains in the Fly delta and Gulf of Papua are fueled primarily by detritus advected from the extensive freshwater marshes and estuarine mangroves up river and within the delta.

A similarly close association has been found between bacteria and phytoplankton in the Changjiang delta and East China Sea.[533] Oxygen consumption rates in the water column are within the same range as those measured within the Papuan shelf; rates are higher in summer than in winter and greater in the delta than offshore. This pattern mirrors those of bacterial numbers and phytoplankton abundance, suggesting a close trophic coupling between both groups.

It remains to be seen whether or not pelagic food chains of other large tidal river deltas in the tropics and subtropics are similar to those on the Brazilian and Papuan shelves being fueled by carbon fixed by phytoplankton and by tidal export of allochthonous detrital carbon.

6.7.3.3 Benthic Processes and Potential Fishery Connections

Work conducted on the Amazon and Papuan shelves show that benthic food chains and microbial decomposition processes are closely linked to sedimentary facies as controlled by factors such as rates of freshwater and sediment discharge and bottom topography.[524] Off the Amazon, lowest bacterial and faunal densities tend to occur within the physically reworked, inshore fluid muds, but with no clear bathymetric patterns. Abundance varies with changes in river discharge, with peak densities occurring during falling and low discharge periods and lowest densities occurring during rising discharge.[534,535] Off the Papuan

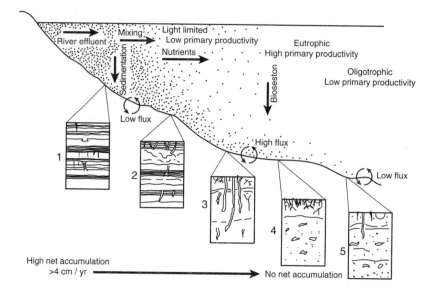

FIGURE 6.16
Idealized model of infaunal responses and sedimentary fabric to discharge from large
tidal rivers. (From Rhoads, D. C. et al., *Cont. Shelf Sci.*, 4, 189, 1985. With permission.)

rivers, highest densities of infauna and bacteria occur within a large
mudbank located southeast of the rivers at mid-shelf. Generally, benthic
faunal and microbial activity are greater in the gulf than within the Fly
delta, where even intertidal muds are too unstable to support benthic
organisms larger than microbes. The benthic infauna of both river
systems share several characteristics:

- Small size (most animals <2 cm in length)
- Numerical dominance of polychaetes over crustaceans and bivalves
- Presence of protobranch mollusc shell layers indicating recent death
  assemblages
- Absence of suspension-feeding bivalves

A similar pattern of low meiofaunal and macrofaunal densities in
rapidly accumulating muds and in oligotrophic sands close to the shelf
break occurs off the Changjiang. This pattern led Rhoads et al.[536] to
propose a generalized model of benthic faunal response to tidal river
discharge (Figure 6.16). Close to the river delta, sediments undergo
episodes of deposition and erosion to the extent that the seabed is too
unstable to sustain benthic life other than bacteria (plate 1). Farther from
the delta, a deepening water column and algal blooms result in a gradient
of conditions increasingly favorable for some infaunal populations (plates

2 to 3). Phytodetritus settles to the sea floor, fueling high densities of surface deposit feeders and microbial decomposition which result in high rates of dissolved nutrient release to the overlying water column. Bioturbation is maximal in this zone, obliterating physical sedimentary structures (plate 4). Farther offshore, primary productivity is lower and rates of sediment accumulation are minimal, resulting in low densities of benthos (plate 5). The location of various sedimentary facies varies in each shelf setting, but the general scheme for the most part applies to the benthos on all three shelves.

Benthic communities on these shelves appear to be regulated by three factors:

- Physical disturbance
- Food limitation
- Predation

All three factors may be operating at any one time. Several lines of evidence point to physical disturbance of the seabed as playing the major role influencing the distribution, abundance, and structure of these tropical and subtropical benthic assemblages. First, X-radiographs of sediment slabs taken from inshore muds on the Amazon and East China Sea shelves and in the interbedded mud and sand facies off the Fly delta, reveal the dominance of primary physical structures. Erosional or burial contacts in subsurface sediments are common, with evidence of truncated biogenic structures suggesting episodes of benthic colonization and elimination. Radiochemical profiles support the radiographic evidence and indicate deep scouring and redepositional episodes. Massive remobilization of sediments to a depth of one meter has been recorded in the inshore fluid muds off the mouth of the Amazon.[523] Laminated deposits form the most common sedimentary profile off all three river systems. Laminated and interbedded deposits likely reflect episodes of erosion and deposition as a result of fluctuations in near-bottom current velocity; high shear stress from strong tidal currents has been observed for all three river areas.

Further evidence of periodic physical disturbance comes from analysis of seasonal and spatial patterns of invertebrate death assemblages[535] buried in deep subsurface deposits off the Amazon. Aller[535] found evidence of cycles of surface exposure and burial of mollusc shells out to a water depth of 40 m, indicating that physical disturbance of the seabed is massive in scale. She also found evidence for large-scale, onshore transport of shells and fragments of relict coral and bryozoans from the outer shelf, followed by deposition and downward mixing in the inshore deposits. The incidence of boring was comparable to other shelves, indicating some biological regulation of the benthos despite the overpowering impact of physical disruption of the seabed.

Predation was initially discounted as being a prime regulatory factor because of the:

- Absence or low abundance of demersal fish or other obvious large predators
- Epibenthos being composed of phyla unlikely to prey on infauna

Examination of fishery data, however, for all three shelves indicates significant commercial landings of finfish (off the Changjiang) and shrimps (off the Amazon and Papuan shelves) in trawling grounds that coincide with highest benthic densities and highly bioturbated facies.[524] This coincidence may be circumstantial but suggests a trophic connection between commercially important fish and shrimp and benthic communities. Alongi and Robertson[524] hypothesized that the main trophic pathway for benthic food chains off the Amazon and Papuan shelves is

$$\text{Detritus + microbes} \Rightarrow \text{shrimp}$$

This is reasonable considering that most of these penaeid species are detritus/microinvertebrate feeders. The larger infauna appears to be a minor trophic pathway to finfish predators. Indeed, compared to shrimp landings, the catch of finfish is minor in the Gulf of Papua[537] and on the Amazon shelf.[538] The scenario of a short food chain is plausible, but clearly more information is needed on the diets of commercially important species. This hypothesis is unlikely to apply to other shelf regions in the tropics and subtropics, because comparatively few other areas regularly experience such massive physical disturbance. There are also large latitudinal differences in the composition of demersal fisheries, even within tropical latitudes.[539]

The dominance of pioneering infaunal assemblages dominated by bacteria and surface detritivores of small size may be an adaptation to a regime of episodic physical disturbance but also suggests food limitation. An analysis of biomass and production estimates of the different benthic trophic groups on the Papuan shelf reveals that infaunal biomass relates positively to pelagic primary production, with bacteria accounting for 75% of total biomass and at least 80% of total benthic production. Compared to temperate benthos, proportionately more production and biomass are vested in bacteria than in the larger size groups on the Papuan shelf. This likely reflects the fact that microbes, being smaller and having shorter generation times than metazoans, can respond more rapidly to, and are less affected by, disturbance.

Studies of bacterial decomposition suggest that re-introduction and episodic remixing of labile organic matter (particularly phytodetritus) may also stimulate rapid bacterial activity in these sediments. The studies of Alongi[540-542] and Aller[543-546] and their colleagues indicate that while rates of decomposition are determined by the amount of available organic matter, the pathways of early diagenesis are influenced more by massive physical reworking. Periodic disturbance and subaerial exposure of anoxic sediments result in vertical elongation of the successive zones of redox reaction sequences.

In most shelf muds, depth scales associated with these zones are typically 0.1 to 1 cm for oxic metabolism, 1 to 10 cm for postoxic diagenesis, and 10 to 100 cm for anoxic processes, with methanogenesis occurring often to a sediment depth of several meters. In muds off the Amazon, Fly, and Changjiang rivers, these zones are vertically expanded. In temperate shelf muds, sulfate reduction normally dominates the reaction sequence, but on all three river-dominated shelves, postoxic diagenesis (iron and manganese reduction) appears to be the major diagenetic pathway. These differences have important consequences for the behavior of trace metals and for burial rates of other elements, such as sulfur.

The best evidence for the diagenetic uniqueness of these shelf deposits comes from the Amazon. On this shelf, net rates of ΣCO_2 production — a measure of total carbon mineralization — are >50 mmol m^{-2} d^{-1} over the upper 1 to 2 m of sediment.[544] Isotopic composition of the dissolved carbon indicates that it is the planktonic carbon rather than the terrestrial material which drives microbial remineralization. Benthic diagenesis is most rapid during periods of low or falling river discharge, although there are no distinct seasonal or spatial trends. Anaerobic mineralization accounts for >75% of total decomposition mostly due to metal recycling. Substantial loss of remineralized carbon to authogenic carbonate formation may occur, but imbalances may also be due to non-steady-state conditions. Little dissolved nitrogen or phosphorus is released from the sediment during stable periods; most release occurs during resuspension events. On average, sediments on the Amazon shelf lose a considerable amount of nitrogen via denitrification (50% of recycled nitrogen) being an inefficient sink for nitrogen and phosphorus derived from upwelling on the outer shelf and from the river. Most phosphorus is lost during resuspension and desorption in the water column.

A model of the sedimentary carbon cycle (Figure 6.17) in the Amazon delta shows that roughly 90% of the carbon remineralized in the sediment (as ΣCO_2) escapes to the overlying water; the remaining 10% is buried as carbonate. Net carbon burial is high despite reoxidation and dissolution of carbonates during disturbance events.

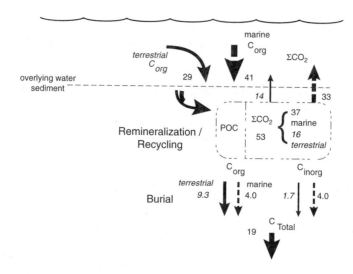

FIGURE 6.17

Carbon fluxes in Amazon deltaic sediments. Terrestrial sources are italicized and fluxes are solid arrows. Sedimentary remineralization and portions of the cycle with dissolved intermediates are enclosed in the dotted rectangle. Units = mmol C m^{-2} d^{-1}. (From Aller, R. C. et al., *Cont. Shelf Res.*, 16, 753, 1996. With permission.)

Nearly 25 to 30% of the buried carbon is in the form of authogenic carbonate, ~20% is marine carbon derived mainly from phytodetritus; the remainder is terrestrial carbon. Nevertheless, nearly 65 to 70% of riverine carbon input onto the shelf is eventually remineralized despite its refractory nature, owing to periodic exposure and re-oxidation of this material during disturbance.

Less extensive information from the East China Sea[547] and Papuan[541] shelves similarly indicates an apparent dominance of postoxic diagenesis. In the Papuan deposits, rates of sulfate reduction are low (4 to 7 mmol S m^{-2} d^{-1}) compared to rates of oxic respiration (18 to 47 mmol O$_2$ m^{-2} d^{-1}), accounting for 21 to 60% of total organic carbon decomposition. Sediments deeper than 20 cm were not examined, but dissolved iron and manganese profiles indicate that metal reduction may be a significant process in these sediments. The data from all three shelves indicate that the spatial patterns of rapid diagenesis mirror the highest rates of phytoplankton production, with close benthic-pelagic coupling.

The biogeochemical and trace element data indicate that diagenesis may not be in steady-state in these shelf deposits. If true, this can have a substantial impact on burial rates and global budgets for many elements. On the Amazon and Papuan shelves, apparently little sulfur is buried, owing to the nature of the source material, which is likely to be highly weathered and composed of reactive oxides. Also, the domination of iron and manganese cycling leaves authigenic minerals poor in sulfur. In most coastal and shelf deposits where sulfate reduction dominates, pyrite (FeS_2) and other reduced sulfides are produced and buried within the sediment. In the Amazon and Papuan settings, repetitive re-oxidation of reduced minerals leads to solubilization and release from sediments, rather than burial. This leaves a C/S signature more reminiscent of freshwater than marine conditions. Re-oxidation during recurring physical disturbance events and the nonsulfidic nature of these deposits lead to very low or undetectable porewater concentrations and flux rates of some trace metals.[542,545] These results may be typical of other high energy coastlines in the wet tropics receiving large amounts of highly weathered, river-borne material, as recently found also on the western Indian shelf.[548]

Michalopoulos and Aller[549] found that reverse weathering (formation of aluminosilicate minerals) occurs within the Amazon deposits, suggesting that a considerable portion of trace elements may be consumed in the process. If such reactions commonly occur in sediments of other wet tropical shelves, global estimates of oceanic element cycles should be revised. Considering that the large proportion of the world's continental shelves are located within the wet tropics, the revised global estimates may be considerably different.

Tropical, river-dominated shelves appear to violate the current paradigm of estuarine systems being highly efficient material traps with little, if any, inorganic or organic matter available for export offshore. These environments are re-visited in the next chapter, in which the role of the coastal zone is examined in conjunction with the rest of the continental shelf.

Chapter 7

THE COASTAL OCEAN II. THE SHELF PROPER AND SHELF EDGE

7.1 INTRODUCTION

The boundaries between the different zones of the coastal ocean can be arbitrary, blurred by the dynamics of an energetic sea. In the previous chapter, nutrient and food web connections between land and river, river and estuary, and estuary and adjacent nearshore zone were explored. This chapter examines the interplay of physical and ecological energetics between the coastal zone and the shelf proper and along- and across-shelf, including processes occurring at the shelf-open ocean boundary.

The flow of energy and nutrients through food webs differs greatly among continental shelves, driven largely by differences in carbon fixation. Rates of fixation are determined ultimately by the confluence of local- and ocean-scale patterns of water circulation, chemistry, and shelf geomorphology. The coastal zone extends across the entire shelf off some of the world's major rivers (Chapter 6), but on most shelves, the estuarine front meanders mainly within the inner shelf. It is on the shelf proper where oceanic and estuarine boundaries frequently intermingle. Tongues of oceanic water regularly or irregularly intrude onto the outer shelf margins often, but not always, to the mid-shelf. At the shelf edge, cold, nutrient-rich oceanic water and associated materials intrude onto the shelf. Such exchanges are often rapid, promoting conditions favorable for higher fertility on the continental shelves than in the open ocean. Higher primary productivity is the main reason why approximately 90% of the world's fish catch is harvested on the continental shelves rather than in the open sea.

Walsh[550] recently itemized the world's shelves based on their location, major rivers and rates of primary production (Table 7.1). The table shows that:

TABLE 7.1

Categorization of the World's Continental Shelves Based on
Location, Major River, and Primary Productivity

Latitude (°)	Region	Major River	Primary Production (g C m^{-2} yr^{-1})
Eastern Boundary Current			
0–30	Ecuador-Chile	—	1000–2000
	Southwest Africa	—	1000–2000
	Northwest Africa	—	200–500
	Baja California	—	600
	Somali coast	Juba	175
	Arabian Sea	Indus	200
30–60	California-Washington	Columbia	150–200
	Portugal-Morocco	Tagus	60–290
Western Boundary Currents			
0–30	Brazil	Amazon	90
	Gulf of Guinea	Congo	130
	Oman/Persian Gulfs	Tigris	80
	Bay of Bengal	Ganges	110
	Andaman Sea	Irrawaddy	50
	Java/Banda Seas	Brantas	110
	Timor Sea	Fitzroy	100
	Coral Sea	Fly	20–175
	Arafura Sea	Mitchell	150
	Red Sea	Awash	35
	Mozambique Channel	Zambesi	100–150
	South China Sea	Mekong	215–317
	Caribbean Sea	Orinoco	66–139
	Central America	Magdalena	180
	West Florida shelf	Appalachicola	30
	South Atlantic Bight	Altamaha	130–350
Mesotrophic Systems			
30–60	Australian Bight	Murray	50–70
	New Zealand	Waikato	115
	Argentina-Uraguay	Parana	70
	Southern Chile	Valdivia	90
	Southern Mediterranean	Nile	30–45
	Gulf of Alaska	Fraser	50
	Nova Scotia-Maine	St. Lawrence	130
	Labrador Sea	Churchill	24–100
	Okhotsk Sea	Amur	75
	Bering Sea	Kuskokwim	170

TABLE 7.1 (continued)

Categorization of the World's Continental Shelves Based on
Location, Major River, and Primary Productivity

Latitude (°)	Region	Major River	Primary Production (g C m⁻² yr⁻¹)
Phototrophic Systems			
60–90	Beaufort Sea	Mackenzie	10–20
	Chukchi Sea	Yukon	40–180
	East Siberian Sea	Kolyma	70
	Laptev Sea	Lena	70
	Kara Sea	Ob	70
	Barents Sea	Pechora	25–96
	Greenland-Norwegian Seas	Tjorsa	40–60
	Weddell-Ross Seas	—	12–86
Eutrophic Systems			
30–60	Mid-Atlantic Bight	Hudson	300–380
	Baltic Sea	Vistula	75–150
	East China Sea	Yangtze	170
	Sea of Japan	Ishikari	100–200
	North-Irish Sea	Rhine	100–250
	Adriatic Sea	Po	68–85
	Caspian Sea	Volga	100
	Black Sea	Danube	50–150
	Bay of Biscay	Loire	120
	Louisiana/Texas shelf	Mississippi	100

Source: Adapted from Walsh,[550] with additional data from Alongi,[12] and Postma and Zijlistra.[551]

- Low latitude upwelling areas on the eastern boundaries of the ocean are usually the most productive shelf ecosystems.

- The subtropical and tropical river-dominated shelves (0 to 30° latitude) on the ocean's western boundaries are nearly as productive as shelves in temperate seas (30 to 60° latitude).

- The shelf systems in many cold temperate and boreal seas (60 to 90° latitude) are comparatively unproductive.

Unfortunately, few of these ecosystems are well understood, despite their importance to humans. Some examples of each of these groups are examined here within the context of coastal-shelf fronts, along- and across-shelf gradients, and shelf-edge processes. The chapter concludes with an examination of the role of continental shelves in fishery yields and carbon budgets of the global ocean.

7.2   SHELF-SEA FRONTS

No clear categorization of fronts exists, although many different types of boundaries or fronts occur in coastal seas:

- Shelf-sea (or tidal) fronts
- Estuarine fronts or plumes
- Shelf-break fronts
- Upwelling fronts
- Island wakes and fronts caused by other land features

Shelf-sea fronts — the boundaries between well-mixed and stratified shelf-sea water masses — occur as a result of changes in tidal mixing. Tidal fronts occur on numerous shelf regions, but the best-studied fronts are those occurring on the northwest European continental shelf, where tidal energy is strong, particularly in the southern North Sea and in the waters around the British Isles. Mann and Lazier[206] have recently discussed tidal fronts in great detail,[50] so I consider here only nutrient cycling and microbial processes.

The persistence of shelf-sea fronts in summer results in the enhancement of primary productivity with, inevitably, an increase in the numbers of potential pelagic consumers. Loder and Platt[552] reviewed phytoplankton activity at tidal fronts and concluded that the process of vertical mixing is the major physical control on phytoplankton growth and production. Vertical mixing exerts control on phytoplankton by enhancing the supply of nutrients and light. Loder and Platt[552] advanced the hypothesis that nutrient (primarily nitrate) transfer across the frontal boundary enhances phytoplankton productivity. How this actually happens is not known.

The Irish Sea[553] scenario indicates distinct ecological characteristics within tidal fronts between surface-stratified (SSW) and bottom-stratified (BSW) waters, as well as between stratified and well-mixed water (MW) masses. This phenomenon evolves when the thermocline begins to form in spring, and the abundant supply of nutrients fuels a rapidly increasing community of photoautotrophs. As the community expands, nutrients are not replenished across the thermocline and they dwindle, but phytoplankton biomass levels off and is maintained during summer. The highest concentrations of phytoplankton lie on the pycnocline within the SSW; numbers of bacteria, protozoans, zooplankton, and their corresponding activities (e.g., nutrient recycling) are enhanced. This enhancement of pelagic food web activity is maintained until surface waters cool and the water column destratifies in autumn. In the bottom-stratified waters, the abundance and activity of the various trophic groups are low, with little recycling of nutrients. In the well-mixed waters, primary productivity is light limited, ameliorating food chain and nutrient cycling activity compared to the pelagic communities in the surface-stratified waters.

Grazing can be intense in the surface-stratified waters. Lochte and Turley[554] performed a series of incubation experiments to determine carbon flux within the SSW of the Irish Sea front. Assuming energetic relationships found in other studies, they were able to estimate the flow of carbon within the pelagic community in the SSW and BSW. In the surface water (Figure 7.1A), the bulk of fixed carbon fluxes through the bacterial community, where most is lost via respiration. Bacterial growth is sustained by a large dissolved organic carbon (DOC) pool maintained by input of DOC from particles and by ambient diffusion. More than one quarter of bacterial biomass is consumed daily by flagellates; ciliates were presumed not to feed on bacteria, which is unlikely as many small ciliates are capable of feeding at the densities observed within the tidal front. Bacterial production and protozoan consumption are lower in the BSW layer, corresponding to lower rates of primary productivity (Figure 7.1B). Further, the low rates of bacterial production may be sustained directly by DOC from ambient particles; in both water masses there appears to be little direct flow of carbon from phytoplankton to the microbial community. The carbon budget (Figure 7.1) is incomplete, as it is likely that the high densities of zooplankton observed within the tidal front play an important trophic role in linking microbial and metazoan communities, including fish and seabirds, in the Irish Sea.[553]

Farther south in the Celtic Sea and across the English Channel to the coast of western Brittany, strong tidal fronts persist in summer, greatly influencing trophic functioning and nutrient flow in pelagic food chains. At the Ushant front in the western English Channel, Holligan et al.[555] and Linley et al.[556] studied the distribution, abundance, and microheterotrophic activity in the pelagic community in relation to water mass structure. Their results indicate peak densities of organisms in the surface waters of the frontal boundary compared to lower numbers within the mixed water masses. Further, they calculated that the heterotrophic demand for fixed carbon, especially in the stratified region, exceeds that produced by the primary producers. They suggested that these needs may be met by transport of materials from adjacent waters, but this has yet to be substantiated.

The size and structure of the phytoplankton are similar to those of communities in the western Irish Sea — diatoms flourishing in the mixed waters, with dinoflagellates replacing them at the front and in the stratified region. Using the observed distribution of dissolved

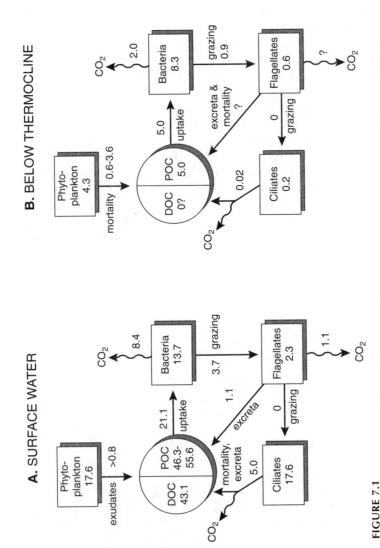

FIGURE 7.1

Carbon flow model within the planktonic microbial loop during summer in (A) surface water and (B) below the thermocline of the stratified waters of the western Irish Sea. Units are mg C l^{-1} and mg C l^{-1} d^{-1} for standing stocks and fluxes, respectively. (Adapted from Lochte, K. and Turley, C. M., in *Proceedings of the Nineteenth European Marine Biology Symposium*, Gibbs, P. E., Eds., Cambridge University Press, London, 1985, 73.)

nutrients, they estimated that in the stratified waters, the flagellates obtain roughly equal amounts of nitrogen from recycled ammonium and upwelled nitrate, whereas the flux of new nitrogen is sufficient to meet nitrogen demand within the frontal boundary. This new nitrogen may actually be advected from adjacent water masses via several possible cross-front mechanisms, such as:

- Baroclinic eddies
- Residual currents
- Vertical transport
- Variations in spring-neap tidal cycles

The ecological importance of these processes remains to be investigated.[205]

River plumes and tidal fronts often co-occur on continental shelves. A good example is the German Bight, located in the southeastern North Sea. Unlike the heat-induced tidal fronts around the British Isles, thermal stratification of water masses in the German Bight is reinforced by persistent salinity gradients. This phenomenon has major consequences for pelagic activity and abundance. The river plumes and tidal fronts show large differences in phytoplankton community composition: autotrophic nanoflagellates with low photosynthetic rates ($0.72\,\mu g\,C\,\mu g\,Chl\,a^{-1}\,l^{-1}$) dominate the tidal front, whereas diatoms with a higher photosynthetic rate ($4.1\,\mu g\,C\,\mu g\,Chl\,a^{-1}\,l^{-1}$) are the major group within the Elbe River plume front.[557] During calm weather in summer, red tides of the dinoflagellates *Ceratium fusus* and *Noctiluca scintillans* occur within weak frontal structures linked to the tidal front in the southwestern part of the German Bight. These red tides cause severe oxygen deficiencies resulting in serious fish mortality near the edge of the Elbe River plume.

7.3   ALONG- AND ACROSS-SHELF GRADIENTS

Here, some of the best-studied continental shelves are viewed from a larger perspective to see how the dynamics of food chains and nutrient cycles vary in response to physical, chemical, and geological forces, from the coastal zone to the shelf edge, and along the longitudinal axis of the continental margin. Because each shelf ecosystem is complex, changing over time and space, any generalizations made about the world's shelves would fail to catch the varied ecologies of individual systems. Many shelf ecosystems, particularly those in the tropics, are very poorly understood. With these caveats in mind, three generalizations about many shelf ecosystems are offered:

- Outwelling of materials from the continents is usually restricted to the coastal zone by complex physical, chemical, and geological processes, but this material greatly influences inshore food chains and nutrient cycles.
- Impingement of nutrient-rich, open-ocean water enhances primary and secondary productivity on the outer shelf and shelf edge, including important fisheries, particularly along boundary currents.
- Little organic matter is exported from the continental margin.

The most important factors determining to what extent a shelf ecosystem validates or violates these generalizations include the:

- Presence or absence of large rivers
- Presence or absence of upwelling
- Location on ocean boundaries
- Shelf width

All of these factors are influenced by climate. Good examples of, and some excellent exceptions to, these generalizations are explored in the remainder of this chapter.

7.3.1   The North Sea

Perhaps no other marine ecosystem has been as long or as well studied as the North Sea. This is not only because it borders the industrialized European nations, but also because it is still one of the world's most important fishing grounds — accounting for roughly 5% of the world's fish catch.[551] However, due to the impact of humans, fish yields and water quality have declined markedly since the late 1960s.

Bordered by the British Isles, the European continent, and the Scandinavian peninsula, the North Sea is connected to the North Atlantic in the north, to the English Channel in the south, and with the Baltic through the Skagerrak in the east. Prominent bottom features are the Fladen Ground, Dogger Bank, Norwegian channel, and the shallower, coastal Southern and German Bights. The hydrography is dominated by strong tidal currents that weaken northwards; the major inflows consist of North Atlantic water of high salinity from the north and south and water of lesser salinity from the Baltic. Numerous smaller inflows from European rivers and the bottom topography help to set up a complex water circulation that also varies seasonally.[557]

Nutrient concentrations in the water column are generally higher in the northern than in the southern North Sea, reflecting input from the North Atlantic, but there have been significant changes over the

TABLE 7.2

Changes in Total Estimated Inputs of Nitrogen and Phosphorus into the
Entire North Sea, Excluding Inputs from the Baltic

Source	Nitrogen		Phosphorus	
	1950	1980	1950	1980
North Atlantic (excluding Norwegian Trench)	5123 (83%)	5123 (74%)	945 (91%)	945 (85%)
Strait of Dover	615 (10%)	–21 (—)	59 (6%)	6 (1%)
Atmosphere	192 (3%)	577 (8%)	9 (1%)	18 (2%)
Rivers, discharges, dumping	264 (4%)	1202 (18%)	22 (2%)	146 (13%)
Total	**6194**	**6881**	**1035**	**1115**

Note: Units are 10^3 metric tons yr^{-1}. Values in parentheses depict percentage of total input.

Source: From Radach, G., *Estuaries*, 15, 477, 1992. With permission.

past few decades. Radach[559] has estimated that the total input of phosphorus and nitrogen has increased by 7.7 and 11.4%, respectively, from 1950 to 1980.

Table 7.2 shows how these changes occurred. For both phosphorus and nitrogen, natural inputs from the North Atlantic have remained unchanged, while those through the Strait of Dover have declined, but inputs from the atmosphere and especially from rivers and other discharge and dumping activities have increased dramatically. Indeed, coastal waters of the Southern and German Bights have experienced a continual, albeit slow, rise in ambient phosphate and nitrate concentrations (Figure 7.2).

Recent analyses imply that these human-induced rises in nutrient concentrations have significantly altered the seasonal and spatial dynamics of the phytoplankton community.[559,560] Trend analyses (1962 to 1984) have shown a strong increase in phytoplankton biomass and a dominance shift from diatoms to flagellates; average diatom biomass has decreased from 6.2 to 4.8 mg C m^{-3}, whereas flagellate biomass has increased tenfold from 2.5 to 25.1 mg C m^{-3}.

There has also been a change in the annual cycle. Phytoplankton activity normally centers around the spring diatom bloom, which is followed by a flagellate bloom often persisting throughout the summer; a second diatom bloom occurs during autumn in some years. In more recent years, the onset of the spring diatom bloom has occurred earlier — in March instead of April — and flagellate biomass remains comparatively high year-round, with dinoflagellate blooms persisting from May until the end of September. On average, phytoplankton production has been increasing incrementally since the 1950s. Long-term averages are 250 g C m^{-2} yr^{-1} in the central North Sea, 150 to 200 g

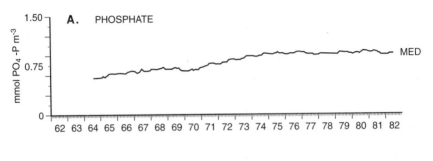

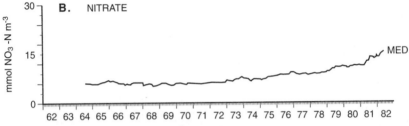

FIGURE 7.2
Long-term trends in water-column phosphate (A) and nitrate (B) concentrations in the
German Bight. The plots depict running 5-year means starting in mid-1964 and ending
in mid-1982. (Adapted from Radach, G., *Estuaries*, 15, 477, 1992.)

C m⁻² yr⁻¹ in the northern North Sea, and 200 g C m⁻² yr⁻¹ in the
South.[560] More recent data indicate that flagellate production in the
South has risen to more than the long-term average of the northern
North Sea.[559]

Dramatic changes have also occurred in the rest of the pelagic
community. Zooplankton biomass and the abundance of fish and seabirds in the North Sea have declined markedly since the first measurements were made in 1948.[558] Ascribing these trends to the effects of
increased eutrophication is difficult and must be considered cautiously,
in concert with long-term climate changes.

A recent series of analyses by Aebischer et al.[12] has shown a remarkable synchrony of long-term changes in abundance of phytoplankton, zooplankton, and herring (*Clupea harengus*) and in the breeding
cycle of the kittiwake (*Rissa tridactyla*), an open-ocean seabird, with the
frequency of dominant westerly weather cells in the North Sea. The
mechanism by which climatic changes might be coupled to community
composition is unclear. Each trophic level may be directly affected by
the weather or there may be a simple regulatory effect up the food
chain. It is equally likely that the reasons are more complicated. Forcing by climatic changes on herring populations could be indirect,
linked to changes in larval transport by wind-driven circulation shifts

or through climate-induced changes in vertical mixing or stratification events. In any case, the climate signal is strong for food webs in the North Sea.

How climate and eutrophication affect the flow of energy and materials through food chains in the North Sea are less well understood. The geographical and seasonal patterns of bacterioplankton abundance and production in the North Sea appear to mirror those for phytoplankton with a time lag of several days.[562] A preliminary model of standing stocks and fluxes (Figure 7.3) indicates that carbon fluxing through the pelagic microbial web in the coastal zone averages roughly 110 g C m^{-2} yr^{-1}, nearly 57% of net primary production; the bulk is lost as respired carbon. How and how much of the assimilated microbial carbon is transferred to metazoans is unclear. The data are insufficient to evaluate the role of the trophic link between phytoplankton and the microbial loop, but the nature of the relationship is likely to be oceanic in character, similar to food web processes in the open waters of the North Atlantic. Herbivorous zooplankton and other planktivorous metazoans are clearly an important trophic link to herring and mackerel (*Scomber scombrus*), but considering the energetic importance of the microbial loop, it is highly unlikely that early energy flow models of the North Sea are still valid. The earlier assumptions of relatively high transfer efficiencies are particularly implausible, considering that the microbial loop acts as a significant shunt for phytoplankton-derived fixed carbon. High fish production and standing stocks may be sustained despite high microbial recycling of carbon if the intermediate trophic levels, including larval and juvenile fish, consume microbial biomass. How, or if, microbial food webs link to fish production remains to be determined.

Benthic food chains contribute to the sustainability of demersal fish stocks, particularly plaice (*Pleuronectes platessa*) and other flatfish. Evidence for a trophic link comes not only from analysis of fish diets, but also from the similarity in the large-scale patterns of macrobenthos and demersal fish abundance. Both macrofaunal abundance and flatfish landings are greatest in the Southern and German Bights and on Dogger Bank in the southern and central North Sea.[558]

Rates of benthic mineralization of organic matter indicate that bacteria recycle the vast bulk of this organic matter in the coastal zone, including the vast expanses of tidal flats within the European estuaries (see Chapter 2). Summarizing available subtidal data, Billen et al.[562] estimate that total sediment carbon metabolism ranges from 50 to 160 g C m^{-2} yr^{-1}. This compares to rates of organic matter sedimentation to the seabed of 20 to 100 g C m^{-2} yr^{-1}. In sandy sediments, recycling of bacterial biomass is required to meet bacterial demand for organic carbon.[563] Obviously, large uncertainties remain, but it is clear that

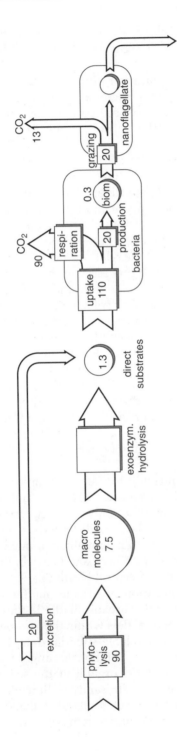

FIGURE 7.3
Estimates of carbon fluxes associated with bacterioplankton in coastal zone waters of the North Sea. Units are g C m^{-2} for mean annual standing stocks and g C m^{-2} yr^{-1} for fluxes. (From Billen, G. et al., *Neth. J. Sea Res.*, 26, 265, 1990. With permission.)

bacteria are responsible for most recycling of organic carbon depositing to the North Sea floor. Stable isotope studies confirm that most of the organic matter settling to the subtidal sea floor is of marine origin.[564]

Recent trophic studies conducted in the North Sea indicate that macrofaunal activities and the sedimentation of organic matter stimulate both bacterial and protozoan abundance; the high densities of benthic nanoflagellates imply that protozoans exert some control over bacterial abundance and production in the surface layers of sandy sediments.[565] In muddy North Sea deposits, however, rates of bacterial production — stimulated by the high levels of dissolved and particulate organic carbon — may outpace consumption by protozoan and metazoan grazers, particularly in anoxic sediments.

An extensive survey[566] of benthic carbon mineralization in the southern North Sea indicates that the greatest rates of microbial activity occur in the rich nursery grounds near the Dogger Bank and in proximity to the Dutch and Belgian coasts. All stations showed moderate rates of oxygen consumption (range: 5.3 to 27.8 mmol O_2 m^{-2} d^{-1}) and sulfate reduction (range: 0.05 to 11.8 mmol SO_4 m^{-2} d^{-1}), with aerobic respiration accounting for 47 to 89% of total organic matter oxidation. Benthic mineralization in the stratified central North Sea accounts for a greater proportion (47%) of net primary production than in the southern North Sea (26%). On average, benthic carbon mineralization in the southern North Sea is equivalent to 17 to 45% of total net primary production, signifying a strong benthic-pelagic link in these waters.

Frontal systems in the southern North Sea also play a strong role in regulating benthic food chains and mineralization rates. The Frisian Front, located southeast of the Dogger Bank, is a remarkable transition zone between the coarse sandy sediments of the Southern Bight and the muddy deposits between the German Bight and Dogger Bank, being characterized by a very large benthic biomass and a tidal front in summer.[567] This transition zone and subsequent benthic enrichment is thought to be the direct result of enhanced sedimentation due to a combination of a sharp decline in tidal current velocities and a steep bottom profile. In summer, increased heat input results in stratification of the water column to the northern part of this area, and residual tidal currents produce a well-mixed water mass to the south, producing a tidal front.

Benthic oxygen consumption measurements indicate that benthic activity is highest in sediments underneath the front (95.1 g C m^{-2} yr^{-1}) compared to deposits north (41.1 g C m^{-2} yr^{-1}) and south (29.4 g C m^{-2} yr^{-1}) of the front. The rates of benthic mineralization were also found to be linked to the spring algal bloom, supporting the idea that benthic activity in the frontal zone is strongly coupled to the deposition of fresh phytodetritus. The oxygen consumption rates indicate that roughly half of the net primary production at the front is mineralized by the benthos, with the remainder presumably mineralized in the water column.

Nitrogen cycling has been measured in numerous North Sea habitats, but only recently have efforts been made to place them within a shelf-scale perspective. Nitrogen release from offshore sandy sediments of the Belgian and Dutch coastal zones ranges from 13 to 30 mg N m^{-2} d^{-1}, capable of sustaining 17 to 50% of phytoplankton requirements in the overlying water column.[562] The significance of nitrogen losses via denitrification are unclear, as few such measurements have been made in connection with nitrogen inputs at a given location. Estimates of nitrogen lost through denitrification range from 1 to 68%, depending upon the method used and the estimated input. The most recent appraisal for the continental coastal zone indicates that of a total nitrogen input of 971,000 metric tons per year, 12,500 metric tons per annum (or only ~1%) are lost by denitrification.[568] If accurate, this estimate implies that either nitrogen is lost by other pathways or the North Sea ecosystem is extraordinarily efficient in conserving nitrogen. Better estimates of nitrogen flux are urgently needed for the North Sea, as anthropogenic inputs are increasing.

7.3.2   Eastern North American Shelf

The eastern seaboard of North America extends from boreal to tropical latitudes and consists of three main biotic provinces (Figure 7.4). The Acadian or boreal province encompasses the coastline from the Scotian shelf out to the Georges Bank-Gulf of Maine region; the Virginian or mid-Atlantic province extends farther south to Cape Hatteras off North Carolina, separating this province from the Carolinian or south Atlantic province, which extends to the very narrow south Florida shelf.

7.3.2.1 The Acadian Province

This ocean-dominated boreal shelf is a complex postglacial environment, characterized by coarse-grained banks intermingled with low-energy basins which trap fine sediments.[569] It is also the location for some of the world's richest fishing grounds: The Grand Banks and Georges Bank (Figure 7.4).

Some attempts have been made to link primary production and fish yields in this province.[570,571] Mills and his colleagues[570,571] estimated rates of primary and secondary production at various depths on the Scotian Shelf region and made some initial attempts to construct

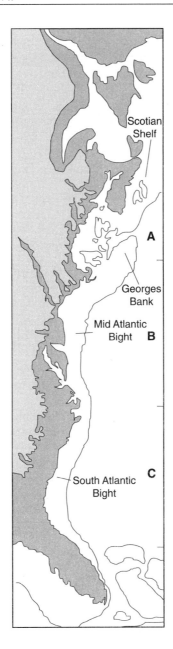

FIGURE 7.4
The continental margin of the eastern seaboard of North America delineating the biotic provinces: (A) Acadian-boreal; (B) Virginian-Atlantic; (C) Carolinian-South Atlantic. (Adapted from Ray, G. C., in *Changing the Global Environment: Perspectives on Human Involvement*, Botkin, D. B., Ed., Academic Press, New York, 1989, chap. 5.)

TABLE 7.3

Comparison of Production (g C m^{-2} yr^{-1}) in Various Regions of the
North American Shelf and the North Sea

	NS	GM	GB	MA	North Sea
Primary production	102–128	162–364	450	150–200	103
Zooplankton	19.3	20.5	20	13.6	12.8
Macrobenthos	8.1	9.7	9.2	17.9	2–5
Fish	3.4	3.2	6.8	3.2	1.4

Note: NS = Nova Scotia, GM = Gulf of Maine, GB = Georges Bank, MA
= Mid-Atlantic Bight.

Source: Data adapted from Mills et al.[571] and Sherman et al.[572]

food web models. These models did not include the microbial loop, as they were constructed before the importance of pelagic microbes was recognized. In fact, an original conclusion that the pelagic food chains are insufficient to support pelagic fish production may be due to the absence of microbial pathways of energy consumption and dissipation, production by autotrophic pico- and nanoplankton and/or incorrect trophic transfer efficiencies in the models. These are understandable oversights considering the lack of data at that time.

The benthos-demersal fish link is clear, but the question of why this region is such a rich nursery ground is still not clear. Table 7.3 shows that the rates of primary productivity are greater on the nursery grounds of Georges Bank as compared to those rates in other areas of the North American Shelf and the North Sea, but these high rates of carbon fixation do not appear to translate into greater rates of pelagic or benthic secondary productivity. Sherman et al.[572] suggest that subtle factors come into play which do not translate into energy or productivity differences, including:

- Temporal changes in fish stocks in relation to their prey (including the timing of the spring bloom)
- Advective losses of pelagic prey by circulation patterns and shelf/slope exchange
- Differences in community structure and species composition

It is likely that no one mechanism is crucial in supporting the large fishery yields from this region. Since the construction of these early models, a better understanding of nutrient cycling and energy flow has arisen which may eventually shed some light on the answer.

A recent study of pelagic food webs in the nearby Gulf of St. Lawrence shows that although food web structure and the main pathways of carbon flow differ greatly between bloom and nonbloom periods, the total flux of carbon does not appear to be significantly altered. During the bloom, herbivorous zooplankton dominate trophic structure, whereas in the postbloom period, the microbial loop dominates; the pelagic food web thus shifts from net autotrophy to net heterotrophy. Despite this trophic shift, the composition, but not the amount of material exported from the euphotic zone to the aphotic zone (and to benthic food chains), is altered. A hypothesis of how this might happen is conceptualized in Figure 7.5. When phytoplankton bloom, the bacteria are fueled via DOC derived from exudates of the large phytoplankton population and by sloppy feeding of protozooplankton and the herbivorous mesozooplankton. A substantial proportion of the large phytoplankton community remains ungrazed and settles from the water column. In nonbloom periods, the mesozooplankton community is omnivorous, feeding mostly on protozooplankton because phytoplankton are scarce. Dissolved nutrients originating from recycling within the microbial loop continue to drive bacterial growth (albeit at a slower rate), but particulate detritus settles out of the euphotic zone in the form of fecal pellets, at a rate roughly equivalent to the rain of phytodetritus during the bloom. Vertical flux and export of detrital carbon continue at a similar rate year-round, although ΣCO_2 flux may be altered. If a similar phenomenon operates on the adjacent Georges Bank and The Grand Banks farther offshore, this phenomenon may explain fish yields on these rich fishing grounds. Unfortunately, overfishing in now taking its toll — as in other regions of the North Atlantic — to the extent that some fish species are now gone and others are in decline.

As in other marine ecosystems, light and nutrients, especially nitrogen, drive phytoplankton growth and production in this province. In the coolest months, light is the principal limiting factor, but with the onset of stratification in the warmer months, nitrogen is thought to become limiting. Cochlan[574] observed a seasonal pattern in the use of ammonium and nitrate in the Scotian shelf waters. In winter, phytoplankton growth is supported mainly by "new" nitrogen advected onto the shelf from the slope. In spring and summer, locally regenerated nitrogen drives the phytoplankton. In all seasons, however, the phytoplankton showed a preference for regenerated ammonium relative to nitrate. High rates of primary production can therefore be sustained as rates of ammonium remineralization exceed uptake rates on the Scotian shelf, even in stratified shelf waters during late spring and summer.

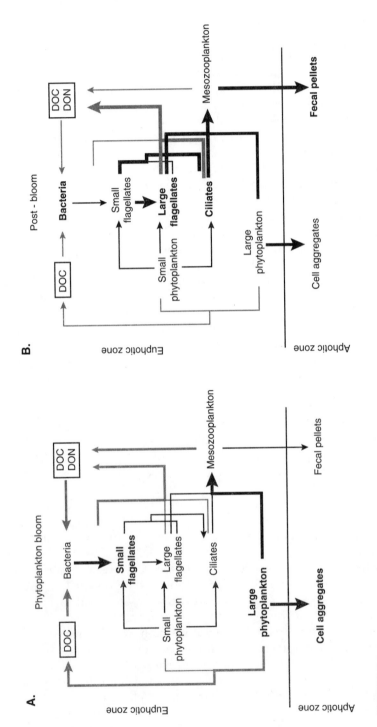

FIGURE 7.5

Pelagic trophic model for the Gulf of St. Lawrence during (A) and after (B) the spring phytoplankton bloom. Wide arrows and bold type depict large fluxes and high standing stocks, respectively. (From Rivkin, R. B. et al., *Science, 272,* 1163, 1996. With permission.)

TABLE 7.4

Estimates of Annual Phosphorus and Nitrogen
Budgets for the Gulf of Maine

Process and Region	Phosphate (10^9 g yr^{-1})	Nitrogen (10^9 g yr^{-1})
Advective inputs		
Northeast channel	6.11	104.1
Scotian shelf	2.52	31.5
Rivers and precipitation	0.06	3.0
Total inputs	8.69	138.6
Advective outputs		
New England Shelf	–8.08	–95.5
Net advective input	0.61	43.1
Nonadvective losses		
Burial	–0.26	–4.4
Organic matter export	–0.35	–5.6
Denitrification	0.0	–33.1
Total nonadvective losses	–0.61	–43.1

Source: Adapted from Christensen, J. P. et al., *Cont. Shelf. Res.*, 16, 489, 1996.

Water advecting from the Scotian shelf and adjacent Northeast Channel is an important source of nutrients for the Gulf of Maine.[575] Nutrient budgets (Table 7.4) reflect the importance of these water masses for nitrogen and phosphorus inputs into the Gulf of Maine; river runoff and precipitation are minor sources. Advection of Gulf of Maine water south to the New England shelf leaves a net import of both nitrogen and phosphorus. These imbalances appear to be met by nonadvective losses such as burial and export of organic matter to the slope. For nitrogen, however, denitrification represents the major pathway through which nitrogen is lost from the system. The export of organic matter constitutes only a small (1.5%) fraction of total net primary productivity in the Gulf of Maine.

These budgets, of course, do not reflect the complexity of nutrient sources for primary productivity. A summary of new and regenerated nitrogen sources (Table 7.5) illustrates the influence of these diverse sources of new nitrogen on primary production in the Gulf of Maine, reflecting inputs from advective and convective water masses, eddies, a plume in the eastern gulf, and coastal and shelf-edge upwelling.

Stable isotope analyses[577] and estimates of benthic oxygen consumption[578] indicate that most of the sediment organic carbon along this boreal shelf is of marine origin and consumed largely by the benthos in the fine sediment areas. Benthic oxygen consumption estimates (Table 7.6) show that in areas containing fine sediments (the

TABLE 7.5

Nitrogen Sources and Resulting Rates of Primary
Production in the Gulf of Maine

Nitrogen Source	Primary Production (g C m^{-2}yr^{-1})
New nitrogen	
Winter convective overturn	25.2
Eastern Gulf plume	36.6
Vertical eddy diffusion	32.2–108
Upwelling	
Coastal Maine	
Estuarine	8.0
Eckman	8.0
Southwest Nova Scotia	36.6
Recycled nitrogen	16–110
Total primary production	162–364

Source: From Townsend, D. W., *Rev. Aquat. Sci.*, 5, 211,
1991. With permission.

Gulf of Maine and Scotian Shelf), benthic respiration equates to a
greater (16 to 19%) proportion of primary production than in areas
(Georges Bank and The Grand Banks) dominated by coarse deposits (5
to 8% of primary production). Grant et al.[578] estimate that the original
food web models may have underestimated benthic consumption by
as much as 50% as well as the magnitude of carbon flow through large
macrofaunal pathways.

TABLE 7.6

Annual Benthic Carbon Consumption in Relation to Coarse and Fine
Sediments in Regions of the North American Shelf

Region	Sediment Texture (%)		Benthic Metabolism (10^6 metric tons yr^{-1})		
	Coarse	Fine	Coarse	Fine	Total
Gulf of Maine	0	100	0.0	4.0	4.0 (19.0%)
Georges Bank	100	0	0.8	0.0	0.8 (5.3%)
Scotian Shelf	66	34	2.2	3.1	5.3 (16.1%)
Newfoundland Shelf	88	12	4.9	1.8	6.7 (8.3%)
Labrador Shelf	68	32	1.0	1.3	2.3 (8.1%)

Note: Values in parentheses are percentage of primary production consumed
     by benthic metabolism.

Source: Data adapted from Grant, J. et al., *Cont. Shelf Res.*, 11, 1083, 1991.

Comparatively little information is available on rates and pathways of organic matter decomposition in these boreal sediments, other than rates of oxygen consumption and denitrification. It is reasonable to believe that rates of anaerobic bacterial activity are rapid in the finer deposits, particularly in spring when phytodetrital remnants from the bloom settle to the bottom. Abundant bioturbating macroinfauna on this shelf also presuppose very active microbial decomposition in subsurface sediments. Rates of sulfate reduction are rapid in sediments of this region.[579] Moreover, profiles of porewater iron and manganese imply that active metal reduction takes place below the zone of denitrification and sulfate reduction.

7.3.2.2 The Virginian Province

South of Cape Cod, the continental shelf widens and is bounded by a coastal zone of extensive freshwater and salt marshes, lagoons, and very large estuaries. The ecosystem on this portion of the shelf is greatly influenced by the mixing of coastal and offshore water masses. It is on this section of the shelf that arguably more research has been conducted on the structure and function of marine ecosystems than on any other. Most of these earlier results and carbon flow models are summarized in Sherman et al.[572] and in Walsh.[550] Here, more recent attempts to integrate the flow of energy and materials from the coastal zone to the shelf edge are discussed.

Despite the myriad economic, social, military, and political uses of this region, it may seem ironic that several recent large-scale programs to investigate shelf processes were stimulated by the question, "Do continental shelves export organic matter?". Posed by Walsh and his colleagues,[580] the hypothesis arose in response to the discrepancy observed between rates of benthic respiration and the deposition rate of ungrazed phytoplankton settling to the shelf floor. It was noted that ingestion of phytoplankton by members of the classical pelagic food web and consumption on the shelf floor were not sufficient to account for the fate of the remaining carbon fixed during the spring bloom. The logical answer was that it was exported either to the central ocean basins or deposited on the continental slope. It was further recognized that if this was the fate of carbon on most continental margins, it would have a significant impact on the global carbon cycle. To test the shelf export hypothesis, the SEEP (Shelf Edge Exchange Processes) and SEEP-II projects were conducted in the central and southern sections, respectively, of the Mid-Atlantic Bight.

The SEEP-I experiment traced the fate of primary production resulting from the spring bloom in 1984. Figure 7.6 depicts the mean carbon flow pathways through pelagic and benthic food webs in the

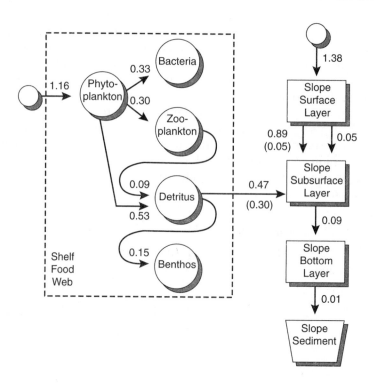

FIGURE 7.6

Summary of carbon flow (g C m^{-2} d^{-1}) within shelf and slope food webs of the Mid-Atlantic Bight during the period of March through April 1984. (From Walsh, J. J., Biscaye, P. E., and Csanady, G. T., *Cont. Shelf Res.*, 8, 435, 1988. With permission.)

Mid-Atlantic Bight. Of the mean carbon fixed daily, roughly half is used by both zooplankton and the microbial loop. The remainder sinks to the shelf floor, where nearly 25% of total organic carbon flux (fecal pellets plus phytodetritus) is consumed by the benthos. The remaining 75% may be transported southwards, where it is either eventually utilized by food webs in the summer or exported to the slope. The protozooplankton and gelatinous zooplankton were not sampled during this experiment, and the above scenario is an oversimplification of very dynamic processes, but it is clear that only a small percentage (10 to 20%) of the original algal carbon ultimately settles onto the adjacent slope.

Significant changes occur in the structure and energetics of pelagic and benthic food webs along the longitudinal and latitudinal axes of the Mid-Atlantic Bight. On average, rates of primary production are highest within the coastal zone, particularly in proximity to the Hudson River, Delaware Bay, and Chesapeake Bay plumes,[572] with a sustained period of high productivity from May to September.

The day-to-day scenario, however, is exceedingly dynamic. Time-series analyses of remote sensing images show large daily shifts in chlorophyll distributions on the shelf, reflecting the synergistic/antagonistic effects of waves, winds, and tides on large-scale, water mass movements.[506] Satellite images taken two days apart (Figure 7.7) with a coastal zone color scanner (CZCS) demonstrate just how rapid are the changes and movements of phytoplankton biomass (and presumably the remaining members of the pelagic food web) in the Mid-Atlantic Bight. Resuspension of near-bottom phytoplankton, offshore and onshore transport, and sinking/downwelling events play a significant role in controlling the spatial and temporal heterogeneity of the plankton. Such variability must be kept in mind when considering any interpretation of static energy and nutrient flow models and budgets.

Factors regulating the benthic communities are not clear. Benthic biomass frequently correlates with changes in sediment composition and grain size, temperature, and food supply. The distribution of benthic biomass and changes in community composition appear to be more a function of benthic responses to the ridge and swale topography persistent in some areas of the shelf (e.g., in the New York Bight) rather than to any simple relationship with bathymetry.[582] In effect, the benthos is ultimately responding to the net effects of currents on the distribution of fine and coarse sediments and to organic matter. How this translates into changes in benthic production, detritus mineralization, and nutrient regeneration is still not clear.

The SEEP-I study by Rowe and his colleagues[583] on the continental shelf south of New England (north of the New York Bight) provides a glimpse into the potential differences in benthic carbon flow between coarse and fine sediments. Comparing organic carbon stocks, fluxes, and turnover times (Table 7.7), they found that the finer, organic-rich deposits are characterized by greater detritus and faunal standing stocks compared to the sandier, organic-poor sediments, but respiration rates were equivalent (0.13 g C m^{-2} d^{-1}). This implies that the turnover of detrital and biotic carbon is faster in the sandy sediments than in the muddier sediments. They attributed these findings to the settling of poorer quality detritus in the finer sediment areas. Overall, these carbon flux estimates indicate that ~25% of the primary production on the shelf is consumed by the benthos. This value is a likely underestimate, as neither sulfate reduction nor iron and manganese reduction were measured directly.

An estimate of the flux of nitrogen on both the New York Shelf and Georges Bank (Table 7.8) shows that recycling by pelagic microbes is the major contributor to nitrogen demand, followed by sediment release and zooplankton excretion. Phytoplankton exudates are a more

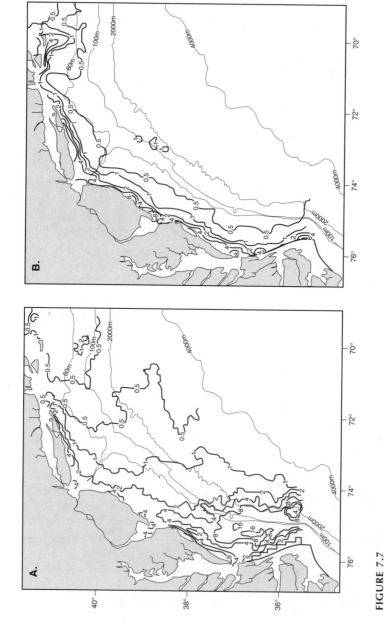

FIGURE 7.7

Short-term changes in chlorophyll concentrations (mg l^{-1}) in the Mid-Atlantic Bight as estimated by the coastal zone color scanner for 19 April 1979 (A) and 21 April 1979 (B). (From Walsh, J. J., *On the Nature of Continental Shelves*, Academic Press, New York, 1988, 520. With permission.)

TABLE 7.7

Inventory of Detrital and Living Organic Carbon
in Shelf Sediments South of New England in
SEEP-I Area

	Organic-Poor Sands	Organic-Rich Fine Sediments
Stocks		
Detritus	78	900
Bacteria	1.03	3
Meiofauna	0.048	0.23
Macrofauna	6.6	7
Turnover times (days)		
Detritus carbon	590	6818
Biotic carbon	58	78

Note: Units are g C m^{-2} for stocks and g C m^{-2} d^{-1} for
fluxes.

Source: Adapted from Rowe, G. T. et al., *Cont. Shelf
Res.*, 8, 511, 1988.

significant pathway on the Georges Bank. Advective and diffusive
inputs of "new" nitrogen in the form of nitrate from the slope are still
required to sustain the nitrogen demand of primary producers, which
on average is ~40% of total requirements. Losses of nitrogen via export
of phytoplankton to the adjacent slope and from removal of harvested
species may be balanced by additional advective inputs from storms or
may be diffused onshore by turbulent motion. In any case, nitrogen
losses via fishing are minuscule compared to other pathways, implying little, if any, direct connection between nitrogen cycling and fishery yields.

As the shelf narrows southwards, there is intense recycling but
little export of phytoplankton-derived carbon to the adjacent slope —
much less than 5% of total carbon production.[584] Even on this narrow
sector, the exchange of material between inner and outer portions of
the shelf is limited. Wirick[585] estimated that phytoplankton production
occurring landward of the 90-m isobath is not exported offshore.
Moreover, although there is significant large-scale, across-shelf transport of plankton biomass produced at the shelf-slope boundary, calculations indicate a small net import. This does not exclude some export
to the slope from zones of high productivity along the 100-m isobath,
but it does place a mimimal upper limit on potential export. Zooplankton blooms are coincident with the algal bloom in spring, but it is
difficult to discern a close trophic coupling between these two groups
owing to large spatial and temporal variations.

TABLE 7.8

Contribution of Recycled Nitrogen to the Annual Nitrogen
Demand of Primary Production on Georges Bank (80 g N m^{-2}
yr^{-1}) and the New York Shelf (60 g N m^{-2} yr^{-1})

Nitrogen Demand	Georges Bank (g N m^{-2} yr^{-1})	New York Shelf (g N m^{-2} yr^{-1})
Recycled sources		
Nitrification	1.5	3.0
Ammonification	0.6	1.2
Sediment organic nitrogen release	4.8	7.8
Zooplankton excretion	5.3	6.8
Microbial plankton remineralization	28.7	14.0
Phytoplankton release	7.5	1.5
Total recycling	48.4	33.9
Required slope input	31.6	26.1
Advective input from storms	34.0	34.0
Diffusive input	42.8	30.6
System losses		
Phytoplankton export	48.0	31.0
Fishing	0.1	0.1

Source: From Walsh, J. J., *On the Nature of Continental Shelves,*
Academic Press, New York, 1988, 520. With permission.

It appears that any trophic coupling between phytoplankton and
zooplankton communities is constrained by the dominance of physical
forces.[586] Zooplankton ingest phytoplankton at rates equivalent to 30
to 35% of primary productivity, with the remainder lost to smaller
pelagic consumers and the shelf floor. Across-shelf, coastal waters
contain 50% more living carbon than do offshore waters, with the
relative proportion of large phytoplankton and grazing plankton decreasing and being replaced by autotrophic nanoplankton and microzooplankton, from inshore to the outer shelf.[587]

Latitudinally, the plankton community becomes increasingly more
heterotrophic towards Cape Hatteras, implying that the carbon advected from the northern sector of the Bight is transferred up through
heterotrophic pathways as it moves southwards. Verity et al.[587] suggest
that what little carbon does get transported to the adjacent shelf does
so in the form of zooplankton fecal pellets. Estimates of total pelagic
and sediment bacterial production[588] indicate that roughly 12 to 24%
and 33 to 66% of daily primary production on the shelf and slope,
respectively, are funneled into microbial carbon consumption.

Rates of particulate flux through the water column are insufficient
to account for the rates of bacterial production in slope sediments and

in the water column, implying that a small carbon subsidy from the shelf, equivalent to 8 to 15% of shelf primary production, would be required. This agrees with other estimates of the amount of shelf carbon potentially available for export. Exactly how much carbon is exported, as well as the location and precise mechanisms of transport, remain to be determined. It is clear, however, that the export subsidy to the adjacent slope is likely to be only a small fraction of total primary production.

7.3.2.3 The Carolinian Province

From south of Cape Hatteras to northern Florida, the continental shelf becomes shallower and widens to form the South Atlantic Bight. More strongly influenced by the Gulf Stream than the northern shelf provinces, this province experiences warmer temperatures. Bordered by numerous coastal barrier islands, behind which extensive salt marshes form, exporting terrestrial material to the coastal zone, the ecosystem of the Bight begins to take on a subtropical character.

Intrusions of nutrient-rich Gulf Stream water, river runoff, and marsh outwelling are the major sources of nutrients to the Bight.[589] Most of the shelf floor is composed of coarse sand with relict coral outcroppings; the exception is the nearshore fringe, which receives terrestrial input from rivers and salt marshes.[589] Outwelled material is retained within 20 km of the coast due to salinity fronts acting as a barrier to the mixing of coastal and shelf waters. Most of the shelf is oligotrophic. Storms may break down these inshore and offshore barriers, but it has been difficult to measure the ecological effects of such episodic events.

Food webs and rates of nutrient cycling are enhanced by outwelling of estuarine materials and intrusions of nutrient-laden waters from the Gulf Stream. In a large-scale survey of benthic biomass and microbial activity, Hanson et al.[589] found highest rates of microbenthic biomass, metabolic activity, and oxygen consumption on the inner shelf, with pockets of intense rates of activity at the shelf break. The irregularity of the latter was attributed to the patchiness of the Gulf Stream waters intruding onto the shelf; eddies and meanders may result in inequitable distribution of nutrients and deposition of particulate organic matter.

The distribution patterns of benthic biomass similarly indicate enrichment from estuarine outwelling and intrusions on the outer shelf, but there are pockets of high biomass at mid-shelf that may reflect sporadic inputs of organic matter from the overlying water-column and from the Gulf Stream intrusions.[590] There are no consistent gradients in benthic structure and biomass with latitude or season.

Tenore[590] showed that species composition changes seasonally, perhaps in response to variations in food inputs and physical forcing.

Subsequent studies reveal that the spatial and temporal patterns of pelagic and benthic respiration, consumption, and nutrient recycling are more complicated than previously believed. Hopkinson's study[591] of pelagic and benthic metabolism in the estuarine plume region of the Bight revealed that the nearshore waters are net heterotrophic, with resuspension regulating the decomposition of organic matter in the sediments and overlying water column. He found that pelagic respiration accounts for 60% of total metabolism, with rapid turnover times of organic carbon in both the sediments and the water column. Seasonality was observed, with metabolic activities relating to water temperature, primary productivity, and the timing of tidal export of detritus. The detritus subsidy has a significant impact on nearshore heterotrophy, although limited to the adjacent coastal zone.[178] Of the total detrital carbon pool available in the nearshore zone, ~31% is derived from tidal outwelling from adjacent estuaries, with the remainder supplied by *in situ* primary production.

That terrestrially derived organic matter has a diminishing role to play in food chain energetics and nutrient processes as distance from the coast increases is supported by evidence from pelagic studies. Moran and her colleagues[592,593] measured the distribution of dissolved lignin-derived compounds across the Bight and found conservative mixing and dilution on the shelf, indicating a large decline in terrestrial material with distance from shore. Spatial and temporal patterns were complex, implying a very variable contribution to the shelf's carbon budget They estimated that, on average, 6 to 36% of DOC in the coastal zone, 5 to 25% of DOC on the inner shelf, and from 3 to 18% of DOC on the mid- and outer shelf, is derived from terrestrial material.

Subsequent analysis of dissolved humic compounds suggests that these substances are not as inert as previously believed. Moran and Hodson[593] observed that coastal bacterioplankton mineralize these substances slowly and inefficiently, with nearly 24% of this material decomposed within several weeks. On the South Atlantic Bight, 11 to 75% of these humic materials are derived from vascular plants, roughly half contributed by coastal marshes and half from rivers. This implies that the influence of vascular plants on the DOC pool of shelf waters is manifested largely in the form of humic compounds.

The conceptual model proposed by Verity et al.[594] links various chemical, physical, and biological observations of coastal and offshore pelagic food web processes in the South Atlantic Bight. Essentially, the estuarine and nearshore waters of the Bight function as a single ecosystem, with plankton production fueled by new nutrients exported from tidal rivers, creeks, and groundwater. These nutrients are utilized and continually recycled within the coastal zone as exchange of biomass and nutrients between nearshore and offshore waters is constrained. The frontal barrier leads to nearshore waters having a long residence time, requiring complete use of available nutrients to sustain the persistent, high rates of primary and secondary productivity inshore. Recycling of nutrients is intense, as nitrogen is recycled through the food web many times during summer; <17% of the nitrogen required to support primary production comes from new inputs.

Most of the carbon fixed by pelagic autotrophs is consumed within the coastal zone, leading to a dynamic equilibrium between producers and consumers in the water column and in the benthos. Beyond the coastal front, seasonal effects of temperature and limiting substrate supply result in a shift in the balance between pelagic production and consumption — net autotrophy in winter and net heterotrophy in summer. This is consistent with earlier observations that pulses of intrusion-enhanced primary production on the outer shelf not transported off the shelf are consumed and respired by pelagic microheterotrophs, implying little export of organic carbon to the open ocean.

The extent to which these processes link to the dynamics of coastal and offshore fisheries is not clear. Low finfish landings in the South Atlantic Bight (compared to the provinces further north) have been attributed to the oligotrophic nature of this region.[594] This is an oversimplification because:

- Lower yields may simply reflect lower catch efforts.
- Finfish abundance may give way to increased landings of crustaceans, particularly penaeid shrimps.
- The Bight is an important nursery ground for species such as the Atlantic menhaden (*Brevoortia tyrannus*), which spawns offshore on the edge of the Gulf Stream and migrates into neighboring southeastern estuaries.

The Bight, therefore, serves as an important nursery area for many species that may not necessarily be caught within the province, suggesting that caution must be applied in the use of fishery statistics in determining the fishery yield of a given shelf region.

7.3.3   The Bering and Chukchi Seas

The broad (~1000 km) interlocking shelves of the Northern Bering and Chukchi Seas (Figure 7.8) are among the world's greatest trawling grounds for pelagic and demersal fish and are a rich nursery area for

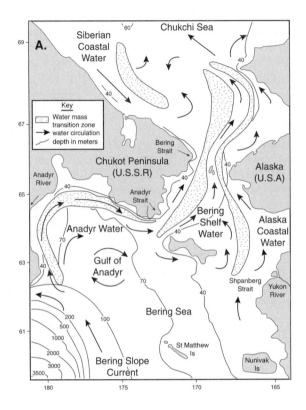

FIGURE 7.8
Charts of the Bering and Chukchi Seas studied during the ISHTAR project. Charts
denote (A) local water circulation, water masses, and bathymetry; (B) phytoplankton
biomass (mg Chl *a* m^{-2}); and (C) macrobenthic biomass. (Adapted from Coachman, L.
K. and Hansell, D. A., *Cont. Shelf Res.*, 13(5/6), 1993 [special issue].)

whales, walrus, seals, and seabirds. Unlike the eastern North American Shelf, the links among fishery yields, primary productivity, and
pelagic and benthic food webs in this polar region are somewhat more
clear, as a result of the recent PROBES (Processes and Resources of the
Bering Sea Shelf) and ISHTAR (Inner Shelf Transfer and Recycling)
programs.[595] These programs were initiated to determine the mechanisms responsible for these rich polar ocean communities.

The original hypothesis for the ISHTAR program was that nutrients draining from the nearby Yukon river are the chief mechanism
sustaining the high rates of primary and secondary production. The
results indicate a more complex scenario in which spatial partitioning
of trophic pathways sustain very different ecosystems between the
eastern and western regions of the shelf. Three water masses (Figure
7.8) entrain upon this broad shelf from the south: the Alaska coastal

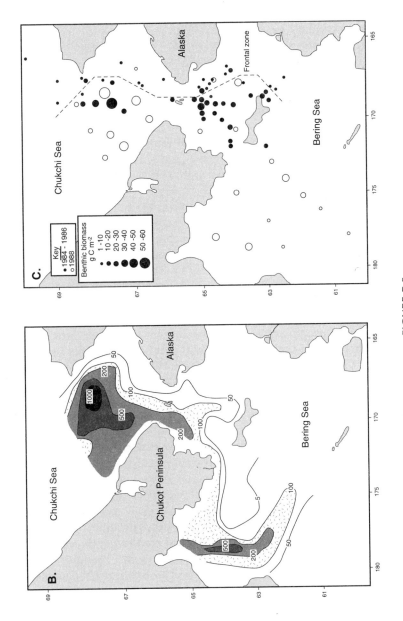

FIGURE 7.8
(continued)

water (ACW), Bering Shelf water (BSW), and Anadyr water (ADW). The eastern portion of the shelf is characterized by the ACW carrying outflow from the Yukon River. The ecosystem of the western shelf is fueled by the ADW, enriched by its nutrient-laden oceanic water. Primary and secondary production are high in the western area, with rates of carbon fixation exceeding 10 g C m^{-2} d^{-1}; an oceanic fauna is upwelled/upsloped onto the shelf by the ADW, resulting in an east-west dichotomy of food webs.[595]

The passage and convergence of the ADW and ACW masses towards the Bering Strait induce a succession of productivity peaks in the Chukchi Sea. Superimposed on the functioning of these separate ecosystems is a cyclic pattern of ice conditions related to warm and cold years, with a 4.7-year cycle between peaks. There is some indication that these interannual weather patterns result in interannual changes in ecosystem functioning.

The spatial distribution of macrobenthic biomass (Figure 7.8C) mirrors that of phytoplankton biomass (Figure 7.8B) and production. The production of phytoplankton (mostly chain-forming diatoms) on the western shelf from the Gulf of Anadyr to the southern Chukchi Sea is high,[49] with daily carbon uptake rates of up to 16 g C m^{-2} d^{-1} and an annual production of 470 g C m^{-2} yr^{-1}. Rates within small plumes northward in summer can be as high as 720 to 840 g C m^{-2} yr^{-1}. To the southeastern Bering Shelf, primary production within the ACW mass is much lower, averaging 80 g C m^{-2} yr^{-1}. Biomass is higher, however, further offshore along the outer shelf edge (Figure 7.8B). This region, coined the "Green Belt" by Springer et al.,[597] is characterized by a predominantly pelagic food web that sustains the pelagic fishery of the Alaskan pollock, now the world's largest single species fishery. Benthic biomass and microbial activity are low in this area and beneath the Alaska Coastal Water Mass compared to the western Bering Shelf, where much of the high biomass of phytoplankton is deposited to the shelf bed to sustain high macrobenthic biomass (Figure 7.8C) and rapid rates of microbial decomposition.

There are differences in the rates of sediment nitrogen flux between the less productive ACW and the eutrophic western Bering Shelf and southeast Chukchi Sea (Figure 7.9). Henriksen et al.[598] measured consistently rapid recycling of nitrogen between benthos and the water column, especially in the southeast Chukchi Sea; rates were significally sloer in Alaskan coastal waters. These differences indicate a direct coupling between pelagic primary production and benthic food chains.[598,599] Across both seas, available nitrate stocks are derived equally from *in situ* recycling and by influx from deep ocean water.

In the eutrophic, shallow areas, enriched benthic communities are a source of food for many benthic predators, such as walrus and,

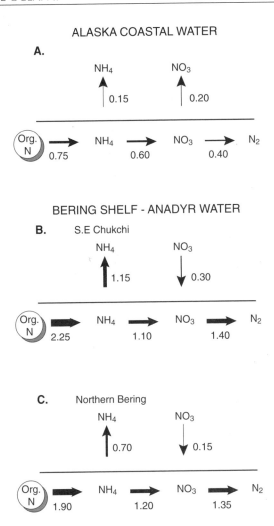

FIGURE 7.9
Nitrogen budgets (mmol N m^{-2} d^{-1}) for Alaska coastal sediment (A), and Bering Shelf-Anadyr sediments in the southeastern Chukchi Sea (B) and the northern Bering Sea (C). (From Henriksen, K. et al., *Cont. Shelf Res.*, 13, 629, 1993. With permission.)

before their fisheries declined, yellowfin sole and king crab. Carbon-flow models of the food webs supporting the outer shelf pelagic (Figure 7.10A) and middle-inner shelf benthic (Figure 7.10B) fisheries show that rates of primary production between these regions are equivalent, but a larger proportion of carbon is shunted through zooplankton and up the food chain to pelagic fishes and seabirds on the outer shelf off the southeastern and northwestern regions of the Bering Sea. Conversely, more carbon flows through benthic food webs to apex demersal

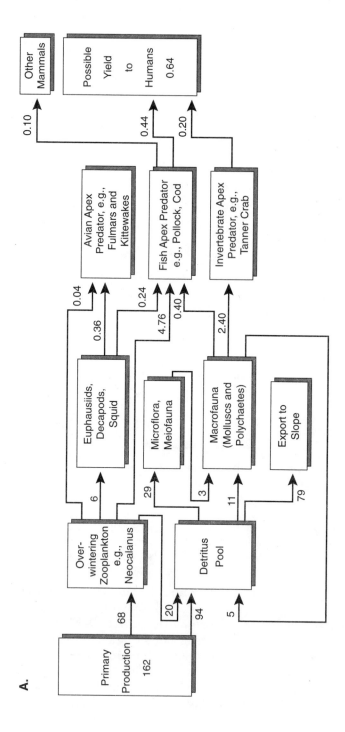

FIGURE 7.10

Annual carbon budgets (g C m^{-2} yr^{-1}) of the food webs on the outer shelf (A) and on the middle shelf (B) of the southeastern Bering Sea. (From Walsh, J. J., *On the Nature of Continental Shelves*, Academic Press, New York, 1988, 520. With permission.)

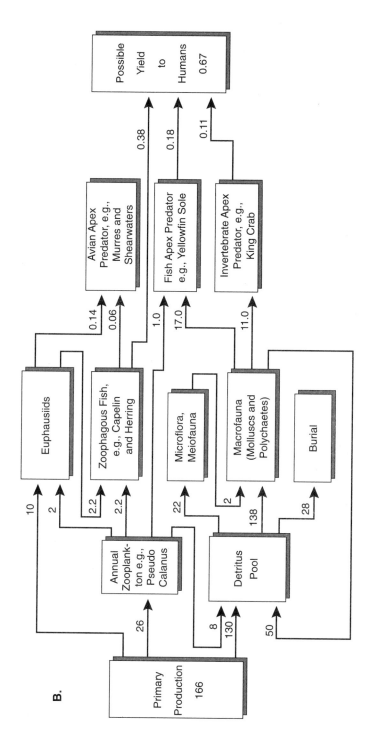

FIGURE 7.10
(continued)

TABLE 7.9

Estimates of Annual Carbon Budgets for the Bering Sea
off Alaska and the Chukchi Sea off Siberia

	Bering Sea $(g\ C\ m^{-2}\ yr^{-1})$	Chukchi Sea $(g\ C\ m^{-2}\ yr^{-1})$
Gross photosynthesis	166.1	246.8
Respiration	136.4	142.8
Net carbon production (gross photosynthesis – respiration)	29.7	104.0
Air-sea (net lateral) exchange	29.7	70.6
Upwelling	0.0	–33.4

Source: Adapted from Walsh, J. J. and Dieterle, D. A., *Progr.
Oceanogr.*, 34, 335, 1994.

predators on the inner (<50 m) and middle (50 to 100 m) shelf regions
where ungrazed phytodetritus accumulates; however, the terminal
yields to humans are similar.

Annual carbon budgets for the entire Bering and Chukchi Seas
(Table 7.9) indicate that these shelves are a sink for atmospheric CO_2
with net storage of carbon (as dissolved inorganic carbon, DIC; DOC;
and particulate organic carbon, POC, not utilized). Differences between shelf seas primarily reflect both overall higher rates of primary
production and air-sea exchange in the Chukchi Sea.[600] As little of the
excess carbon appears to be buried in shelf sediments, it is presumably
exported to the continental slope, where it may sustain shelf-edge and
slope food webs throughout the winter. Wind-driven currents are
favorable for off-shelf transport of excess production throughout summer.

The importance of terrestrial matter to carbon flow on this shelf is
limited; only within the Alaskan Coastal Water mass does terrestrially
derived carbon from the Yukon River contribute significantly (~30%)
to total input. This is supported by extensive measurements[601] of $\delta\ ^{13}C$
and atomic C:N ratios in surface sediments. These measurements show
a progression of increasing $\delta\ ^{13}C$ and decreasing C:N values from east
to west, implying that inputs from land are limited to Alaskan coastal
waters and that most of the sedimentary carbon on this shelf is of
marine origin.[601]

This ecosystem is highly productive because of the intense physical processes that occur at the shelf edge. Tidal mixing and transverse
circulation and eddies in the Bering Slope Current transport nutrients
to the ocean surface, where they are rapidly and efficiently utilized by
the pelagic food web. Enhanced standing stocks of both pelagic and
benthic secondary producers result in favorable feeding conditions

and increased concentrations of fishes and squids, which, in turn, attract large populations of marine mammals and birds. Springer et al.[597] suggest that three factors account for the high productivity of this ecosystem:

- Sustained primary production
- Intense food web exchanges
- High trophic transfer efficiencies at the shelf edge

7.3.4   The Great Barrier Reef Shelf

Continental shelves dominated by large tropical rivers exhibit characteristics different from those of temperate and boreal latitudes. The occurrence of large tracts of carbonate and mixed carbonate-terrigenous sedimentary facies[12] on some continental margins receiving considerably less, or less continuous, rainfall is another characteristic of the tropics. Examples of such shelves include areas of the eastern Gulf of Mexico, the Seychelles, the Brazilian shelf, the Red Sea, the inner shelf off Northwest Africa, and the Puerto Rico shelf.

The exemplar is the continental margin off the Queensland coast of northeastern Australia (Figure 7.11). The most obvious feature of this shelf is the Great Barrier Reef (hereafter GBR), a matrix of 2500 individual reefs extending along the outer shelf rim for roughly 2600 km from 25 to 9.2°S latitude. Most of the shelf between the GBR and the coast is shallow (≤60 m) and forms an extensive lagoon connected to the Coral Sea by numerous channels between reefs and by the Capricorn Channel at the southern terminus.

The GBR shelf is commonly divided into three sections: northern, central, and southern. The across-shelf distance between the reef matrix and the coast increases from north to south (Figure 7.11). Southeasterly trades are the dominant winds from April to October; in summer, winds are more variable and tropical cyclones are common. Water circulation is complex, particularly in proximity to reefs, but is controlled mainly by diurnal and semidiurnal tides, the southeasterly tradewinds, and wind-driven waves and swells. The dominant low-frequency water motion is northward longshore, generated by the East Australian Current flowing south in the Coral Sea and currents partly driven by the southeasterly tradewinds.[602] High-frequency water motion occurs throughout the shelf, especially in relation to internal seiches, as well as trapping of waves in embayments and lagoons; forcing by tropical cyclones can be intense in summer. Subsurface intrusions of Coral Sea water onto the outer shelf occur most frequently in summer.

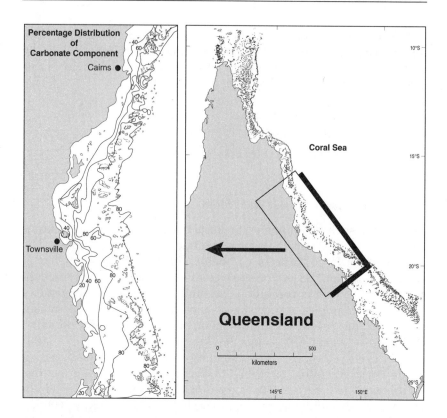

FIGURE 7.11
Chart of the Great Barrier Reef shelf (right panel) showing a portion of the across-shelf carbonate gradient (left panel) in sediments from the coastal zone to the outer shelf. (Adapted from Belperio, A. P. and Searle, D. E., in *Carbonate-Clastic Transitions, Developments in Sedimentology 42*, Doyle, L. J. and Roberts, H. H., Eds., Elsevier, Amsterdam, 1988, chap. 5.)

The predominance of longshore water movement is partly responsible for an across-shelf gradient from well-mixed coastal waters overlying terrigenous silts, clays, quartz, and silica sands to an abrupt shift to clear, nutrient-poor waters overlying sedimentary deposits increasing in carbonate content seaward.[603] Sea-level fluctuations facilitate limited intermixing of terrigenous and carbonate sediments.

During the summer wet season, coastal boundary fronts may form between low-salinity, turbid plumes and shelf water masses. Mixing between coastal and offshore waters may be inhibited at other times of the year during some wind conditions.[602] The largest source of modern terrigenous sediment and freshwater to the shelf is direct pulses of fluvial input mainly in discrete flood events during the wet season. Total freshwater and sediment discharge into the GBR shelf averages

~100 km^3 and ~28 × 10^6 metric tons, respectively.[602,603] With the exception of extreme climatic events, most of this material mixes within the coastal zone, transported longshore in a northerly direction. The major external sources of nutrients to the shelf include:

- Upwelled Coral Sea water
- Precipitation
- Nitrogen fixation
- River runoff

The shift from a terrigenous nearshore region to a reef and ocean-dominated middle and outer shelf results in a clear progression of change in the structure and function of pelagic and benthic communities. Phytoplankton production[604] across the central GBR shelf ranges from 0.3 to 1.8 g C m^{-2} d^{-1}; lowest rates (<0.25 g C m^{-2} d^{-1}) are most frequently measured within the coastal zone. When algal blooms occur in response to either wet season river runoff or summer stratification (*Trichodesmium*) events, rates of carbon fixation can exceed 4.0 g C m^{-2} d^{-1}.

Most of the phytoplankton biomass and productivity occur within the picoplankton- and nanoplankton-size classes, except during algal blooms when microplankton-sized diatoms prevail. Seasonality has not been observed on the mid- or outer shelf; responses to summertime upwelling on the outer shelf are indirect, with increases in phytoplankton biomass occurring in discrete layers or patches. Phytoplankton productivity increases with latitude, with average annual rates ranging from 354 to 424 g C m^{-2}.

Studies of zooplankton community structure have identified distinct inner- and outer-lagoon assemblages with transitional mid-lagoon communities.[605] Recent evidence suggests that zooplankton production is also greatly influenced by the boundary between trapped coastal water and offshore water. For instance, peak abundances of some fish larvae occur at this front, probably in response to enhanced abundance of prey.[605] McKinnon and Thorrold[606] investigated zooplankton community structure and production across this transition zone. They observed peak zooplankton abundance during the summer flood period, particularly in the coastal zone, but little difference in species composition across the shelf. Zooplankton abundance and copepod egg production increased in response not only to wet season flooding, but also to intensity of upwelling events.[606] These results imply that some members of the coastal and offshore zooplankton communities are opportunistic, poised to respond quickly to climatological and hydrographical events.

TABLE 7.10

Estimates of Biomass and Production of Benthic
Trophic Groups in Missionary Bay within the
Central Great Barrier Reef Shelf

	Biomass[a] (g C m^{-2})	Production[a] (g C m^{-2} yr^{-1})
Bacteria	8.2	350
Microalgae	0.1	11
Protozoa	0.05	1
Meiofauna	0.5	7.5
Macroinfauna	0.9	2.7
Epifauna, including fish	0.2	0.4

[a] To a depth of 20 cm.

Source: Data from Daniel and Robertson,[607] Alongi,[451,608-612]
Alongi and Christoffersen,[613] and Alongi et al.[614,615]

Benthic communities may be similarly adapted to respond rapidly
to land-ocean processes in the coastal zone. In some inshore areas of
the shelf, significant outwelling of detritus from adjacent mangrove
forests occurs. Benthic studies[451,607-615] show that outwelling is limited
to the coastal zone, except during summer cyclones and other major
storms, when pulses of terrigenous material are transported offshore
and some offshore sediments may be transported landwards.

Benthic biomass in the coastal zone is dominated by highly productive bacterial assemblages; abundance and diversity of other benthos
are often very low. The macroinfaunal communities are dominated by
small-sized, surface-dwelling polychaetes and crustaceans and are
devoid of large equilibrium, bioturbating species. These results were
ascribed to the poor nutritional quality of outwelled riverine and
mangrove litter and to low rates of phytodetrital input and *in situ*
microalgal production. Moreover, a variety of physical disturbances
such as smothering by detritus, resuspension of sediments by monsoons, and large tides during intervals of low sea level are important
in regulating the benthic communities. Under physically disturbed
conditions, bacteria can respond and recover more rapidly than larger
benthic organisms. The size and trophic structure of these inshore
communities are reminiscent of those inhabiting other shallow tropical
shelf habitats that are frequently disturbed.[12]

The distribution of benthic biomass and secondary production of
the different size groups (Table 7.10) shows that bacteria are the dominant members of the benthos. These estimates are crude but are supported by empirical estimates of the rates and pathways of carbon flow

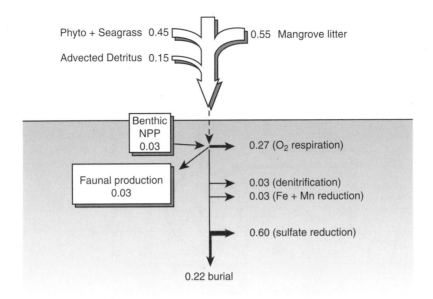

FIGURE 7.12

Preliminary carbon budget (g C m^{-2} d^{-1}) for sediments of the coastal zone within the central Great Barrier Reef lagoon. Compiled from data in Torgersen and Chivas[616] and references provided in footnote of Table 7.10.

at one station within the coastal zone of the central GBR shelf (Figure 7.12). This preliminary budget shows that of an average supply rate of 1.15 g C m^{-2} d^{-1} (mangrove-derived carbon), approximately 81% is mineralized by aerobic and anaerobic bacterial communities within the upper 20 cm of sediment, with the remaining 19% buried in deeper sediment layers. Nearly 45% of the carbon taken up by bacteria is respired. The burial rate is high but presumably declines along with sedimentation rates across the shelf and is lower in coastal areas with less extensive mangrove forests or smaller riverine input, or both.

The fate of this large stock of bacterial biomass is unknown, but given the low abundance and productivity estimates of potential benthic predators, most is probably recycled by the next generation of bacteria or consumed by pelagic microheterotrophs during resuspension events. This model is crude as it does not encompass other coastal sites receiving lesser amounts of mangrove detritus nor does it reflect the shift from anaerobic to aerobic decomposition pathways when disturbance events erode, resuspend, and oxidize the subsurface sediments.[609]

Additional measurements (see references in Table 7.10) point to the dominance of sediment bacteria on the central GBR shelf. A preliminary nitrogen budget (Figure 7.13) indicates that low standing

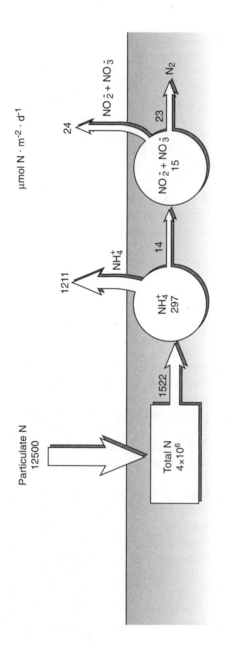

FIGURE 7.13
Preliminary nitrogen budget (mmol N m^{-2} d^{-1}) for mid-shelf sediments of the central Great Barrier Reef lagoon. (Compiled from data in Alongi.[608,617])

amounts of nitrogen are rapidly mineralized via ammonification, nitrification, and denitrification. Sources of nitrogen in the sediment are not known, but sources other than phytodetritus are required to balance the rates of microbial nitrogen mineralization.

A crude estimate of nutrient inputs to the central GBR shelf points to nitrogen fixation by *Trichodesmium* as the single largest source of nitrogen, followed in decreasing order by rivers, upwelling, rainfall, and sewage; the prime phosphorus sources are upwelling and river discharge and, to a much lesser extent, precipitation and sewage.[610] Total external inputs of nitrogen and phosphorus are small, however, compared with microbially recycled nutrients. Furnas et al.[618] suggest that vertical exchange processes (including resuspension events) between sediments and the overlying water are crucial in regulating the availability of nitrogen and phosphorus for pelagic and benthic food webs. Far too little is known of benthic-pelagic coupling, denitrification losses, and interconnections between the lagoon and the Great Barrier Reef to make any sophisticated food web models or nutrient budgets for this tropical shelf.

7.3.5   Other Tropical Continental Margins: The West African and Sunda Shelves

It is enticing to speculate that the generally lower fishery yields on tropical compared to temperate shelves[619] is partly the result of greater sequestering and trophic loss of carbon to pelagic and benthic, microbial food webs. Unfortunately, except for upwelling areas, estimates of ecosystem-level processes are few for tropical continental margins. Here, two contrasting tropical shelves — greatly different from the GBR shelf — are discussed: the West African Shelf off Sierra Leone and the Sunda Shelf in Southeast Asia.

The continental shelf off Sierra Leone lies between the rich upwelling ecosystems off northwest Africa to the north and Guinea to the south. This shelf had an intensive commercial fishery until fish stocks became depleted in the late 1970s and early 1980s.[620] A permanent thermocline exists at mid-shelf, but on the inner shelf, the waters are well mixed by high river discharges during the monsoons which occur from May to September. Longhurst[620] estimated the flow of carbon on this shelf using data obtained from within the Sierra Leone estuary and adjacent coastal zone, as well as coarse sand and shell-gravel deposits offshore. Developed before the importance of pelagic microbial food webs was acknowledged and highly simplified, the budgets nevertheless demonstrate a significant difference in carbon flow and trophic links between muddy inshore and sandy offshore communities. The major difference is more abundant and productive benthic communities and phytophagous fish assemblages (mostly the clupeid *Ethmalosa fimbriata*) inshore. Longhurst[620] estimated that most fixed carbon is not consumed by benthic and pelagic food webs but rather is utilized by microbes, buried, or exported to the slope or inshore mudbanks. It is likely that much of the net carbon is respired by bacteria. These discrepancies can only be resolved by further study, but Longhurst's analysis[620] demonstrates clear energetic and trophic linkages between primary producers and their consumers, including fish, on this shelf.

In contrast, the waters of most of Southeast Asia overlie an impressively continuous, broad, and shallow (<100 m) continental margin called the Sunda Shelf. This shelf is bisected by many submerged river channels, reflecting the high rates of precipitation in this region. Even in areas distal to the large rivers, such as the Mekong, numerous smaller rivers feed large quantities of terrigenous material and freshwater to the shelf, leading to low surface salinities and relatively little exchange with the deep ocean. Lack of exchange greatly affects the structure and energy flow of pelagic and benthic food webs on this shelf.

ECOPATH models constructed by Pauly and Christensen[621] for areas bordering the northern rim of the Sunda Shelf shed some light on the major flows of carbon to demersal and pelagic fishes. Comparison of the models for the Gulf of Thailand, the Vietnam coast, southern China, and off of peninsular Malaysia and southeastern Sumatra suggests major differences in trophic structure and flow of carbon to higher trophic levels, including humans. In the Gulf of Thailand, fish yields are intermediate to those of the other two shelf areas: roughly 50% of fish production is lost to humans, with 37% consumed by apex fish predators and the remainder lost to invertebrate predators. Phytoplankton production is equivalent among the three areas, but biomass and productivity of the other trophic compartments differ considerably. In the Gulf of Thailand, the detrital and benthic pools are large, reflecting the fact that demersals are the largest catch group. In contrast, zooplankton and detritus are the largest trophic groups off the coasts of Vietnam and South China: zooplankton-feeding fish and other pelagics are proportionally the largest catches in this area of the shelf. The high catch yield off peninsular Malaysia and Sumatra consists mainly of small pelagics and diverse demersal species. The higher rates of fish production in this area may be ascribed to the generally high rates of secondary production of pelagic crustaceans and benthic communities.

The models imply that fish production is closely related to the mean rates of trophic transfer efficiency: the mean value for the ecosystem off Malaysia/Sumatra is higher (11.5%) than for the Gulf of

Thailand (8.9%) and the least productive Vietnamese/Chinese ecosystem (6.4%). These results must be considered cautiously, as they encompass a large number of unverified assumptions, but they are provocative in implying that fish production is a direct consequence of the structure, trophic functioning, and efficiency of energy transferred within tropical pelagic and benthic food webs.

High pelagic production and fish catches in other areas of the Sunda Shelf are linked to alternating downwelling and upwelling events in response to monsoons, as observed off eastern Indonesia at the interface between the Banda and Arafura Seas.[622] During upwelling events caused by the southwest monsoon, primary production can exceed 400 to 500 g C m^{-2} yr^{-1} and may be linked to large standing stocks of small pelagic fishes. Upwelling and other events occurring at the shelf edge are of considerable importance in understanding why some shelf ecosystems are more productive than others.

7.4 PROCESSES AT THE SHELF EDGE

Exchange of materials between the coastal zone and the shelf proper is limited on most continental margins, separated by various fronts and boundaries into well-mixed, turbid coastal waters and clear, stratified mid- and outer-shelf waters which break down episodically. Material exchange between continental shelves and the open ocean is also limited, when considered in proportion to the total primary production on the shelf. This is not the case, however, for narrow, active margins bordering eastern boundary currents, where upwelling drives shelf food webs, or for several continental margins in the wet tropics receiving massive riverine discharge

7.4.1   Wind-Driven, Coastal Upwelling

Coastal upwelling is a complex process, driven mainly by winds causing surface waters to diverge from the coast and to be replaced by the upwelling of cold, nutrient-rich waters from the adjacent deep ocean.[623,624] Other forms of upwelling at the shelf edge were discussed earlier, including buoyancy-driven upwelling (e.g., off the Amazon shelf), tidally driven upwelling (e.g., Celtic Sea, Georges Bank), and intrusions driven by Western boundary current instability (e.g., South Atlantic Bight). Our discussion here is restricted to wind-driven upwelling.

Two books have recently provided integrated views of the world's open ocean and coastal upwelling systems, so this section

TABLE 7.11

Estimates of New Phytoplankton Production Derived
from Various Regions Including Upwelling Systems

Region	Area (10^{12} m^2)		New Production (10^{15} g C yr^{-1})	
Tropical open ocean	90	(25%)	—	
Upwelling	—		1.5	(21%)
Turbulent mixing	—		0.7	(9.7%)
Southern ocean	77	(22%)	1.1	(15.3%)
Subarctic gyres	22	(6%)	0.3	(2%)
Coastal upwelling	4	(1%)	0.8	(11.1%)
Monsoonal	5	(1%)	0.4	(5.5%)
Subtropical gyre	114	(32%)	0.5	(6.9%)
Continental margins	45	(13%)	—	
Western boundaries	—		0.7	(9.7 %)
Estuarine exchange	—		1.2	(16.6 %)

Note: Values in parentheses are percentages of total ocean
estimates.

Source: Adapted from Summerhayes, C. P. et al., Eds., *Upwelling in the Ocean: Modern Processes and Ancient Records,* John
Wiley & Sons, Chichester, 1995, chap. 15.

only summarizes how wind-driven coastal upwelling impacts energy
and material flows. Much of the discussion which follows is drawn
from individual chapters in these two books. I refer the reader to these
excellent volumes for a more extensive analysis of physical and chemical processes associated with upwelling ecosystems and their effect on
the geological record.

Five great coastal current-upwelling systems are known:

- The Canary Current off Northwest Africa
- The Benguela Current off Southwest Africa
- The Peru Current off of Peru and Chile
- The Somali Current in the northwestern Indian Ocean
- The California Current off Oregon and California

These systems, plus upwelling that occurs in other ocean regions,
by area account for a disproportionate amount of new production in
the global ocean (Table 7.11). Upwelling ecosystems are productive
because high concentrations of nutrients (mainly nitrate) brought up
from the deep ocean are injected into the euphotic zone, stimulating
primary production. Most studies point to a significant time-lag between the onset of an upwelling event and the inevitable response of

increased phytoplankton production. Pelagic grazers are overwhelmed by the rapid increase in algal biomass (composed mostly of diatoms), leading to excess production sedimenting to the sea floor.

How much of this carbon eventually deposits to the benthos or is transported as CO_2 to the atmosphere depends upon the extent of recycling efficiency within the pelagic food web. The factors regulating changes in the structure and function of the pelagic food web within these systems are still not understood. For instance, in the upwelling events of the eastern tropical Pacific, phytoplankton production is lower than expected considering the abundant nutrients available, implying that other limiting factors such as available iron play a pivotal role in controlling food chain production.

A new view has emerged of coastal upwelling systems as sites in which tongues or filaments of upwelled and downwelled water masses intermingle and extend out to hundreds of kilometers from the shelf edge. This latter scenario has been recently observed and measured in the coastal transition zone off the northern coast of California,[625] where the highest concentrations of plankton occur as narrow filaments in response to jet-induced upwelling. Such processes, rather than the classical view of nutrient-laden, cold water being advected along the shelf edge, may be responsible for the high productivity.

Temporal and spatial oscillations in food web dynamics within upwelled and downwelled waters are still poorly understood, because very few studies have attempted to relate ecological processes to such physical forces. It is not even clear whether the elevated algal biomass observed in upwelling systems is the result of enhanced *in situ* production or the net effect of entrainment and aggregation of phytoplankton induced by converging water masses.

It is not possible to predict the occurrence of upwelling events, even though the key physical forces (wind variability, bottom topography, surface heating, latitude, stratification) that induce upwelling are known. A good example of this unpredictability is the Leeuwin Current off Western Australia.[626] The basic pattern of this major current stream is favorable for coastal upwelling, but no such events occur. Despite these uncertainties, most coastal upwelling ecosystems exhibit a series of common characteristics:[623,624]

- Closed circulation cells
- Strong lateral and benthic advection
- High carbon deposition and burial
- Low rates of carbon recycling
- Large stocks of top predators
- Low biodiversity

Energy-flow analysis[627] of the two major eastern boundary up-welling regions — the Peruvian and Benguela systems — suggest that these ecosystems are several times more efficient in producing fish compared with coastal ecosystems not influenced by upwelling (e.g., Chesapeake Bay, Baltic Sea). Models indicate that the higher yields of planktivorous (in the Benguela system) and carnivorous (in the Peruvian system) fish may be due to higher production-to-biomass ratios and shorter food chains. Fewer transfers of energy and materials would result in more carbon available to sustain high fish production. Of course, fish production is very variable in both of these ecosystems, being dependent upon climate (e.g., El Nino-southern oscillation [ENSO] events) and oceanographic factors that are not considered in such models.

How upwelling impinges on shallow, inshore areas is just begin-ning to be understood. Within the nearshore zone off central Chile, Peterson et al.[628] observed both temporal and alongshore spatial sepa-ration of pelagic primary and secondary production, controlled by distinct upwelling phases, each eliciting a different biological response. Advection of newly upwelled water into the coastal zone results in seaward transport of plankton communities, replaced by different assemblages transported laterally. As the upwelling events wane, shear-ing of vertical currents separates this new assemblage from offshore assemblages advected into the nearshore from deeper waters. This results in limited separation of zooplankton from primary producers nearshore, while their offshore counterparts are more closely coupled on the shelf edge.

In contrast, the Banc d'Arguin ecosystem, a shallow inshore ex-panse of tidal flats and seagrass beds covering an area of 10,000 km^2 off the coast of Mauritania, is clearly dependent upon input of upwelled water.[629] Studies conducted during the Dutch-Mauritanian "Banc d'Arguin 88" project show that nutrient-rich water and associated plankton from the Canary Current intrude onto the Mauritanian shelf and penetrate deep into this shallow sea to bathe and fertilize exten-sive beds of the seagrasses, *Zostera noltii, Cymodocea nodosa,* and *Halodule wrightii.* The imported plankton support dense populations of filter-feeding molluscs inhabiting a seaward sill. This filter-feeding zone is several kilometers wide. Fish yield is low, partly due to the extremely dense populations of piscivorous birds that come to feed and breed within the Banc d'Arguin. The seagrass beds sustain a detritus-based benthic food web characterized by low-density benthic communities, certain species of which may be consumed by several species of wad-ing birds. The influx of upwelled water results in a negative estuary with higher salinity waters nearshore. This leads to little organic mat-ter being exported from the Banc d'Arguin.[629]

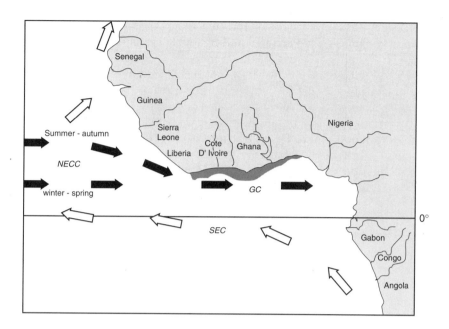

FIGURE 7.14
Chart of surface circulation in the eastern tropical Atlantic and hatched area along the Ghana coast depicting the seasonal upwelling and *Sardinella aurita* fishery (shaded area). NECC = North Equatorial Counter Current; GC = Guinea Current; SEC = South Equatorial Current. (Adapted from Binet, D. and Marchal, E., in *Large Marine Ecosystem: Stress, Mitigation and Sustainability*, Sherman, K., Alexander, L. M., and Gold, B. D., Eds., American Academy for the Advancement of Science, Washington, D.C., 1993, chap. 2.)

Further south, the upwelling system off the Gulf of Guinea (Figure 7.14) differs from other eastern boundary current systems in being climatically forced by oscillations of the inter-tropical convergence zone (ITCZ), the doldrums area separating northern and southern tradewinds at the equator. Crossing of the ITCZ north of the equator initiates the onset of monsoonal rains in June. Subsequent river runoff drains into the tropical surface waters (TSWs) which are warm and dilute. The TSW is displaced during the dry season (July to September) by cold, nutrient-rich water upwelling from the South Atlantic central water (SACW) mass. Movement of the ITCZ back south of the equator triggers a second monsoon season extending from September to November, ending the short period of upwelling.

The upwelling region extends only a short distance longshore and close to the coast, due to the convergence of several water masses; these large-scale circulation patterns, particularly the relationship between upwellings and river inputs, are not well understood. The upwelling sustains a pelagic fishery.[630] The most abundant species are

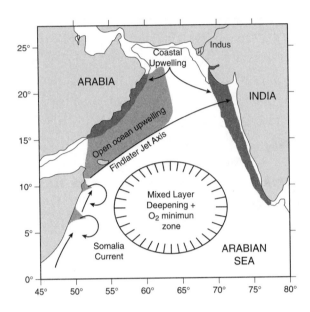

FIGURE 7.15

Chart depicting open ocean and coastal upwelling in the northern Arabian Sea during the southwest monsoon resulting from the influence of the Findlater jet and upward Ekman pumping. (Adapted from Summerhayes, C. P. et al., Eds., *Upwelling in the Ocean: Modern Processes and Ancient Records,* John Wiley & Sons, Chichester, 1995, 422.)

Sardinella aurita, S. maderensis, Engraulis encrasicolus, Brachydeuterus auritus, and, occasionally, *Scomber japonicus.* The collapse and rapid recovery of the *Sardinella aurita* fishery in the late 1970s and early 1980s were not caused solely by overfishing, suggesting that the fish stocks may be susceptible to changes of equatorial climate.[631]

Monsoons trigger upwelling regimes along other tropical continental margins, the best-studied example being in the northern Arabian Sea where equatorial, open ocean, and coastal upwelling are all important.[632-634] All three types of upwelling systems occur in the Arabian Sea from off the coast of Oman around to the southern tip of India; the Somali upwelling regime farther south in the equatorial western Indian Ocean is also generated by the onset of monsoons (Figure 7.15). During the southwest summer monsoons, both coastal and open-ocean upwellings intrude cold, nutrient-laden water onto and along the continental margin bordering the northern Arabian Sea. These upwellings are driven by a combination of atmospheric, oceanic, and continental phenomena, such as strong coastal winds which increase out to the Findlater Jet stream.[634] The strong southwest monsoon promotes an intense oxygen minimum zone between 200 and 1500 m. This zone is formed by microbial decay of the high production induced by upwelling, facilitated by deep Indian Ocean water of low oxygen content.

These monsoon-driven events have a significant impact on pelagic food webs and nutrient cycles in coastal and offshore waters. Measurements suggest high rates of primary productivity in the western Arabian Sea during the southwest monsoon; rates of production and standing stocks decline eastwards until coastal influences, such as from the Indus Delta and smaller rivers along the west coast of India, and coastal upwelling facilitate an increase in plankton and benthic production.[635] Off the Indian coast, red tides (*Noctiluca* in summer and *Trichodesmium* in winter) follow phytoplankton blooms that occur during both monsoon seasons. The intrusions of upwelled water during the southwest monsoon may penetrate into river mouths, stimulating high abundance of phytoplankton and zooplankton.[633] Lack of data precludes any objective assessment of the effects of upwelling on benthic food chains in this region. The few data that exist imply that benthic biomass is extremely variable beyond the coastal zone due to low oxygen conditions in near-bottom shelf waters and physical disturbance.[12]

Penetration of cold upwelled water into rivers, estuaries, and bays on narrow continental margins can enhance primary and secondary productivity and, in turn, stimulate fishery production. The best-studied example of this phenomenon is the Finisterre coastal upwelling system off Galicia in northwest Spain, where intrusions of North Atlantic central water into the interior of the rias (i.e., rivers) stimulate high rates of primary production (rates average $\geq$700 mg C m^{-2} d^{-1}) which support intensive raft cultivation of the mussel, *Mytilus edulis*.[636] These cultures have a significant impact on the food web structure and nutrient recycling within the rias. Tenore et al.[636] contrasted the benthos of two rias: the Ria de Arosa, where mussel cultivation is intense, and the Ria de Muros, where cultivation is much less intensive. Benthic carbon flow budgets for the Ria de Arosa (Figure 7.16A) and the Ria de Muros (Figure 7.16B) illustrate how the higher sedimentation rates of biodeposits from the mussel rafts and associated epifauna in the Ria de Arosa lead to higher rates of anaerobic microbial oxidation and low macrobenthic biomass in the sediments beneath the rafts. Burial rates were not measured but are presumably rapid.

Subsequent studies on the adjacent shelf suggest that significant outwelling of detritus does occur[637] in concert with the upwelling of cold North Atlantic Central Water. The upwelling and outwelling phenomena enrich the benthic regime of the outer and inner shelf zones, respectively. These enriched benthic communities sustain an important demersal fishery for hake (*Merluccius merluccius*) and blue whiting (*Micomesistius poutassou*). Also, the greater abundance of the

A. Ria de Arosa

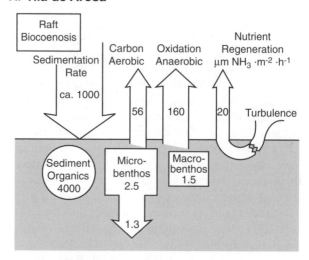

B. Ria de Muros

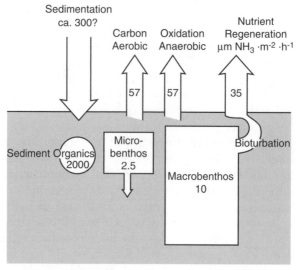

FIGURE 7.16

Benthic carbon budget in rafted areas of the Ria de Arosa (A) and nonrafted areas of the Ria de Muros (B) in northwest Spain. Standing stocks are g C m⁻² and fluxes are g C m⁻² d⁻¹. (From Tenore, K. R. et al., *J. Mar. Res.*, 40, 701, 1982. With permission.)

Norwegian lobster, *Nephrops norvegicus*, in the adjacent coastal zone suggests direct support from detritus outwelling from the Rias Bajas. [637]

Biogeochemical studies[638-640] in the Ria de Vigo, a ria with no mussel cultures and located south of the Rias de Muros and de Arosa, indicate that significant differences in the fluxes of carbon, nitrogen,

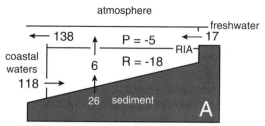

WINTER (vigorous circulation and resuspension)

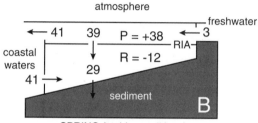

SPRING (stable conditions)

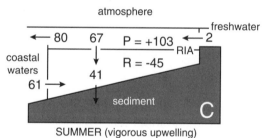

SUMMER (vigorous upwelling)

FIGURE 7.17
Mass balance for organic carbon (mol C per season) in the Ria de Vigo (northwest Spain) in (A) winter, (B) spring, and (C) summer. P = photosynthesis; R = respiration. (From Prego, R., *Geochim. Cosmochim. Acta,* 57, 2041, 1993. With permission.)

and phosphorus are linked to the seasonal pulses of upwelling. Mass balance estimates for carbon show that in winter (Figure 7.17A) there is significant net export of organic carbon despite low gross primary production; carbon is supplied to the ria via import, resuspension, and freshwater input. After the spring phytoplankton bloom (Figure 7.17B), the ria is stable and at steady-state with respect to carbon. Vigorous upwelling in summer stimulates production and mineralization, with the ria behaving as a carbon trap when organic material settles to the sea floor (Figure 7.17C). Organic nitrogen, in contrast, is exported from the inner to the outer ria year-round, but the total nitrogen balance suggests that the ria acts as a nitrogen trap during spring and summer.

The pattern for phosphorus is similar; in winter, the ria flushes more phosphorus into the adjacent ocean than it receives, but during the rest of the year, the opposite occurs. The net exchange is an accumulation of phosphorus (mainly in organic form), equivalent to 38% of annual inputs into the ria.

7.4.2   Export to the Deep Ocean: The Tropics Revisited

Rejection of the export hypothesis formulated by Walsh and his colleagues[580] applies to many continental margins, but not to all. Lack of consideration for the uniqueness of other continental shelves can have serious consequences for attempts to estimate nutrient budgets for the coastal ocean worldwide. Here, the export of materials from wet tropical shelves and the significance of export from temperate shelves to adjacent slopes are examined.

That a relatively small amount of the carbon fixed on the majority of temperate shelves is exported to the adjacent slopes does not mean that such inputs are unimportant for slope ecosystems, or that some equilibrium does not exist between the shelf proper and adjacent deep ocean. The cycles of nitrogen, phosphorus, and other elements between shelf and slopes may be similarly dynamic. Recent calculations of carbon and nitrogen flux by Walsh[641] imply that a close biogeochemical interdependence exists between shelf and slope ecosystems. Based on data collected from several systems, Walsh's calculations[641] suggest that, on average, upwelled nitrogen (mostly nitrate) is the largest source of new nitrogen to continental shelves. Moreover, this import is roughly balanced by the amount available for new production and export to the adjacent deep ocean. Perhaps as much as one third of this total available nitrogen pool is exported, with most being utilized within the deep water column, leaving only a small fraction (<10%) to deposit on the sea floor in particulate form. Converted to carbon, these estimates imply that 20 to 30% of shelf carbon is mineralized or stored in slope sediments. These shelf subsidies may account for a proportionally larger fraction of slope energy flow, as rates of carbon oxidation and nutrient recycling are generally slower in the deep ocean.

The presence of deep-sea fans hundreds of kilometers offshore from some of the world's largest tropical rivers (such as the Indus, Ganges, Brahamaputra, and the Irrawady) is well known and indicates significant export of terrestrial material to the tropical deep ocean. Until recently, it was thought that these fans are relict, cut off from their original sources of sediment supply by the Holocene transgression. This explanation was supported by deep-sea seismic and tectonic observations, as well as by evidence suggesting little significant seaward growth of subaerial deltas for the past two centuries. Indeed, until recently, lack of geological data from most of these tropical shelves precluded any attempts to suggest mechanisms for offshore transport.

A variety of sedimentological, geochemical, and biogeochemical studies[642-644] on tropical shelves and their adjoining slopes and on deep-sea fans now indicates modern export and deposition of organic and inorganic materials to the deep ocean. The rates of material flux remain unquantified but cast doubt on the wisdom of excluding consideration of these fluxes in global ocean flux models (see next section). Geological evidence from the Bay of Bengal and the Bengal fan[642] suggest that sediments from the Ganges-Brahmaputra river system are accumulating most rapidly on the shelf in close proximity to a steep channel (called the "Swatch of No Ground") that leads into the adjacent deep sea.

Stable carbon isotope signatures in long cores taken from the Bengal fan[643] generally support the sedimentological data but indicate that the rate of accumulation increased rapidly during the early Miocene period and decreased thereafter to a rate consistent with present delivery estimates. Moreover, the isotope data suggests that the organic matter deposited to the Bengal fan is composed of a mixture of terrestrial and marine-derived organic carbon varying with past changes in climate and hydrological conditions on the shelf. This interpretation is supported by sediment trap studies[644] in the Bay of Bengal revealing variations in particulate flux in relation to seasonal monsoons. Peak flux rates coincide with maximum discharge from the Ganges and Brahmaputra rivers during the southwest monsoon. Offshore advection of the river plumes stimulates primary production and subsequent deposition of phytodetritus and zooplankton fecal pellets, which mixes with lithogenic and organic material derived from land. Aggregation of these mixed organic phases with inorganic particles enhances sinking of this material into the deep ocean. Sediment traps concurrently deployed in the adjacent Arabian Sea[644] show that particle fluxes are related mainly to wind-induced mixing of deeper water layers, as well as nutrient enrichment of surface waters during the southwest and northeast monsoons.

Storage of this material in Arabian Sea sediments is controlled not only by factors influencing rates of supply (e.g., river discharge, primary production, monsoons), but also by factors enhancing preservation of the organic matter.[645] The oxygen minimum zone (see page 304) persists on the slope bordering the northern Arabian Sea, but higher concentrations of organic carbon off the west coast of India compared with on the Arabian and Pakistan slopes suggests that enhanced storage of organic carbon is due to several factors other than anoxia, including sediment texture and clay composition, shelf width and slope gradient, currents, and the extent of terrigenous sedimentation.[645]

Stable isotope studies of organic material within the Indus[464] and Zaire rivers and in adjacent slope sediments[647] indicate significant dilution and transport of high-quality riverine organic matter with inorganic material. The sources of organic matter and the rate of transport play a significant role in the quality of the POC that eventually is deposited on the slope.[645] Ittekott and Arain[646] estimate the rivers on the Indian subcontinent can contribute from 8 to 11% of the POC buried globally in continental margin sediments. The Zaire river study[647] supports the scenario for the Bengal fan in that organic carbon derived mainly from C_3 plants is diluted seawards by inorganic particles, with the proportion varying in relation to wet season vs. dry season discharge and to year-to-year differences. Deposits laid down during drier years show proportionately more organic carbon of marine planktonic origin.

Rivers draining the island of New Guinea similarly export significant, if unquantified, amounts of marine and terrestrial carbon to the adjacent deep ocean. The distribution of particulate elements and stable carbon signatures in sediments on the Gulf of Papua, and adjacent slope and rise off southern Papua New Guinea imply significant transport of material to the deep sea.[648,649] The carbon isotope data, extrapolated to the entire region of Oceania, suggest the potential for as much as 90×10^6 metric tons yr^{-1} of POC being transported via rivers to the global ocean — roughly twice the amount of organic carbon exported by the rivers of North and South America and Africa combined. This is a crude estimate but, if accurate, suggests that Oceania may be the major source of POC to the global ocean — a source not previously considered in carbon budgets of the world ocean.

The carbon isotope data of Bird et al.[648] implies that 40% of the Gulf of Papua sediments contain ≥75% terrestrially derived POC, with 50% of the shelf containing ≥40% terrigenous POC. The origin of this terrestrial POC can be gleaned from the mass balance estimates for organic carbon in the Fly delta. Using a simple two-box model, Robertson and Alongi[650] estimated total organic carbon input from the river to the delta to be on the order of 1.7×10^{12} g C yr^{-1}, composed nearly equally of DOC and POC. Pelagic and benthic respiration in the delta accounts for the loss of ~60% of total organic carbon input, and benthic microbial uptake of DOC accounts for 33% of DOC input from the river. This leaves more than 3.0×10^{11} g C yr^{-1} available to be exported from the delta to the adjacent shelf and deep sea. Most of this exported carbon appears to be of mangrove origin as little of the riverine organic supply reaches the Gulf of Papua. Indeed, benthic trawls found significant quantities of macrodetritus of mangrove origin on the adjacent Coral Sea floor. This is in agreement with earlier observations[651] of large quantities of branches, seeds, leaves, and whole mangrove logs lying on the adjacent slope and continental rise. Transport must be rapid (perhaps during storms), considering that more than 80% of the plant seeds recovered appeared to be still viable. How much of the total net production of mangroves and other vegetation that is eventually exported to the deep-sea is unknown.

From our examination of some of the best-studied shelf ecosystems, it is reasonably clear that continental shelves in the wet tropics most often violate the generalizations offered at the beginning of this section. The exceptions are due mostly to the presence of large rivers, particularly those bordering narrow shelf margins. It is highly probable that many large and many smaller, mountainous rivers in tropical latitudes, for instance, export significant quantities of organic matter to the deep ocean. What impact wet tropical shelf ecosystems have on present estimates in nutrient budgets of the global ocean is unknown.

7.5 NUTRIENT CYCLES
AND GLOBAL CHANGE IN
THE COASTAL OCEAN

A few mass balance estimates of carbon and nitrogen flow for the world's coastal ocean exist, summarizing our present knowledge of sources, fluxes, and sinks of organic matter. These exercises help us to assess the significance of the coastal ocean in global nutrient budgets and how coastal ecosystems may be affected by global climate change.

7.5.1   Mass Balance Estimates and
Ocean-Atmospheric Exchanges

Wollast[652] constructed a preliminary global mass balance of organic carbon on the continental shelf and slope (Table 7.12). The carbon model implies that more than 82% of carbon fixed by shelf phytoplankton is respired by shelf benthic and pelagic consumers. The remainder (~14%) is exported to the continental slope where most carbon is respired by slope consumers below the thermocline. Of the original carbon fixed on the shelf, only 4% is eventually buried in shelf and slope sediments. This export pathway is more significant than previously believed, reflecting the more recent estimates of shelf export in tropical and boreal latitudes. This also implies less importance of carbon burial in coastal sediments, but Wollast[652] cautioned that recent recognition of increases in river particulate loads and higher sedimentation rates on some shelves (as we have seen for shelves of the wet tropics) make this value uncertain.

TABLE 7.12

Global Mass Balance of Organic Carbon Fluxes (10^{15} g C yr^{-1}) on the Continental Shelf and Continental Slope

---

Continental Shelf

Primary production	6.9
	⇓
Lost as pelagic respiration	3.7
Deposited to sediments	2.2 ⇒ 2.0 respired, 0.2 buried
Available for export	1.0
	⇓

Continental Slope

Import from shelf	1.0
Lost as pelagic respiration	0.8
Deposited to sediments	0.2 ⇒ 0.18 respired, 0.02 buried

Source: Data adapted from Wollast, R., in *Ocean Margin Processes in Global Change*, Mantoura, R. F. C., Martin, J.-M., and Wollast, R., Eds., John Wiley & Sons, Chichester, 1991, 365.

A more extensive and detailed carbon budget of the global ocean constructed by Smith and Hollibaugh[653] (Figure 7.18) reveals some significant differences with Wollast's carbon budget (Table 7.12). First, the Smith and Hollibaugh (SH) budget indicates that respiration exceeds primary production by 1.4%, whereas Wollast (W) calculated that coastal waters are slightly net autotrophic (production-to-respiration ratio [P:R] = 1.03). The net heterotrophy of the coastal ocean in the SH model is compensated by terrigenous loading of DOC and POC which was not considered by Wollast. Second, less carbon settles to the sediment in the SH model (~15% of net primary production, or NPP) compared to the W model (~32%). Both models agree that proportionately little carbon is available for preservation in sediments (SH model: ≤2% of NPP; W model: 3% of NPP) in contrast to Walsh's[641] estimates (see page 308). The sediment carbon cycle in the SH model is balanced by the return of carbon (mainly as DOC efflux) to the water column. Third, the SH model indicates that carbon equivalent to ~4% of total NPP is exported to the open ocean, three and one half times less than calculated in Wollast's budget The fate of this material is not distinguished from that of the carbon fixed in the open ocean, but both models generally agree that only a small proportion is preserved in deep-sea sediments. Although the contribution from tropical rivers was included in the SH model, it is likely that more recent estimates of carbon flux from tropical river-dominated shelves[648] would result in higher global average rates of carbon input and more absolute amounts of carbon being remineralized or buried in slope and continental rise sediments.

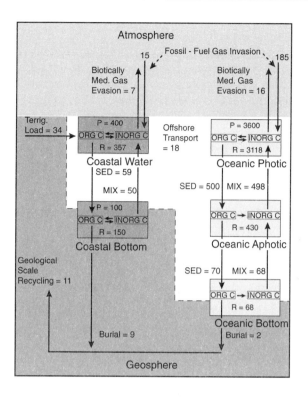

FIGURE 7.18

Steady-state model of the organic carbon cycle in the coastal ocean and the open ocean. Units are 10^{12} mol C yr^{-1}. Dashed arrows are approximate rates of penetration of fossil fuel CO_2 into the marine biosphere. (From Smith, S. V. and Hollibaugh, J. T., *Rev. Geophys.*, 31, 75, 1993. With permission.)

The atmosphere plays a significant role in coastal and open-ocean carbon cycles, but there has been considerable disagreement and confusion over whether or not the ocean is a net sink or source of CO_2. The Smith and Hollibaugh[653] model indicates that the global ocean is net heterotrophic and therefore a net source for atmospheric CO_2. However, their estimate of net export of biologically mediated carbon gases from the ocean (total: $\sim23 \times 10^{12}$ mol yr^{-1}) to the atmosphere is much less than present atmospheric inputs of carbon from the burning of fossil fuels to the global ocean, estimated at about 200×10^{12} mol yr^{-1}. A more recent analysis[600] of CO_2 cycling in the coastal ocean indicates that the ocean is presently a sink for CO_2 due to fossil fuel burning and changes in land use (deforestation, etc.). Walsh and Dieterle[600] suggest that significant differences in atmosphere-ocean exchange of CO_2 occur with latitude because of natural changes in partial pressure, upwelling, and warming of shelf waters resulting in net release of CO_2. Temperate and polar shelves sequester from 8.3 to 9.7×10^{13} mol C yr^{-1}, but tropical shelves (for the reasons above) release 3.1×10^{13} mol C yr^{-1} of CO_2 to the atmosphere, for a net input on the order of 5.2 to 6.6×10^{13} mol C yr^{-1}. These estimates are critically dependent on the assumptions that the POC is buried, the DOC is advected to below the aphotic zone, and the ΣCO_2 crosses the lysocline of adjacent slope and basin waters.

Various scenarios and models have been constructed to assess the impact of increases in fossil fuel inputs to the ocean which are expected to continue well into the next century. Further discussion is beyond the scope of this book, but it suffices to say that better understanding of processes such as carbon preservation in sediments is urgently required. Present models[654] predict that much of the increased inputs of atmospheric CO_2 to the ocean will translate into increased carbon fixation by primary producers; much of this organic matter will presumably sink below the euphotic zone, where it will either be respired or deposited to the deep-sea floor. This scenario is complicated by forecasted increases in the severity of eutrophication and of river runoff into the coastal zone (see next chapter). It is here that the carbon cycle is most closely linked to the nitrogen and phosphorus cycles, as increased discharge of nitrogen and phosphorus compounds increase primary production and fluxes of organic carbon in the water column and sediments.

Wollast[655] constructed preliminary global budgets of nitrogen in the coastal ocean (Figure 7.19A) and in the world ocean (Figure 7.19B). The same caution should apply to these budgets as for the carbon estimates. Nevertheless, on average, the largest exchanges of nitrogen in the world's coastal zones in decreasing order of importance are

- Sedimentation and regeneration in sediments
- Upwelling
- Export to the open ocean
- Groundwater
- River inputs from land
- Coastal ocean-atmospheric exchanges

By far, most nitrogen leaves the water column as particulate matter settling to the benthos and as either particulate or dissolved nitrogen exported to the adjacent open ocean. Denitrification is the next largest loss of nitrogen from the coastal ocean, with a net loss of nitrogen to the atmosphere and a much smaller loss as buried nitrogen in sediments. Denitrification losses roughly equate with river inputs.

The total flux of recycled nitrogen accounts for ~78% of the nitrogen required for primary productivity; the remainder is new nitrogen

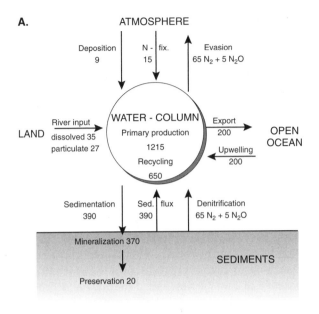

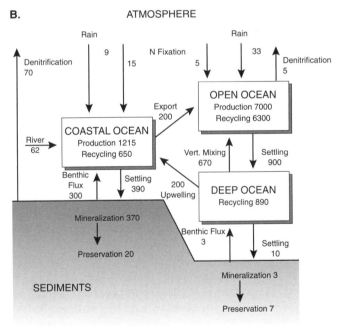

FIGURE 7.19
Preliminary global budget of nitrogen in the coastal zone (A) and for the entire ocean
(B). Fluxes are 10^{12} g N yr^{-1}. (Adapted from Wollast, R., in *Interactions of C, N, P, and S
Biogeochemical Cycles and Global Change*, Wollast, R., MacKenzie, F. T., and Chou, L.,
Eds., Springer-Verlag, Berlin, 1993, 195.)

contributed mainly from the open ocean, with much smaller contributions from land and the atmosphere. The generic fluxes of nitrogen in the global ocean (Figure 7.19B) support Walsh's contention[641] that the export of nitrogen from the coastal ocean is an important flux to the open ocean, where recycled nitrogen accounts for a larger share (90%) of primary production requirements than in the coastal ocean. In fact, only ~13% of primary production as nitrogen leaves the upper layers of the open ocean, with most settling through the deep water column to be recycled; little (0.1%) deposits to the deep-sea floor. Denitrification is of minor importance in the open ocean, as most occurs in coastal sediments.

The small contribution of nitrogen from river inputs does not diminish the impact of human discharge, as many estuaries and other coastal zones have been severely impacted by eutrophication. Even comparatively small shifts in the ratio of nitrogen to phosphorus input can lead to uncoupling of primary production to nutrient limitation, causing an imbalance of coastal nutrient dynamics and food webs.[656] More recent studies[657,658] also suggest that present estimates of global losses via denitrification are severe underestimates, by as much as 50%. A recent modeling exercise by Christensen[659] implies that an increase in denitrification rates would lead not only to a decline in POC flux to the deep ocean, but also a shift in carbon flux between the atmosphere and the ocean. Any deviation from presumed steady-state would clearly have an enormous impact on coastal and oceanic nutrient cycles and food web energetics.

Such a change is already happening, but these models still do not adequately reflect many natural processes, particularly those operating in the tropical ocean, such as the effect of inputs of materials carried by the large and small rivers of Oceania and Southeast Asia. Accounting for the inherent differences between high-latitude and tropical coastal seas would have a dramatic impact on model predictions and estimates, as noted for example in Walsh and Dieterle's estimates[600] of latitudinal differences in ocean-atmospheric exchanges of carbon dioxide.

Recent data of river inputs to coastal zones in the tropics and other regions have resulted in reappraisals of the phosphorus[660] and silicon[661] cycles. Treguer et al.[661] have now revised the net input of dissolved silica to the ocean to be 6.1×10^{12} mol yr^{-1} with rivers being the major contributors (Table 7.13). Inputs are close to being balanced with outputs, but most of the biogenic silicon is internally recycled. The bulk of the silicon comes from land via rivers, but little is buried within coastal sediments, as most is rapidly regenerated to the water column during early diagenesis of inorganic and organic detritus in sediments.

TABLE 7.13

Revised Budget of Silica (10^{12} mol yr^{-1})
in the World's Ocean

Inputs	
Rivers	5.0 ± 1.1
Eolian	0.5 ± 0.5
Sea floor weathering	0.4 ± 0.3
Hydrothermal	0.2 ± 0.1
Total	6.1 ± 2.0
Outputs	
Coastal	1.2 ± 0.7
Deep sea	5.9 ± 1.1
Total	7.1 ± 1.8
Biological cycle	
Production	240 ± 40
Ratio of internal cycling to inputs	23:53

Source: From Treguer, P. et al., *Science,*
268, 375, 1995. With permission.

7.5.2 Accumulation and Preservation of Organic Carbon

Little organic matter is stored in the water column, so understanding how much carbon accumulates and is preserved in sediments and the factors controlling these processes is crucial in determining the ultimate fate of this material. Until recently, it was thought that organic carbon is preferentially preserved under anaerobic conditions, particularly in sediments beneath waters of high productivity and in anoxic basins (see Section 6.5.2.3). Figure 7.20 shows current estimates of the distribution of organic carbon burial in various marine environments. It shows that most organic matter is buried in marine deltaic and shelf sediments; little is buried in anoxic basins, on slopes, in upwelling zones, or in the deep sea. Hedges and Keil[16] estimated a global burial rate of 0.16×10^{15} g C yr^{-1}, slightly higher than previous estimates, reflecting recent assumptions of more extensive burial in upper slope deposits but not including more recent studies of export from tropical shelves. Ultimately, less than 0.5% of global ocean productivity and less than 10% of organic material depositing to the sediments are preserved in marine sediments. The main factors controlling organic carbon preservation include:

- Primary productivity
- Sediment accumulation rate
- Source of organic matter
- Oxygen content of near-bottom waters

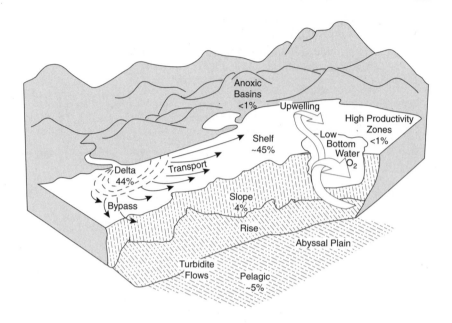

FIGURE 7.20

Graphic depicting estimates of the percentage of organic matter burial within various regimes of the continental margin. (From Hedges, J. I. and Keil, R. G., *Mar. Chem.*, 49, 81, 1995. With permission.)

Despite problems in separating organic matter from sediment mineral matrices and our ignorance of the biochemical composition of sediment organic matter, advances have been made in understanding factors controlling carbon preservation in marine sediments. Recent studies[447,662,663] show that rates of organic matter decomposition under anoxic conditions are often equal to, or greater than, aerobic decay. The balance between aerobic and anaerobic decomposition depends upon several factors:[450]

- Oscillations in redox status
- Degree of bioturbation
- Age and nutritional quality of organic matter
- Benthic community composition and structure

Aller's original hypothesis[450] that oscillation in redox status affects organic matter decomposition was tested in a series of laboratory experiments conducted by Kristensen et al.[663] Incubating labile (aged diatoms) and poor-quality (fresh barley hay) detritus under oxic, then under anoxic, conditions in separate chambers, they found that decay rates of the diatom detritus were fastest under anoxic conditions, but no significant difference was observed in the decomposition of the hay, either under aerobic and anaerobic conditions. They suggested that the slower rates of anaerobic decay of the diatom detritus were limited by the initial hydrolytic and fermentative attack on the pre-decomposed material. The limiting step in the decay of the poor-quality detritus was the slow release of DOC; after depletion of the DOC pool, rates of decomposition declined. These results indicate that rates of decomposition of easily hydrolyzable and leachable compounds in fresh detritus are equivalent under oxic and anoxic conditions, but anaerobic mineralization of detritus (composed mostly of structural macromolecules) is inhibited by inefficient and slow hydrolysis of structurally complex compounds.

Recently, Lee[664,665] suggested that some organic matter may be preserved in sediments by the production of new bacterial biomass derived from the decomposition and acquisition of organic carbon from the original material. This idea assumes that bacteria are not heavily grazed in anoxic sediments and that bacteriovores, such an anaerobic protozoans, are themselves not heavily consumed. This is an intriguing idea, similar to the trophic dead-end argument[5,7] (see Section 2.2) formulated to explain the fate of microbes in sediments. Organic geochemical[666,667] and trophic[5,7,665] data support the scenario that organic matter in sediments is converted into bacterial biomass and that some fraction of this material may eventually accumulate in anoxic sediments. Research is needed to test the idea that organic carbon preservation is linked to the fate of bacterial and protozoan biomass in anoxic sediments. Such research would be a clear instance where biogeochemical and trophic paths would cross to unravel the fate of organic matter in the coastal ocean.

7.5.3   Are Coastal Ecosystems Net Heterotrophic or Net Autotrophic?

Photoautotrophs (e.g., phytoplankton, macrophytes) and bacteria are, respectively, the main producers and consumers of fixed carbon in the coastal biosphere. What is less understood is whether or not a net balance exists between production and consumption of carbon in coastal ecosystems.

Is the coastal ocean in steady-state? The budgets discussed in Section 7.5.1 suggest that pristine coastal ecosystems are a source of carbon (in the form of CO_2) for the atmosphere. In reality, increases in the concentration of carbon gases in the atmosphere since the beginning of the Industrial Revolution have tipped the balance in favor of the ocean now being a sink for CO_2. The mechanisms of ocean-atmosphere exchange and predictions of future trends are uncertain and

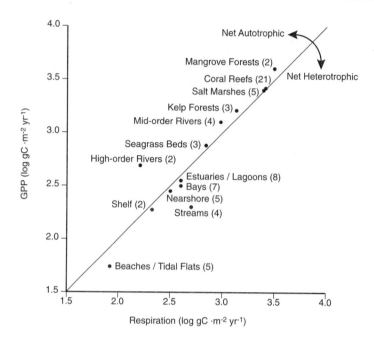

FIGURE 7.21

Plot of the balance between gross primary production and total respiration in various aquatic ecosystems, including streams and rivers. Note that data are $\log_{10}$ transformed. Each point is the arithmetic mean of data obtained from several systems (numbers in parentheses). (Data from references in Heip et al.,[421] Smith and Hollibaugh,[653] and Heath[668] and from data summarized in Chapters 2, 3, 4, 5, and 6).

controversial. It is therefore important to understand whether or not a balance exists between organic carbon production and metabolism, if predictions of impacts of further climate change on coastal ecosystems are to be accurate.

A plot (Figure 7.21) of mean rates of gross primary production vs. total system respiration for various coastal ecosystems (rivers, estuaries, lagoons, bays, nearshore, and shelf zones) reveals that most ecosystems are either slightly autotrophic (rivers, seagrass beds, kelp and mangrove forests, coral reefs, and salt marshes) or slightly heterotrophic (streams, beaches and intertidal flats, estuaries and lagoons, bays, and nearshore and open shelf regions).

As nearshore and shelf regions by area overwhelmingly dominate the coastal ocean, the conclusions of Smith and Hollibaugh[653] and Heip et al.[421] that the coastal ocean is slightly heterotrophic and a source of CO_2 appear correct. If one considers the inherent variability in rates of gross primary production and respiration within these ecosystems and the very sparse data (obtained mostly from pristine or slightly polluted systems), it seems best to be conservative and conclude that the steady-state scenario is one of overall carbon balance. Expected increases in eutrophication, deforestation, and habitat destruction (Chapter 8) will result in an overall shift to net heterotrophy in some habitats (e.g., sandy beaches) and to net autotrophy in others (e.g., coral reefs) depending upon the nature of the disturbance. In any case, a severe imbalance will result in the global ocean because of non-steady-state invasion of CO_2 from the burning of fossil fuels.

7.6 GLOBAL ESTIMATES OF FISHERY YIELDS TO HUMANS

Fisheries production in estuaries, bays, lagoons, and other coastal habitats correlate with primary productivity (see Section 6.5.3.3), but other factors play important roles in sustaining coastal fisheries. These include high structural diversity, abundant food, predation limited by turbidity, and enhanced production in the wake of river discharge.

Nixon[669] and Mann[670] have speculated that more intensive yields of harvestable organisms in the marine biosphere compared with freshwater ecosystems, are indirectly caused by greater spatial and temporal variations of physical energy inputs, leading to such phenomena as upwelling-downwelling, vertical mixing, tidal exchanges, tidal fronts, and stratification-destratification of water masses. These physical events may enhance transfer of materials and energy up the food chain. A simple example would be when vertical mixing, caused by wind and tidal currents, erodes the benthic boundary layer, permitting more rapid release of nutrients to the water column to enhance phytoplankton production, eventually making more food available to fish. Variations in physical forces, such as solar heating, freshwater input, temperature, and wind strength, may lead to variations in fish production.

Events affecting fish stocks at one ocean boundary can affect fish populations at the opposing boundary. The best example of such synchrony are the events generated by the ENSO phenomena when periods of El Nino cause sardine catches off Peru to decline, but leads to the advection of subantarctic waters off Tasmania favoring a pelagic food web dominated by large zooplankton (mainly the euphasiid *Nyctiphanes australis*). These euphasiids are, in turn, the main food for the jack mackerel, *Trachurus declivis*. Mann[670] provides additional examples for the argument that physical forces operating on and through food webs, are prime factors determining variations in fish stocks and production.

The extent to which physical factors are a dominant force in controlling fish stocks in the tropical ocean is less understood, although several studies have indicated enhanced fish production in proximity to river discharge,[670,671] monsoonal upwelling,[672] or a combination of

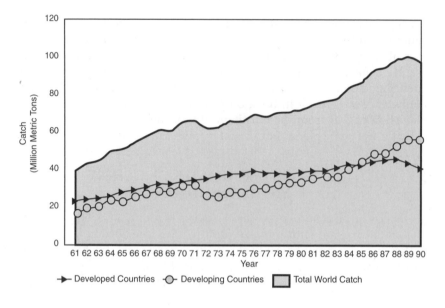

FIGURE 7.22
Changes in total global marine fish catch (including aquaculture) by developed and developing countries from 1961 to 1990. (Data from FAO, *Fish and Fishery Products,* Circ. No. 821, Rev. 2, Food and Agricultural Organization of the United Nations, Rome, 1992, 3.)

both.[673] Longhurst and Pauly[619] observed that fish yields in the tropics are lower than expected (~16% of global production) given that roughly half the world's ocean lies within the tropics. They emphasize that the low yields of fish in the tropics is an enigma but speculated that tropical fishes may be less productive because life in warmer water results in higher metabolic costs and less energy available for growth. Other workers[674,675] have also hypothesized lower fish yields, particularly for demersal species, in the tropics compared with catch estimates from colder waters.

A comparison of productivity and fish yields to humans for some boreal, temperate, and tropical shelf ecosystems revealed no clear latitudinal trends, as differences within a climatic zone are as great as they are among them.[12] Latitudinal comparisons cannot be conclusive because information on tropical fisheries lags behind data on higher latitude fisheries which have been worked more extensively. Alongi[12] points out that differences in fish production among ocean regions are likely to be constrained to a narrow range because of attenuation of energy transfer up the food chain; the basic laws of thermodynamics must apply to all ecosystems regardless of climate and latitude.

Regardless of cause, recent data (Figure 7.22) and predictions[676] indicate that maximum sustainable fish yields have been reached.

Technological limitations may prevent maximum yields from being attained in the open ocean, but fish stocks of the continental margins are already fully exploited. The proportion of the total catch (Figure 7.22) attributed to developing nations has now surpassed that of the developed countries.[677] This bodes poorly for the future, as population growth continues to increase in developing countries, nearly all of which border tropical seas. The growth of human populations in the tropics not only places additional strains on already fragile, coastal ocean resources, but invariably also leads to escalation of pollution and destruction of habitats — problems which are discussed in the next and final chapter.

Chapter 8

DEGRADATION AND CONSERVATION

8.1 A GLIMPSE AT THE GLOBAL PROBLEM

Situated at the confluence of land, sea, and air, the coastal ocean — its structure and processes composed of terrestrial, freshwater, marine, and atmospheric elements — is exploited by humans for food, recreation, transport, waste disposal, and other needs. For these reasons and others (Table 8.1), roughly 60% of the world's human population resides close to a coast.[678] The projected increases in population; greater urbanization, industrialization, and affluence; and rapid transportation foreshadow greater pressure on the living and non-living resources of the coastal ocean.

Estimates vary, but most experts[678-682] agree that a large proportion of the world's coastal habitats are in various stages of degradation. Figure 8.1 shows the global progression of humankind's begrimed handprints on coastal resources and habitats. In particular, the map indicates that large expanses of coastline bordering the industrialized nations of the northern hemisphere are experiencing considerable pollution. Exacerbation of the problem in the future, however, is expected mostly in the developing nations of the tropics, where population growth and attendant urbanization are fastest.[683]

Regardless of biogeographic province, the litany of abuse of coastal habitats is similar. Recent reports[681,682] by GESAMP* and NRC** identified the same key problems of coastal degradation and pollution (Table 8.2). The greatest concerns are coastal development and accompanying modification and destruction of habitats, eutrophication, microbial

---

* The Joint Group of Experts on the Scientific Aspects of Marine Pollution (GESAMP) is an advisory body to the heads of eight organizations of the United Nations (UN, UNEP, FAO, UNESCO, WHO, WMO, IMO, and IAEA).

** The Committee To Identify High-Priority Science To Meet National Coastal Needs, Ocean Studies Board Commission on Geosciences, Environment, and Resources, National Research Council of the National Academy of Science, Washington, D.C.

TABLE 8.1

Major Uses of Coastal Resources by Humans

Fishing	Timber
Sand and gravel dredging	Harbors, marinas
Aquaculture	Housing and reclamation
Oil, gas, and mineral exploration	Tourism
Recreation	Waste disposal (including sewage)
Collecting and hunting (e.g., shells)	Canals and other diversions
Fuel	

pathogens, fouling by plastics, and persistent increases in chlorinated hydrocarbons. Despite aroused public concern, the GESAMP found that radionuclides, most trace elements (particularly lead), and crude and refined oil now present less of a problem for a variety of reasons, including national and international environmental regulations, declining use of leaded fuels, and improvements in technology. These bodies do have their critics[684] in that other urgent problems — such as the decline of marine mammal populations, overfishing and the collapse of fisheries, fish and shellfish diseases, and crashes of seabird populations — have not been addressed. Some critics simply do not agree that radionuclide and oil pollution, for instance, are of lesser concern.

One of the most difficult problems in discerning a pollution problem is being able to distinguish a deleterious effect from natural variability. This inherent difficulty has become most evident in arguments

TABLE 8.2

Major Problems in the World's Coastal Ocean

Eutrophication
Coastal development (including aquaculture)
Habitat modification/destruction/alteration
Disruption of coastal hydrological cycles (including river discharge)
Point and nonpoint source release of toxins and pathogens
  (including synthetic organic and petroleum-related chemicals)
Introduction of exotic species
Fouling by plastic litter
Buildup of chlorinated hydrocarbons
Shoreline erosion/siltation accelerated by deforestation,
  desertification, and other poor land-use practices
Unsustainable/uncontrolled exploitation of resources
Global climate change and variability
Noise pollution

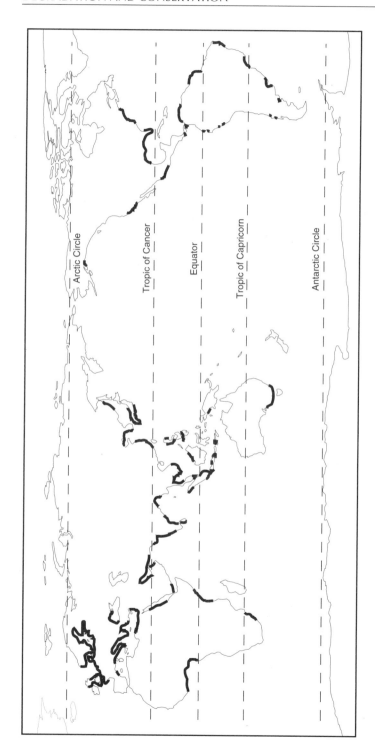

FIGURE 8.1
World chart highlighting coastal areas significantly impacted by man. Based on information found in References 679 to 683, 692, 695, 697, 698, 701, 703, 705, 707, and 708.

TABLE 8.3

Sources of Pollutants in the Coastal Ocean

Source	All Pollutants (% contribution)
Runoff and other land discharges	44
Atmosphere	33
Maritime transportation	12
Dumping	10
Offshore production sites	1

Source: From GESAMP, *The State of the Marine Environment*, Blackwell Scientific, Oxford, 1990. With permission.

concerning the possible effects of the buildup of greenhouse gases on global climate change. For instance, reconstruction of summer temperatures from ring widths of pine trees from northern Europe indicates a strong periodicity of 30 to 40 years which may reflect natural oscillations of the thermohaline circulation in the North Atlantic. Such statistical links suggest that caution must be applied before unequivocal judgments can be made of harmful effects. The danger lies in being able to discern an anthropogenic effect before it becomes irreversible.

Most pollutants enter the coastal ocean from land and from the atmosphere (Table 8.3). Both of these sources account for roughly two thirds of total pollutant inputs, with the remainder coming from maritime accidents and discharges, dumping, and from offshore petroleum production. The large contributions from land and atmospheric inputs reinforce the importance of the connections of land and atmosphere with the coastal zone and the urgent need to understand better the nature of these relationships. There is increasing recognition that coastal pollution problems are closely linked to contamination of attendant watersheds and aerosols in the lower atmosphere.

Most of these problems have been chronic in the developed nations since industrialization. As the problems have been thoroughly reviewed,[680-682,686-688] this chapter focuses mostly on eutrophication and coastal development, cited by GESAMP and NSR as the most serious problems facing the coastal ocean globally.

8.2 EUTROPHICATION

Eutrophication — an increased rate of supply of nutrients leading to enhanced primary production — is caused mainly by loading of nitrates and phosphates (and secondarily by organic matter) derived from fertilizers, livestock waste, and fossil fuel emissions. Perhaps more than any other coastal pollution problem, eutrophication has an immediate and demonstrable impact on coastal ecosystems. The most heavily affected areas are the Baltic and North Seas, the Mediterranean, and the Northwestern Atlantic.

Enhanced growth of primary producers responding to the increased nutrient supply initially leads to an alteration of species composition and, secondarily, to toxic or nuisance blooms, behavioral effects, and shading. The rates of autotrophic production then exceed the rate of consumption, resulting in the settling of the excess organic material, the decomposition of which leads to extreme effects such as oxygen depletion, buildup of toxins (e.g., sulfides), smothering, and mortality of benthic and some pelagic species. Ultimately, anaerobic conditions develop in normally oxidized surface sediments and overlying waters, resulting in mass death of many assemblages.

Many such episodes are seasonal, accentuated by peak summer temperatures and mild climate. Present evidence suggests that such eutrophic episodes are increasing in both frequency and severity in many coastal areas.[689-691] As described in Chapters 6 and 7, the effects of eutrophication are best described for the Baltic and Adriatic regions, but other areas (for example, off the coasts of India and Pakistan, Japan, and Korea and in the Caribbean) are becoming increasingly eutrophic. In the coming decades, coastal waters of Asia, Latin America, and Africa will experience such events with greater frequency. In all these areas, most attention has focused on the most obvious symptoms of coastal eutrophication, such as enhanced hypoxia, harmful algal blooms, and fish kills.

Much less obvious is the encroachment into rivers and their watersheds of subtle increases in inputs of nutrients, sediments, and organic matter caused by:

- Deforestation and land-clearing for crops and grazing
- Accelerated use and loss of fertilizers and pesticides
- Increased residential and industrial waste discharge

There are considerable long-term data from the industrialized nations to show that nutrient concentrations and loads are increasing in coastal rivers and estuaries (see Chapters 6 and 7).

A recent analysis[692] of nutrient concentrations and fluxes from 42 of the world's major rivers has established that the mean annual concentrations of both nitrate and nitrate exports to the coastal ocean are positively related to human population size within a river's watershed. These are very strong relationships considering the interplay between anthropogenic, biological, and abiotic factors that influence the behavior of nitrate. Other analyses, including watershed area and rate of river-water flow were not significant, underscoring the overriding impact of human numbers on nutrient flux from land to the coastal ocean.

These results support other evidence that atmospheric deposition, deforestation, and agricultural inputs have significant impacts on the amounts of nitrate exported to the coastal zone. Paerl's analysis[691] of the contribution of atmospheric nitrogen deposition to coastal eutrophication indicates that from 10 to over 50% of coastal nitrogen is derived from wet and dry deposition from the atmosphere. This atmospheric material is an important source of "new" nitrogen to many coastal ecosystems and is especially critical considering that estuarine and coastal primary producers are generally nitrogen limited and thus sensitive to nitrogen enrichment. In the Baltic, for instance, atmospheric deposition contributes 29.3% to total nitrogen loading, compared with 58.5% from terrestrial runoff and 12.2% from nitrogen fixation.[691]

Figure 8.2 shows that most of the atmospheric inputs of nitrogen to the ocean occur in areas in proximity to the heavily industrialized nations in the northern hemisphere. Most of this material is dissolved in rainwater in the form of nitrite and nitrate; there is considerable dry deposition of nitrogen in the form of aerosols and particle-associated NH_3/NH_4. Organic nitrogen deposition can be locally important with most in the form of amino acids and urea, the source of which is unknown. All of these inorganic and organic species can be readily assimilated by phytoplankton and other primary producers in estuarine and coastal waters.

Ecosystem models offer some insight into the contribution of anthropogenic inputs and their fate in watersheds and associated rivers and estuaries. A good example is the model constructed for the Westerchelde estuary in the southwest of The Netherlands by Soetaert and Herman.[693] The nitrogen model (Figure 8.3A) shows that anthropogenic inputs from farm land and industry (30×10^3 metric tons yr^{-1}) to the estuary are intimately intertwined with natural fluxes, approximating the annual loading from the river (36×10^3 metric tons yr^{-1}). Roughly 85% of this combined land-river input is exported to the adjacent coastal sea, with some net organic import. Most of the inorganic nitrogen within the estuary is nitrified, denitrified, or buried (Figure 8.3A). Compared to previous budgets constructed prior to increases in human inputs, the present model indicates that the net effect of eutrophication is a small net increase in denitrification, but a large increase in export of inorganic nitrogen to the coastal zone.[693]

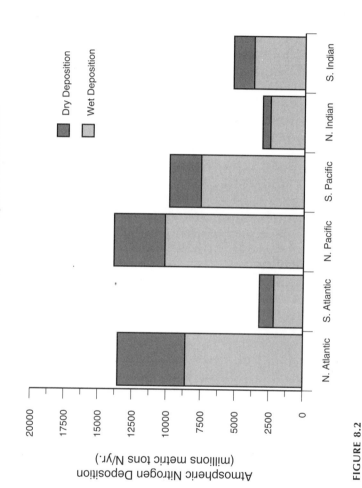

FIGURE 8.2

Annual wet and dry deposition of atmospheric nitrogen to the global ocean classified by northern and southern hemispheres. (From Paerl, H. W., *Ophelia*, 41, 237, 1995. With permission.)

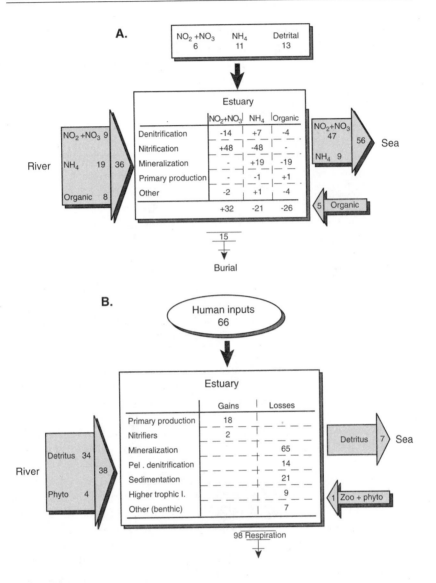

FIGURE 8.3

Nitrogen (A) and carbon (B) budgets of the Westerschelde estuary (1000 metric tons yr⁻¹) indicating net gain (positive values) and net loss (negative values). Left arrow represents inputs from the river to the estuary proper, top arrows depict anthropogenic inputs, and arrows at right represent exchange between estuary and the North Sea. (Adapted from Soetaert and Herman.[693,694])

A companion model of carbon in the Westschelde estuary (Figure 8.3B) indicates that, in contrast to nitrogen, little excess carbon is exported to the coastal zone compared to when the estuary was less polluted. The increases in anthropogenic carbon have resulted in an attendant rise in microheterotrophic activity within the water column and sediments of the estuary to the extent that only refractory carbon is exported. Roughly 64% of the carbon input into the estuary is derived from human sources; *in situ* production is minor. Net carbon loss from the estuary equates to more than two thirds of total carbon input. Most (~68%) of this carbon loss is via microbial mineralization. The differential quality of organic detritus within the land-estuary-sea system results in the estuary importing organic nitrogen and exporting refractory carbon, but acting as a net trap for labile organic matter.

No such models or similarly comprehensive empirical data exist for tropical polluted rivers and estuaries.[695] Indeed, as noted in earlier chapters, few data are extant for tropical estuaries in general. The best-studied tropical estuaries are arguably those in India,[696] where evidence indicates that very productive systems in proximity to large cities such as Bombay have undergone considerable deterioration since industrialization and urbanization (and increases in sewage discharge) began increasing at mid-century.

Most efforts to investigate the impacts of pollution in tropical coastal waters have naturally focused on structural changes in planktonic and benthic communities rather than on ecosystem processes. One recent exception is the study by Ramaiah et al.[696] of the estuary draining into Bombay harbor. They found that microbial activity in the water column was not obviously stimulated by river inputs or organic wastes draining into the estuary but are lower than rates reported for other tropical estuaries. They suggest that the industrial wastes dumped into this system are toxic, inhibiting the growth of most microbial assemblages.

Reviews of the status of tropical rivers in Asia by Lean et al.[697] and Dudgeon[695] suggest that the river systems of this region are among the most polluted on Earth, ascribed directly to human population increases and resultant growth of domestic and industrial waste discharge and agricultural runoff. Untreated sewage is the biggest problem. On average, roughly 70% of the large rivers of Asia are polluted.[695,697,698] Dudgeon[695] identified two other major threats to Asian rivers in addition to organic pollutants:

- Degradation of the watershed/drainage basins, mainly as a result of deforestation and overgrazing by livestock
- Uncoordinated river/drainage basin control and regulation

The first problem leads directly to sediment loss, siltation, and excessive flooding during the monsoon seasons. As for the second problem, dams and other massive projects have been widely constructed for decades, but the scale and number of such projects are increasing

exponentially in pace with human population growth, greater affluence, industrialization, urbanization, and domestication. The clearest
examples of such a problem are the massive projects underway and
planned for the Yangtze and Mekong Rivers. These projects will alter
the natural tide of events, such as monsoonal flooding, to which many
tropical organisms are adapted, including migrations for breeding and
feeding. All three problems — pollution, watershed degradation, and
modification of river flow — have led to severe declines in the number
of aquatic species (reptiles, river dolphins, macrophytes, fishes, invertebrates) and terrestrial animals associated with the watersheds/drain-
age basins.

Tropical habitats, such as seagrass beds and coral reefs, are especially sensitive to eutrophication,[14,683,698,703,707] considering their need for
clear water and low ambient concentrations of dissolved and particulate nutrients. For coral reefs, four effects of eutrophication are now
recognized:[700]

- Weakening of coral skeletons
- Stimulation of overgrowth by benthic micro- and macroalgae smoth-
  ering living corals
- Reduction in water clarity by algal (including phytoplankton) blooms
- Settlement and mass death of primary producers, leading to oxygen
  depletion

Figure 8.4 is a good example of a coral reef severely impacted by
accelerated runoff from mining activities, resulting in a shift in dominance from living corals to macroalgae. Clearly such changes in the
structure of reef food chains translate into alterations of energy and
nutrient flow and exchange with overlying waters and adjacent habitats, not to mention destruction of some of the most beautiful and
delicate ecosystems on Earth. Eutrophication is increasing rapidly in
the tropics, but, as illustrated in the next section, it is not necessarily the
worst problem.

8.3   HABITAT MODIFICATION AND DESTRUCTION

The most blatant form of ecosystem alteration is physical removal and
destruction of habitat. Such destruction is associated with:[680,683]

- Agriculture and industry
- Dredging and channelization or diversion of water
- Filling for solid waste disposal

FIGURE 8.4

The net result of a shift from a coral- to an algal-dominated trophic structure on a coral reef receiving excess nutrients in the South Pacific. (Photograph courtesy of T. Done).

- Roads
- Mariculture
- Commercial and residential development
- Construction of dams, dikes, levees, and sea walls
- Spraying of pesticides and herbicides
- Mining

Often overlooked are natural causes, such as subsidence, drought, storms, erosion, and sea-level variations, but these are outpaced by human activities. Estuarine habitats such as salt marshes, mangrove forests, and seagrass beds are the most afflicted. In the U.S., for instance, it is estimated that more than half of all freshwater and estuarine wetlands have been destroyed since European colonization.[701] Worldwide, the loss figure varies widely among biogeographic provinces, but the global average is thought to approach 30%.[702]

In tropical Asian rivers and others such as the Nile, Yangste, Mekong, and Jordan rivers, the modification of river discharge basins has become one of the most contentious issues of our time. Changes in river hydrology can occur by natural causes, but clearly humans have greatly altered the course and outflow of many of the world's rivers. The best documented cases are the Colorado and Nile Rivers, where sediment discharge has dropped from 135 to 0.1 metric tons yr^{-1} and from 150 metric tons yr^{-1} to nearly zero after damming.[681] Efforts are now being made to restore the natural hydrology of these and other rivers, but it is not known if the damage is irreversible. GESAMP[681] estimates that, since the 1950s, the discharge of many rivers has declined because of various damming projects: a 20% decline in river runoff in Africa and North America, 15% in Europe, 14% in Asia, 4% in Australia, and 6% in South America.

Reduced river discharge has resulted in a reduction in groundwater flow with resulting encroachment of salt into drainage water inland as well as increases in subsidence. Such problems have been recorded in the southeastern U.S., Australia, Europe, and Asia.[680,681] Intrusion of salt water has even affected rain forests and has caused die-back of mangrove forests in the Indus and Niger deltas. Clearly, the manipulation of watersheds and river hydrology have resulted in severe impacts on the structure and function of coastal ecosystems, with attendant loss of fishery yields and coastal habitats.

Much attention has been paid to habitat losses in temperate regions, but the pace of destruction has accelerated rapidly in the developing world. Hatcher et al.[14] argue that tropical habitats such as mangrove forests and coral reefs are more fragile and less resistant and resilient to disturbance than are many temperate ecosystems. If true, this presages severe restoration problems (see Section 8.4). Coral reefs and seagrass beds are threatened mostly by:[14]

- Sewage
- Industrial wastes
- Agricultural chemicals
- Thermal effluents
- Turbidity
- Sedimentation
- Dredging
- Construction
- Mining
- Blasting, trampling, and anchoring by fishermen and tourists
- Trawling
- Removal of organisms and living and dead coral structures

Wilkinson[703] estimates that nearly 10% of the world's coral reefs are beyond recovery, and an additional 30% of reefs are in danger of collapse within the next decade. The reefs of Southeast Asia, East Africa, and the Caribbean are at greatest risk.

The best-documented degradation of coral reefs is that found in the Caribbean. Through a combination of natural and human disturbances, many coral reefs in this region have seriously degraded over the past 20 to 30 years. In northern Jamaica, most corals have been destroyed by disease, overfishing, and hurricanes, to the extent that barely 5% still exist today. Hughes[704] has documented a shift in community structure of the Jamaican reefs from the coral dominants, *Acropora palmata, A. cervicornis, Montastrea annularis,* and *Agaricia agaricites,* to reefs dominated by fleshy macroalgae, including species of *Sargassum, Lobophora, Dictyota,* and *Halimeda.* This phase-shift was preceded by hurricane damage, overfishing, and disease problems that resulted in mass mortality of the echinoid, *Diadema antillarum,* an important grazer controlling benthic algal communities. The decline of *Diadema* and herbivorous fish from over-trapping was followed by a long and extensive benthic algal bloom which began in 1983 and continues today. This scenario provides a clear warning of how rapidly coral reef communities can undergo change by a variety of disturbances and how interdependent various communities are to one another in maintaining stability on a coral reef. As pointed out by Hughes[704] management practices must focus on restoring control of benthic algae by herbivores, most easily done by regulating overfishing.

TABLE 8.4

Status of Coral Reefs in Five of the ASEAN Countries as
Percentage of Live Coral Cover

Country	Status			
	Excellent	Good	Fair	Poor
Thailand	2.6	24.2	31.6	41.6
Malaysia	11.4	52.8	27.5	8.3
Philippines	1.3	7.5	49.2	42.0
Singapore	2.8	9.2	20.4	67.6
Thailand	16.9	42.1	34.8	6.2

Source: Data adapted from Chou, L. M. et al., in *Living Coastal
Resources of Southeast Asia: Status and Management,* Wilkinson,
C. R., Ed., Australian Institute for Marine Science, Townsville,
1994, 8.

In Southeast Asia, the ASEAN*-Australia Living Coastal Resources
project has been monitoring coral reefs, mangroves, and seagrass beds
for the past decade and has documented the degradation of these
ecosystems. Chou et al.[705] have summarized the extent of degradation
and the loss of reef productivity (Table 8.4). At present, only 1 to 17%
of coral reefs in Southeast Asia can be described as being in excellent
condition, with most categorized in fair to poor condition in Indonesia,
the Philippines, and Singapore. Coral reefs in all ASEAN countries
have undergone significant degradation and loss of productivity over
the past 50 years (Table 8.4). The major causes are sewage, wastes from
agriculture and industry, overfishing, and sedimentation.[706] Natural
disturbances do occur, but reefs can recover after several years only if
not under severe stress from human activities.

The degradation of mangrove forests and seagrass beds, although
more severe, has until recently received less attention than the decline
of coral reefs. In Southeast Asia, more than 50% of mangrove forests
have been cleared, mostly for timber and mariculture;[707] nearly 20 to
60% of seagrass beds have been lost to pollution, overfishing, and
industrial development.[708] Nearly 43% of the world's 17 million hectares of mangrove forests are found in Southeast Asia, where losses
have been greatest in the Philippines and in Thailand.

The greatest threat to mangroves worldwide is conversion of forest areas to finfish and shrimp ponds. In the Philippines, roughly 80%
of mangrove forests have been lost to shifting aquaculture, primarily
the cultivation of milkfish.[707] In Thailand, similar conversion for aquaculture has resulted in losses of 50 to 80%.

---

* Association of South-East Asian Nations includes Brunei Darussalam, Indonesia,
Malaysia, Phillippines, Singapore, Thailand, and Vietnam.

Another threat is clear-felling for woodchips for the rayon industry in Japan. Losses from this activity are entirely unnecessary because mangrove forests can be sustainably harvested. Many mangrove forests have been lost by a combination of human activities. For instance, the mangrove forests of Ecuador have been extensively damaged or lost mainly for construction of shrimp ponds and, to a lesser extent, for logging for charcoal.[709] Estimates of loss vary by province from 6 to 35%, with a national average of about 14% as of 1987. Some of these losses may have been natural.

The third major threat is chronic pollution by oils, heavy metals, herbicides and pesticides, nutrients, and pathogens. This latter threat has remained poorly documented, despite a few comprehensive studies in Australia and Panama.[14]

Many threats to living resources are difficult to estimate, as a large proportion of the tropical coastlines remain unstudied. This is best understood for the soft-bottom resources of Southeast Asia which have been wholly undersampled until 1985 because of a lack of appreciation, funding, and training. Even then, most benthic studies have focused on community structure rather than on function. As noted in earlier chapters, the seabeds of these tropical areas receive most of the freshwater and sediment (and pollutants) draining from massive and small rivers. Trawling is also destroying most of the soft-bottom biocoenoses before they can even be evaluated for their diversity and levels of fishery stocks. This is unfortunate as benthic communities can also be used as indicators of other forms of environmental change.

Other coastal habitats, such as salt marshes, have undergone significant alteration or loss worldwide. Many such problems have come about as a result of nearsighted management decisions. Other than gross pollution, such difficulties are perhaps the greatest threat to many temperate marshes and wetlands. For instance, on the southeast coast of the U.S., roughly 11% of coastal marshes are diked and flooded to control mosquitoes and sandflies and to attract wintering waterfowl. These impoundments are successful but are of great concern as they significantly alter marsh water level, decrease salinity, increase nutrient retention, decrease nutrient and material export, and result in alteration of species composition and numbers of inhabitants. Montague et al.[710] reviewed the problems of marsh impoundments and concluded that the largest concern may be reduced access by estuarine fish, shellfish, and other organisms that may be able to take advantage of the available food and high rates of algal production and standing stocks in the impoundments. Many of the above problems can be ameliorated by periodic opening, especially during episodes when the impounded water shows signs of stagnation. It is possible to ameliorate the effects of many human alterations to coastal habitats, but data are lacking in this regard and with respect to restoring degraded habitats. Progress, however, is slowly being made in some localities.

8.4 RESTORATION ATTEMPTS: PROBLEMS AND PROGRESS

Concerted attempts to restore damaged coastal ecosystems to their natural state have been ongoing since pollution became a major social and political issue in the 1960s. Some estuaries, such as the Thames, the Hudson, and the Rhine, are considerably cleaner now than when pollutant discharge into these estuaries peaked in the late 1960s to early 1970s. It is likely that some ecosystems can revive naturally given sufficient time, but recovery is greatly dependent upon both the type and toxicity of the pollutant(s) as well as the frequency and areal extent of discharge. Some ecosystems may be somewhat more resilient and resistant than others. Mangroves and salt marshes, for instance, can tolerate considerably greater inputs of nutrients than coral reefs or seagrass beds.[14,711] This does not mean that they are unaffected by pollutants, but are merely more resistant.

Efforts to restore mangrove forests and seagrass beds have been ongoing for the past two decades. Failures have been common, but notable successful efforts have been made in the Caribbean and, more recently, in Southeast Asia.[712] Fool-proof criteria do not exist. Good preparation, a careful choice of a resilient species, and adequate anchoring and planting method techniques are crucial for success.

8.4.1 Seagrass Beds

For restoring seagrasses, the most successful methods involve:

- Use of transplanting entire 10- to 20-cm wide plugs of sediment, roots, rhizomes, and blades from a natural site to the damaged bed
- Hand-planting of seeds or seedlings
- Transplanting turfs, sods, and grids that are shallow plugs from a natural bed and do not necessarily include the entire root rhizome system

Thorhaug[712] reviewed these efforts and concluded that of more than 165 attempts at restoration, 75 were deemed successful, giving a success rate of about 45%. Most replanting efforts have been made in the U.S., mostly in southeast Florida and the Gulf of Mexico and, to a lesser extent, in North and South Carolina and along the west coast. Other efforts have been successfully made in Europe, Central and South America, the Philippines, and Australia.

The plugging technique has been the most successful, probably because the plant and root rhizome system are least disturbed during transplanting. In the tropics, the planting of seeds and seedlings appears to be the most common method, but less successful than planting whole plugs. Seeds and seedlings are more susceptible to destruction by wave action, cyclones, and other disturbances that resuspend the sediment. The success rate for seeds and seedlings is slightly greater than 50%.

The success rate also appears to be low for attempts to plant shoots, but some species survive better than others. Thorhaug et al.[713] found that *Thalassia* exhibited better survival than *Halodule* in exposed subtropical and tropical beds, but *Syringodium* was the least successful transplant species. Obviously, species differ greatly in nutrient and light requirements and resilience to disturbance. Better understanding to these differences would enhance the success rate of restoration efforts.

Success in transplanting seagrasses is only the first step on the road toward full recovery. The return of animals to the beds is crucial, because the role of these systems as nursery grounds for invertebrates, reptiles, mammals, fish and other commercially important species is most often cited to justify restoration efforts. To date, only a handful of studies have documented the rate of recovery of the animal communities in restored seagrass beds.[714,715] The results from these efforts suggest that threshold shoot densities are required to sustain recovery, which may take several years; even partial recovery of the diversity of species is better than none at all.

8.4.2 Mangrove Forests

Mangroves appear to be more amenable to transplantation and restoration than most other halophytes, with several hardy species exhibiting wide salinity and temperature tolerances.[716] Mangroves can be restored by transplanting propagules, young trees, or more mature trees. The least expensive method is to plant propagules, because growing mangroves to the young- or mature-tree stage requires a nursery, nutrients, seawater system, personnel, money, and time. The success rate for propagules improves if seeds that have been nursed for up to a year are transplanted, rather than seeds being picked off a tree and planted immediately.

Trees grown to no more than one meter in height are most tolerant to transplantation, because mortality from disease and disturbance lessens as mangroves survive past the young-shoot stage. Trees at this stage of development are at their fastest rate of growth and are capable to laying down roots to stabilize the substrate. Transplanting young trees is more costly, but the survival rates are much higher than planting propagules.

Restoration of mangrove forests has been practiced for many years in India, Malaysia, Thailand, Bangladesh, and Burma, where successful and profitable forestry practices have been established. Silvicultural techniques are well known for mangroves. The most common result of such practices is the development of successful monocultures, most often species of *Rhizophora*. Perhaps the most extensive and best-documented example of mangrove restoration is the afforestation program in Bangladesh, which began in response to the damage caused by a particularly destructive cyclone in 1970.[717] The program began earlier in 1966 but gained speed after the cyclone. To date, more than 120,000 hectares have been planted. The dominant species have been *Sonneratia apetala* and *Avicennia officinalis*. Many efforts have been by trial and error, but, on the whole, the program has been successful. These restored forests now offer better protection, jobs, food, and a range of products for the inhabitants of the extensive, low-lying coastal deltas of Bangladesh.

Another major effort to restore mangroves has been taking place in Indonesia, where the conversation of mangrove forests to *tambak* (brackish water mariculture ponds) has resulted in extensive loss of forests, particularly on the north and east coasts of Java.[716] Forests have also been unsustainably exploited in Sumatra, Kalimantan, and Irian Jaya for wood chips. The losses have been so extensive that fish yields have declined precipitously and coastal erosion is a major problem on most islands.[718] Since 1964, nearly 40,000 hectares of mangrove have been replanted, mostly with *Rhizophora*, *Avicennia*, and *Sonneratia* species. Most of the rehabilitation has occurred in areas of abandoned fish ponds and in logged and degraded intertidal land.

Vietnam has also been the site of extensive restoration of mangrove forests, lost by defoliants used during the war and, more recently, by logging and mariculture enterprises. Minh Hai province in southern Vietnam has been the site of most of the country's restoration efforts, where over 36,000 hectares have been replanted. As in Indonesia, *Rhizophora*, *Avicennia*, and *Sonneratia* have been the most widely replanted species, but there have been severe problems such as slow growth, rotting of propagules, infestations, and attacks by wildlife. The greatest threat continues to be the conversion of forests for shrimp ponds. While this practice is now regulated to some extent, recent evidence from many mixed shrimp pond-mangrove forestry enterprises in the Mekong delta region (Figure 8.5) indicate that yields of

FIGURE 8.5

A typical mixed shrimp pond-mangrove (*Rhizophora apiculata*) enterprise in the Minh Hai province in the Mekong Delta, Vietnam. Trees are approximately 6 years old. (Photograph by author.)

both shrimp and mangrove timber are declining. The reasons are not known but may include:

- Low pH and oxygen concentrations in tidal waters
- Poor hydrology in the ponds
- Residual herbicides
- Lack of sufficient mangrove area to sustain seed populations of shrimp postlarvae

Vast expanses of replanted forests are monocultures, suggesting that the lack of biodiversity may have some role in the poor sustainability of these plantations.

8.4.3 Salt Marshes

Similar restoration efforts have been attempted for other estuarine wetlands, particularly salt marshes in the U.S. These restoration measures have been accompanied by extensive analyses on the effects of tidal elevation, nutrient additions, hydrology, and geomorphology — and their synergistic/antagonistic effects — on transplanted marshes.[719-721] For transplanted marshes of southern California, Gibson et al.[719] hypothesized that the addition of nutrients to newly constructed marshes inhabiting organic-poor soils would stimulate plant growth and production. They found that less than 5% of nitrogen added to the soil was recovered in aerial plant tissue. The bulk of the nitrogen was lost via microbial decomposition and leaching through the sandy soil. After 2 years, continued additions of organic matter and nitrogen had no effect on nutrient pools or plant growth, both of which remained poor compared to adjacent natural marshes.

Another study[720] suggests that one factor can either greatly enhance or impede the effect of another factor. In transplanted salt marshes of Louisiana, Wilsey et al.[720] observed that the effects of macro- and micronutrient additions were enhanced in marshes transplanted to elevated areas. Growth and biomass of transplanted *Spartina alterniflora* were more than twice that of *Spartina* plants in nonelevated plots, suggesting that adding nutrients alone is not sufficient for success of transplanted marsh grasses.

Constructed wetlands may take many years to approximate ecological conditions of natural marshes or may never do so, depending upon whether or not best management practices are used to replicate natural wetland conditions. A good example is the tidal marshes of coastal Virginia where even after 7 years many ecological conditions in constructed marshes do not approximate natural conditions.[721] Havens et al.[721] recommend that more effort should have been made to replicate several conditions better, including more use of natural soil, higher stem densities, planting of more mature scrubs, mixing habitats and marsh microtopography, and more complex routing of drainage ditches to enhance freshwater outflow and nutrient retention and to minimize soil erosion.

8.5 SUSTAINABILITY: IMPLICATIONS FOR MANAGEMENT

It is clear that the rate of exploitation by humans of many natural coastal ecosystems for food, fuel, and other items is unsustainable. Sustainable exploitation of renewable resources depends on managing the resources to ensure that the rate of extraction does not exceed the rate of replacement — in the case of living resources, the rate of reproduction. Moreover, the rate of reproductive output must be maintained and preferably exceeded to result in a net reproductive surplus.[722] Sustainable yields can be ascertained in a variety of ways, but none are foolproof or necessarily accurate. Continual change is the rule for living resources, not constancy. This makes it difficult to estimate whether or not renewable resources are being sustainably exploited. Hilborn et al.[722] note that the only certain way to determine maximum sustainable yield of some large-scale resources is to first overexploit. This has been the case for various coastal fisheries.

The Matang Mangrove Forest Reserve (40,711 hectares), located on the western coast of peninsular Malaysia, is a prime example of a coastal ecosystem that has been well managed as a forestry reserve on a sustainable basis since 1902,[723] but it now shows some signs of declining yield.[724] Up to the present day, minimal management efforts have been the apparent key to successful sustainability. Forest regeneration (mostly *Rhizophora apiculata*) is largely natural on the reserve, the number of forestry personnel is very small, and there are minimal regulations. Forest blocks of 1000 hectares are harvested annually for charcoal production and poles, with a 30-year rotation cycle. Plots are thinned every 15 years and all of the remaining leaves, branches, and roots are retained within the plots to minimize erosion and maximize nutrient retention.

As seen in the photograph (Figure 8.6), there is still extensive loss of biomass and disturbance to sediments in the clear-felled plots even 2 years after harvesting. Growth and regeneration are rapid, but the loss of tree diversity and nutrients stored in the extracted timber may now be taking its toll on ecosystem sustainability. There may be other causes for the declining timber

FIGURE 8.6
The remains of a *Rhizophora apiculata* forest 2 years after clear-felling in the Matang Mangrove Forest Reserve in northwestern peninsular Malaysia. (Photograph by author.)

yields, such as too-frequent disturbance of the biogeochemical cycles, but these factors are only now being examined. The humid climate in Malaysia is conducive to rapid rates of *R. apiculata* growth and regeneration, but clearly there are temporal and spatial limits to sustainable exploitation. Moreover, the uniqueness of the growth conditions which have allowed these forests to be exploited for nearly a century implies that the management scheme used in the Matang Forest Reserve is not necessarily applicable for sustainable development of other mangrove forests. Unfortunately, not enough is known about energy flow and nutrient cycles in exploited and unexploited mangrove forests to make definitive rules for successful management of such systems.

The ecosystem approach, however, has long been used to estimate potential yields. Analyses of energy and materials flow often set obvious upper limits on production of harvestable species, and many ecosystem simulation models (e.g., ECOPATH) have been used for this purpose. Such trophic analyses are only as good as the empirical data (and assumptions) on which they are based; consider, for example, the lack of reproductive and life-history data for many mammal, fish, and plant species. There is a particular lack of life history information for many commercially harvested species.[722]

Further, many assumptions valid for a temperate coastal zone may be invalid for a coral reef or a mangrove forest. The modeling approach for estimating sustainability also does not adequately account for the temporal and spatial complexity inherent in every food web and ecosystem. So, in effect, models are good heuristic tools that play an important role in helping to identify gaps in our knowledge rather than as a predictive tool.

Presently, there are two schools of thought about the best direction to take in researching management of living resources.[722] One school suggests that the best way to tackle the problem of providing research relevant for management is to perform intense, detailed research on the basic ecological conditions of the systems to provide a better level of understanding and better management. The other school states that traditional research does not provide the relevant information necessary for better management and that research should be targeted to specific problems and to better monitoring. In truth, there is no universal best solution for all cases. This does not mean that scientists and organizations have not tried to push optimal solutions — often with predictable disasters.

Methods based on experimentation and observation are still the most reliable to estimate potential yields, and the ecosystem energetics approach can play a useful role in some instances. Good examples of how tracing the flow of energy and nutrients can be an important environmental management tool are the studies currently in progress to assess the sustainability of semi-intensive shrimp ponds on the Caribbean coast of Colombia. Larsson and his colleagues[725] have been estimating the input of natural and manmade resources required from the adjacent coastal ecosystems to sustain shrimp farms. To date, they estimate that a semi-intensive shrimp farm requires enormous support from the adjacent ecosystem — resources from an area equivalent to 35 to 190 times the size of an average farm. The energy subsidy is on the order of 295 to 1 for each energy unit of shrimp protein produced. More than 80% of coastal primary production is required to sustain the operation. They estimate that about 20 to 50% of the country's total mangrove area is required to supply sufficient postlarvae to the farms. These calculations suggest that such semi-intensive farms are ecologically unsustainable.[725]

One of the keys to successful management of sustainable exploitation is controlling the exploiters in such a way as to maximize yield.[726] Another key problem is focusing research to address questions directly related to sustainability. For instance, in assessing how to minimize the impact of shrimp pond effluents on receiving mangrove forests and to sustain yields of both shrimp and timber in Southeast Asia, Robertson and Philips[727] indicated that most mangrove research to date does not help in assessing the ability of mangroves to tolerate nutrients from these ponds. They suggested the following research requirements:

- What is the assimilative capacity of mangrove trees and sediments for nitrogen and phosphorus?
- What are the tolerance limits of mangrove species to particulate and dissolved organic matter?
- How much nutrient loading results in complete anaerobiosis of mangrove sediments, tidal water, and tree mortality?
- What is the role of sediment type on the assimilative capacity to sorb excess nutrients?
- What silviculture practices will maximize forest yields and rates of nutrient uptake?
- What are the indirect effects on food webs?
- What combinations of hydrological regimes and effluent loads can minimize impacts on mangroves and maximize shrimp and timber yields?

From this example, it is apparent that such practical questions are also of academic interest, and it is likely that this is the case for sustainability and management issues for other coastal ecosystems.

8.6   CONSERVATION: TOOLS AND IMPEDIMENTS

Attempts to conserve marine resources and habitats have been ongoing for many decades in many countries, but not until recently have there been concerted international efforts to do so. A significant breakthrough was achieved in July 1992 with the adoption by most maritime nations of coastal conservation recommendations at the United Nations Conference on Environment and Development (UNCED). The declaration states that nations will and must:

- Prevent, reduce, and control marine coastal degradation
- Develop ways to enhance the potential for sustained use of living marine resources to meet nutritional, social, economic, and development needs and goals
- Promote the integrated management and sustainable development of coastal regions of the global ocean

Related efforts have been made by the Intergovernmental Oceanographic Commission (IOC) of the United Nations Educational, Scientific, and Cultural Organization (UNESCO), fostering greater efforts for coastal countries to develop monitoring and remedial programs to improve degraded habitats. Additional United Nations efforts include the Food and Agricultural Organization (FAO) and UNESCO's Man and the Biosphere and Coastal Marine (COMAR) programs. Other organizations that make considerable coastal conservation efforts include the World Bank, World Wildlife Fund, International Whaling Commission, and the World Conservation Union (IUCN). Many international laws and treaties limit coastal exploitation: The Law of The Sea Treaty, The MARPOL Convention, the London Dumping Convention, and various Arctic, Antarctic, and fisheries agreements. Regional agreements, such as in the ASEAN region, have been ongoing for the past decade.

Perhaps of greatest consequence has been an emphasis on a holistic, integrated management approach rather than on piecemeal, habitat-by-habitat attempts to come to grips with the problem. This reflects frustration with the failure of many small-scale efforts in the past, as well as recent recognition that the coastal ocean is an integrated system with no one habitat ever being completely isolated from another.

One outgrowth of this idea is the concept of large marine ecosystems (LMEs). Several international organizations have begun to support large ecosystem-scale programs focusing on resource sustainability, economically viable development, and ecosystem "health". Sherman[728] lists 29 large marine ecosystems and subsystems, within which the major research efforts are

- Fish spawning strategies
- Productivity and trophic studies
- Analyses of stock fluctuations
- Recruitment and mortality
- Natural variability
- Human perturbations

The strategy is that these efforts are funneled into four management streams:

- Bioenvironmental and socioeconomic modeling
- Optimization of sustainable fisheries yields
- Amelioration of human-induced stress
- Improvement of ecosystem "health"

It is too early to predict the success of this strategy, although initial efforts in developed countries are encouraging.

The oldest, yet still most reliable, conservation method is the concept of protected areas, in which specified tracts of land or coastal margin are protected from various forms of human exploitation.[729] Two current programs that are based on the protected areas concept and are being applied to coastal marine areas are the Biosphere Reserve Concept within the Man and the Biosphere Program and the IUCN program promoting marine protected areas (MPAs).

The Biosphere Reserve Concept consists of designating specific areas as reserve territory composed of:

- Inner core area in which all marine life is strictly protected
- Buffer zone in which any human interference is strictly controlled
- Outer transition area where cooperation with the neighboring populace is sought in developing sustainable management practices

Both terrestrial and coastal reserve efforts have had severe problems, many because earlier guidelines were unrealistic and vague. More recent attempts[729] to clarify their role and list priorities may improve additional use of the biosphere reserve concept.

The long-term vision of the IUCN program, to establish and provide for a global system of marine protected areas, has met with greater organizational success due to the coordinated efforts of IUCN (particularly its Commission on National Parks and Protected Areas), the World Bank, the World Conservation Monitoring Centre, and the Great Barrier Reef Marine Park Authority (GBRMPA). The first step

TABLE 8.5

Inventory of Subtidal Marine Protected Areas
(MPAs) Classified under Marine Regions, as
Determined by Working Groups Sponsored by
the IUCN, GBRMPA, and the World Bank

Marine Region	Number of MPAs	Percentage of Total
Antarctic	17	1.3
Arctic	16	1.2
Mediterranean	53	4.0
Northwest Atlantic	89	6.8
Northeast Atlantic	41	3.1
Baltic	43	3.2
Caribbean	104	7.9
West Africa	42	3.2
South Atlantic	19	1.4
Central Indian Ocean	15	1.1
Arabian Sea	19	1.4
East Africa	54	4.1
East Asian Seas	92	7.0
South Pacific	66	5.0
Northeast Pacific	168	12.8
Northwest Pacific	190	14.5
Southeast Pacific	18	1.3
Australia/New Zealand	260	19.9
Total	**1306**	

Source: Adapted from Kelleher, G. et al., *A Global
Representative System of Marine Protected Areas,* Vol. 1,
GBRMPA, World Bank, and IUCN, Washington, D.C.,
1995.

was to organize the world's coastal areas into 18 biogeographical
regions (Table 8.5), to develop an inventory of the status of existing
MPAs in each of these regions and setting management priorities for
zones without MPAs. The global survey[730] identified a total of 1306
MPAs (Table 8.5). Most marine regions have few MPAs — more than
55% are concentrated in the Caribbean, Northeast Pacific, Northwest
Pacific, and Australia/New Zealand. The median size of existing MPAs
is 1584 hectares, the largest being the Great Barrier Reef Marine Park
(34.4 million hectares), the Galapagos Islands Marine Resources Reserve and Whale Sanctuary (8 million hectares), and the Milieuzone
Noordzee reserve (2 million hectares) in The Netherlands. Tropical
coastal areas generally have the fewest MPAs. The management effec-
tiveness of about 70% of these MPAs is unknown, with equal numbers
of the remaining MPAs categorized at high, moderate, and low levels
of management. 155 MPAs have been identified as being of high

conservation priority. Hope of developing and enforcing current and new MPAs depends upon funding, greater application of science to management, garnering community support, training people to manage properly, and striking a balance among implementation, monitoring, and review. At present, this concept offers the greatest hope of securing adequate conservation of coastal ocean resources.

The areas of greatest concern are in the developing world where marine reserves appear as an economic luxury and are subject to intense over-exploitation. Polunin[731] argues that different, more pragmatic approaches are necessary concerning conflicts between modern and traditional uses, over-fishing, and human over-population. This may be ameliorated in the short term by modeling the whole ecosystem to establish upper and lower limits of exploitation.[731]

Insufficient information is not the only impediment to marine conservation. Other impediments include:[732]

- Cultural habits
- Economic valuation and differing costs to benefits of harvesting resources
- Lack of funds
- Lack of political interest
- Legal gaps and overlaps of jurisdiction
- Territorial sovereignty
- Gap between rich and poor nations
- Lack of coordination among international and national agencies
- Poor communication between scientists and managers
- Military use and restrictions

The supposed incompatibility between economic and ecological theory is surmountable, if efforts are made to make environmental exploiters understand that the depletion of one resource cascades upon a number of other exploitable resources — to their ultimate detriment. Some recent economic pragmatists have focused on multiple-use aspects of exploitation, based on the ecological reality that resources between and even within habitats are interlinked and that wise use and selective exploitation of multiple resources make good economic sense.

An example of a combined economic/ecological strategy promoting conservation is the mangrove woodchip industry of the Bintuni Bay area of Indonesia. A cost-benefit analysis by Ruitenbeck[733] incorporating linkages among mangrove clearing, fish production offshore, traditional uses, and the benefits of retaining sufficient mangroves to control erosion and sustain tree growth and replenishment, indicates that the optimal strategy is to cut selectively 25% of the available mangroves — $35 million (U.S. dollars) greater profit value than clear cutting. Further cutting would not be cost effective owing to the need to outlay additional money for manual replanting, for relocating to other mangrove sites, and losses in fisheries yield and crops such as sago. Many conservationists decry such economic rationalization, but considering current rates of habitat loss, the melding of economic and ecological strategies may, in the end, be the only practical means of conserving some precious coastal resources.

8.7 GLOBAL CLIMATE CHANGE: COASTAL IMPLICATIONS

The continual growth of industrial and transport emissions into Earth's atmosphere is considered to be the greatest threat to the survival of terrestrial and marine ecosystems. Model predictions vary as to the actual amounts of greenhouse gases that will interact with the global ocean (see previous chapter), but all agree that the net exchange has already reversed from the ocean being a source to a sink, for CO_2. Many consequences as a result of climate change are predicted for the global ocean:

- Increased ultraviolet radiation and water temperatures
- Altered ocean circulation and upwelling due to increased vertical stability
- Weakening poleward and seasonal alteration of storms, surface currents, and precipitation
- Decrease in polar and glacier ice
- Rising sea level

All of these changes will have attendant impacts on life on Earth.[732,734-739] Some scientists have warned that these predictions are highly uncertain, given the enormous natural variation of Earth's climate and our limited ability to distinguish natural from human-induced variability.[734,736] These uncertainties are reflected in some conflicting predictions.[737,738]

A detailed analysis of global climate change is beyond the scope of this book, but some recent reports[739,740] provide some insights on how some specific ecosystems, such as mangrove forests and coral reefs, may be impacted. Table 8.6 summarizes the major predicted impacts of global climate change on these tropical ecosystems. We can presume

TABLE 8.6

Predicted Responses of Mangrove and Coral Reef Ecosystems to Global
Climate Change

Rise in Mean Sea Level (Prediction: 6.0-cm Rise per Decade)

Mangroves
    Landward progression of mangroves; rate determined by rate of sea-level rise
    Erosion at seaward front
    Altered zonation patterns
Coral reefs
    Will not threaten most coral reefs; advantageous for some, permitting
      vertical growth
    Will devastate habitability of low coral islands and corals of other
      low-lying areas

Increase in Atmospheric CO_2 Concentration (Prediction: 0.5% Rise per Year)

Mangroves
    Enhanced water
    May not increase canopy photosynthesis, growth, and litterfall
    Increased rate of vertical accretion
    Species responses will vary
Coral reefs
    Lower carbonate production due to surface water changes in alkalinity and ΣCO_2
    Increased algal growth and increased competition with corals
    Enhance the effect of nutrient pollution and over-fishing

Increase in Atmospheric Temperature (Prediction: 0.3 °C per Decade)

Mangroves
    Expansion of mangroves into higher latitudes
    Increase in mangrove photosynthesis and respiration, litterfall
    Increase in rates of microbial decomposition
    Increase in plant and animal biodiversity
    Change in plant and animal densities, community composition, growth,
      and reproduction
    Reduced rate of vertical accretion
Coral reefs
    Will not threaten coral survival
    Decreased coral cover
    Increased bleaching of corals and other calcifiers
    Temperature extremes will endanger corals by increasing vulnerability
      to other stresses

Changes in Precipitation

Mangroves
    Decreased soil salinity and increased water content with increased rainfall
      will increase plant growth and vice versa
    Changes in animal composition and zonation dependent on salinity tolerances
Coral reefs
    Decreased coral cover on coastal reefs
    Increased vulnerability to other stresses
    Increased rate of sedimentation, nutrients, and other contaminants onto reefs

TABLE 8.6 (continued)

Predicted Responses of Mangrove and Coral Reef Ecosystems to Global
Climate Change

Climate Change Impacts on Human Use and Exploitation

Mangroves
    Increased erosion
    Increased flooding in low-lying areas
    Increased salt intrusion
    Increased damage from storms
Coral reefs
    Increased localized damage
    Endangered species pressed to extinction (e.g., sea turtles)
    Shift in current patterns impeding reproduction and larval colonization
    Damaged ozone layer disrupting recruitment

Source: Based on Field,[716] Norse,[732] Pernetta et al.,[734] Bardach,[735] UNEP,[739] and Wilkinson and Buddemeier.[740]

that other coastal and oceanic ecosystems may be similarly affected to some extent.

The predicted rise in sea level will not threaten most mangrove forests and coral reefs and may even be beneficial for some systems, but for other systems a sea-level rise may be devastating. For instance, coral reefs fringing low islands or coastal reefs would be particularly sensitive to sea-level changes. It is, therefore, difficult to predict the impacts of sea-level rise given the diversity of physical settings of many coastal systems.

The impacts of predicted increases in atmospheric CO_2 concentration and temperature are somewhat easier to envision, as some empirical data exist on the effects of altering temperatures and carbon dioxide concentrations on coral and mangroves. Increases in CO_2 concentration are not expected to greatly affect mangrove growth or photosynthesis, but species responses may vary; a temperature rise, however, would result in an increase in mangrove gross and net primary production and the productivity and respiration of most mangrove-associated organisms, including microbes.[739] What long-term impact such an accelerated pace of ecosystem functioning will have is unknown. Coral reef ecosystems may experience a dominance shift from corals to macroalgae, similar to the impacts of excess nutrients and thermal pollution.[740] Bleaching of photosynthesizing organisms is a likely impact with a decrease in calcium carbonate production due to changes in surface water chemistry.

Changes in the precipitation patterns and frequency and intensity of storms will directly affect soil salinity and plant water-use efficiency in mangroves. Generally, increases in precipitation will be beneficial, whereas drier conditions will be deleterious. The opposite is true for coastal reefs, as increased rainfall would result in greater sedimentation, nutrient, and contaminant fluxes from land threatening coral survival.

Changes in land use, utilization, and exploitation by humans as a response to global climate change would also greatly impact coastal ecosystems, increasing local and regional damage, further endangering fragile species and their reproductive efforts, and increasing flooding and erosion in low-lying areas.

Management and conservation practices will have to be altered as our climate changes. These new practices must include:

- Increased afforestation of wetlands in suitable areas
- Construction of protective barriers
- Relocation of people and agricultural and maricultural activities
- Close monitoring for changes in coastal systems using satellite technology (e.g., digital geographical information systems)

Given the current endangered status of many of the world's coastal ecosystems, an important prelude to ameliorating possible impacts of global climate change would be to minimize the current pace of destruction and anthropogenic inputs into the coastal ocean.

8.8 A FINAL REMARK

Coastal ecosystems are not divorced from processes occurring on land or in the atmosphere; changes to terrestrial or atmospheric processes can greatly affect the structure and function of coastal food webs. The food webs that support life in the coastal ocean are productive and support us in a variety of ways, but they are delicate mosaics, intricately and beautifully adapted to the physics, chemistry, and geology of the sea.

Life in the global ocean is more reactive and thus arguably more vulnerable than life on land. Marine primary producers and supporting food webs are smaller, with a higher rate of turnover. Marine autotrophs are also more readily decomposable than their terrestrial counterparts, who must expend significantly greater amounts of energy to construct and maintain support structures. The viscosity and buoyancy of salt water permit a variety of lifestyles, rapid motion, and exchange of energy and materials among and between pelagic and benthic food chains; the pace of life and death in the coastal ocean is correspondingly quick.

The coastal ocean is more variable than the open sea and therefore subject to more frequent disturbances by humans and nature. At present, the production and consumption of organic matter in the coastal ocean are in rough balance, but this may change, given the present rates of carbon input from anthropogenic sources and alteration of natural inputs. We must do more to protect and preserve our coastal life, for having already done so much to degrade our ecosystems, we can do no less.

REFERENCES

1. Pomeroy, L. R., The ocean's food web, a changing paradigm, *Bioscience*, 24, 499, 1974.
2. Azam, F., Fenchel, T., Field, J. G., Gray, J. S., Meyer-Reil, L. A., and Thingstad, F., The ecological role of water-column microbes in the sea, *Mar. Ecol. Prog. Ser.*, 10, 257, 1983.
3. Wiegert, R. G., The past, present, and future of ecological energetics, in *Concepts of Ecosystem Ecology: A Comparative View*, Pomeroy, L. R. and Alberts, J. J., Eds., Springer-Verlag, New York, 1988, chap. 3.
4. Henrichs, S. M., Early diagenesis of organic matter: the dynamics (rates) of cycling of organic compounds, in *Organic Geochemistry*, Engel, M. H. and Macko, S. A., Plenum Press, New York, 1993, chap. 4.
5. Alongi, D. M., The fate of bacterial biomass and production in marine benthic food chains, in *Recent Advances in Microbial Ecology*, Hattori, T., Ishida, Y., Maruyama, Y., Morita, R. Y., and Uchida, A., Japan Science Society Press, Tokyo, 1989, 355.
6. Blackburn, T. H., Microbial food webs in sediments, in *Microbes in the Sea*, Sleigh, M. A., Ed., Ellis Horwood, Chichester, 1987, 39.
7. Kemp, P., The fate of benthic bacterial production, *Rev. Aquat. Sci.*, 2, 109, 1990.
8. O'Neill, R. V., DeAngelis, D. L., Waide, J. B., and Allen, T. F. H., *A Hierarchical Concept of Ecosystems*, Monographs in Population Biology 23, Princeton University Press, Princeton, 1986, 253.
9. DeAngelis, D. L., *Dynamics of Nutrient Cycling and Food Webs*, Population and Community Biology Series 9, Chapman & Hall, London, 1992, 270 pp.
10. Jones, C. G. and Lawton, J. H., Eds., *Linking Species and Ecosystems*, Chapman & Hall, New York, 1995, 387 pp.
11. Giblin, A. E., Foreman, K. H., and Banta, G. T., Biogeochemical processes and marine benthic community structure: which follows which?, in *Linking Species and Ecosystems*, Jones, C. G. and Lawton, J. H., Eds., Chapman & Hall, New York, 1995, chap. 4.
12. Alongi, D. M., The ecology of tropical soft-bottom benthic ecosystems, *Oceanogr. Mar. Biol. Ann. Rev.*, 28, 381, 1990.
13. Saenger, P. and Holmes, N., Physiological, temperature tolerance, and behavioral differences between tropical and temperate organisms, in *Pollution in Tropical Aquatic Systems*, Connell, D. W. and Hawker, A., Eds., CRC Press, Boca Raton, FL, 1992, 69.
14. Hatcher, B. G., Johannes, R. E., and Robertson, A. I., Review of research relevant to the conservation of shallow tropical marine ecosystems, *Oceanogr. Mar. Biol. Annu. Rev.*, 27, 337, 1989.
15. Bird, M. I., Chivas, A. R., and Head, J., A latitudinal gradient in carbon turnover times in forest soils, *Nature*, 381, 143, 1996.

16. Hedges, J. I. and Keil, R. G., Sedimentary organic matter preservation: an assessment and speculative synthesis, *Mar. Chem.*, 49, 81, 1995.
17. Short, A. D., Physical variability of sandy beaches, in *Sandy Beaches as Ecosystems*, McLachlan, A. and Eramus, T., Eds., Junk, The Hague, 1983, 133.
18. McLachlan, A. and Turner, I., The interstitial environment of sandy beaches, *Mar. Ecol.*, 15, 177, 1994.
19. Pearse, A. S., Humm, H. J., and Wharton, G.W., Ecology of sandy beaches at Beaufort, *Ecol. Monogr.*, 12, 135, 1942.
20. Mare, M. F., A study of a marine benthic community with special reference to the micro-organisms., *J. Mar. Biol. Assoc. U.K.*, 25, 517, 1942.
21. Postma, H., Tidal flat areas, in *Coastal -Offshore Ecosystem Interactions*, Jansson, B., Ed., Springer-Verlag, Berlin, 1988, 102.
22. Brown, A. C. and McLachlan, A., *Ecology of Sandy Shores*, Elsevier, Amsterdam, 1990, 328 pp.
23. Robertson, A. I., Sandy beaches and intertidal flats, in *Marine Biology*, Hammond, L. and Synnot, R. N., Eds., Longman Cheshire, Melbourne, 1994, chap. 16.
24. Griffiths, C. L., Stenton-Dozey, J. M. E., and Koop, K., Kelp wrack and the flow of energy through a sand beach ecosystem, in *Sandy Beaches as Ecosystems*, McLachlan, A. and Erasmus, T., Eds., Junk, The Hague, 1983, 547.
25. Robertson, A. I. and Hansen, J. A., Decomposing seaweed: a nuisance or a vital link in coastal food chains?, *CSIRO Mar. Lab. Res. Rep.*, 75, 1982.
26. Robertson, A. I. and Lucus, J. S., Food choice, feeding rates, and the turnover of macrophyte biomass by a surf-zone-inhabiting amphipod, *J. Exp. Mar. Biol. Ecol.*, 72, 99, 1983.
27. Robertson, A. I. and Lenanton, R. C. J., Fish community structure and food chain dynamics in the surf-zone of sandy beaches: the role of detached macrophyte detritus, *Exp. Mar. Biol. Ecol.*, 84, 265, 1984.
28. Poovachiranon, S., Boto, K. G., and Duke, N. C., Food preference studies and ingestion rate measurements of the mangrove amphipod *Parhyale hawaiensis* (Dana), *J. Exp. Mar. Biol. Ecol.*, 98, 19, 1986.
29. Alongi, D. M., Microbial-meiofaunal interrelationships in some tropical intertidal sediments. *J. Mar. Res.*, 46, 349, 1988.
30. Koop, K. and Lucas, M. I., Carbon flow and nutrient regeneration from the decomposition of macrophyte debris in a sandy beach microcosm, in *Sandy Beaches as Ecosystems*, McLachlan, A. and Erasmus, T., Eds, Junk, The Hague, 1983, 249.
31. Kemp, P. F., Deposition of organic matter on a high energy sand beach by a mass stranding of the cnidarian *Velella velella* (L.), *Estuarine Coastal Shelf Sci.*, 23, 575, 1986.
32. Hutchings, L., Nelson, G., Horstman, D. A., and Tart, R., Interaction between coastal plankton and sand mussels along the Cape coast, South Africa, in *Sandy Beaches as Ecosystems*, McLachlan, A. and Erasmus, T., Eds., Junk, The Hague, 1983, 481.
33. Steele, J. H., Comparative studies of beaches, *Phil. Trans. Roy. Soc. Lond. B.*, 274, 401, 1976.
34. Munro, A. L. S., Wells, J. B. J., and McIntyre, A. D., Energy flow in the flora and meiofauna of sandy beaches, *Proc. Roy. Soc. Edin.*, 76B, 297, 1978.
35. Alongi, D. M., The role of intertidal mudbanks in the diagenesis and export of dissolved and particulate material from the Fly Delta, Papua New Guinea, *J. Exp. Mar. Biol. Ecol.*, 149, 81, 1991.
36. Warwick, R. M., Joint, I. R., and Radford, P. J., Secondary production of the benthos in an estuarine environment, in *Ecological Processes in Coastal Environments*, Jeffries, R. L. and Davey, A. J., Eds., Blackwell, Oxford, 1979, 429.
37. Ellison, R. L., Foraminifera and meiofauna on an intertidal mudflat, Cornwall, England: populations, respiration and secondary production; and energy budget, *Hydrobiologia*, 109, 131, 1984.

38. Kuipers, B. R., de Wilde, P. A. W., and Creutzberg, F., Energy flow in a tidal flat ecosystem, *Mar. Ecol. Prog. Ser.*, 5, 215, 1981.

39. Asmus, H. and Asmus, R., The importance of grazing food chain for energy flow and production in three intertidal sand bottom communities of the northern Wadden Sea, *Helgo. Meersunters.*, 39, 273, 1985.

40. Schwinghamer, P., Hargrave, B., Peer, D., and Hawkins, C. M., Partitioning of production and respiration among size groups of organisms in an intertidal benthic community, *Mar. Ecol. Prog. Ser.*, 31, 131, 1986.

41. Duyl, van F. C. and Kop, A. J., Seasonal patterns of bacterial production and biomass in intertidal sediments of the western Dutch Wadden Sea, *Mar. Ecol. Prog. Ser.*, 59, 249, 1990.

42. Cammen, L., Annual bacterial production in relation to benthic microalgal production and sediment oxygen uptake in the intertidal sandflat and an intertidal mudflat, *Mar. Ecol. Prog. Ser.*, 71, 13, 1991.

43. Kemp, P. F., Bacterivory by benthic ciliates: significance as a carbon source and impact on sediment bacteria, *Mar. Ecol. Prog. Ser.*, 49, 163, 1988.

44. Hondeveld, B. J. M., Bak, R. P. M., and van Duyl, F. C., Bacterivory by heterotrophic nanoflagellates in marine sediments measured by uptake of fluorescently labeled bacteria, *Mar. Ecol. Prog. Ser.*, 89, 63, 1992.

45. Epstein, S. S. and Shiaris, M. P., Rates of microbenthic and meiobenthic bacterivory in a temperate muddy tidal flat community, *Appl. Environ. Microbiol.*, 58, 2426, 1992.

46. Epstein, S. S., Burkovsky, I. V., and Shiaris, M. P., Ciliate grazing on bacteria, flagellates, and microalgae in a temperate zone sandy tidal flat: ingestion rates and food niche partitioning, *J. Exp. Mar. Biol. Ecol.*, 165, 103, 1992.

47. Montagna, P. A., Blanchard, G. F., and Dinet, A., Effect of production and biomass of intertidal microphytobenthos on meiofaunal grazing rates, *J. Exp. Mar. Biol. Ecol.*, 185, 149, 1995.

48. Epstein, S. S. and Gallagher, E. D., Evidence for facilitation and inhibition of ciliate population growth by meiofauna and macrofauna on a temperate zone sandflat, *J. Exp. Mar. Biol. Ecol.*, 155, 27, 1992.

49. Middelburg, J. J., Klaver, G., Nieuwenhuize, J., and Vlug, T., Carbon and nitrogen cycling in intertidal sediments near Doel, Scheldt Estuary, *Hydrobiologia*, 311, 57, 1995.

50. Middelburg, J. J., Klaver, G., Nieuwenhuize, J., Wielemaker, A., de Haas, W., and Vlug van der Nat, J. F. W. A., Organic matter mineralization in intertidal sediments along an estuarine gradient, *Mar. Ecol. Prog. Ser.*, 132, 157, 1996.

51. Talbot, M. M., Bate, G. C., and Campbell, E. E., A review of the ecology of surf-zone diatoms, with special reference to *Anaulus australis*, *Oceanogr. Mar. Biol. Ann. Rev.*, 28, 155, 1990.

52. Lewin, J., Schaefer, C. T., and Winter, D. F., Surf-zone ecology and dynamics, in *Coastal Oceanography of Washington, and Oregon*, Landry, M. R. and Hickers, B., Eds., Elsevier, Amsterdam, 1989, chap. 12.

53. McLachlan, A. and Romer, G. S., Trophic relations in a high energy beach and surf-zone ecosystem, in *Trophic Relationships in the Marine Environment*, Barnes, M. and Gibson, R. N., Eds., Aberdeen University Press, Aberdeen, 1990, 356.

54. Heymans, J. J. and McLachlan, A., Carbon budget and network analysis of a high-energy beach/surf-zone ecosystem, *Estuarine Coastal Shelf Sci.*, 43, 485, 1996.

55. Blackburn, T. H. and Sorensen, J., Eds., *Nitrogen Cycling in Coastal Marine Environments*, John Wiley & Sons, Chichester, 1988, 451.

56. Cohen, Y. and Rosenberg, H., Eds., *Microbial Mats: Physiological Ecology of Benthic Communities*, American Society of Microbiology, Washington, D.C., 1989.

57. Stal, L. J. and Caumette, P., Eds., *Microbial Mats*, Springer-Verlag, Berlin, 1994.

58. Joye, S. B. and Paerl, H. W., Contemporaneous nitrogen fixation and denitrification in marine microbial mats: rapid response to runoff events, *Mar. Ecol. Prog. Ser.*, 94, 267, 1993.

59. Joye, S. B. and Paerl, H. W., Nitrogen cycling in microbial mats: rates and patterns of denitrification and nitrogen fixation, *Mar. Biol.*, 119, 285, 1994.

60. McLachlan, A. and McGwynne, L., Do sandy beaches accumulate nitrogen?, *Mar. Ecol. Prog. Ser.*, 34, 191, 1986.

61. Cockcroft, A. C., and McLachlan, A., Nitrogen budget for a high-energy ecosystem, *Mar. Ecol. Prog. Ser.*, 100, 287, 1993.

62. Kerner, M. and Wallmann, K., Remobilization events involving Cd and Zn from intertidal flat sediments in the Elbe Estuary during the tidal cycle, *Estuarine Coastal Shelf Sci.*, 35, 371, 1992.

63. Kerner, M., Coupling of microbial fermentation and respiration processes in an intertidal mudflat of the Elbe estuary, *Limnol. Oceanogr.*, 38, 314, 1993.

64. Bock, M. J. and Miller, D. C., Storm effects on particulate food resources on an intertidal sandflat, *J. Exp. Mar. Biol. Ecol.*, 187, 81, 1995.

65. de Jonge, V. N. and van Beusekom, J. E. E., Wind- and tide-induced resuspension of sediment and microbenthos from tidal flats in the Ems estuary, *Limnol. Oceanogr.*, 40, 766, 1995.

66. Campbell, E. E. and Bare, G. C., Ground water in the Alexandria dune field and its potential influence on the adjacent surf-zone, *Water SA*, 17, 155, 1991.

67. Harrison, S. J. and Phizacklea, A. P., Seasonal changes in heat flux and heat storage in the intertidal mudflats of the Forth estuary, *J. Climatol.*, 5, 473, 1985.

68. Ridd, P., Sandstrom, M. W., and Wolanski, E., Outwelling from tropical tidal salt flats, *Estuarine Coastal Shelf Sci.*, 26, 243, 1988.

69. Asmus, R., Nutrient flux in short-term enclosures of intertidal sand communities, *Ophelia*, 26, 1, 1986.

70. Turner, R. E., Geographic variation in salt marsh macrophyte production: a review, *Contr. Mar. Sci.*, 20, 47, 1976.

71. de Leeuw, J. and Buth, G. J., Spatial and temporal variation in peak standing crop of European tidal marshes, in *Estuaries and Coasts: Spatial and Temporal Intercomparisons*, Elliott, M. and Ducrotoy, J. P., Eds., Olsen and Olsen, Fredensborg, 1991, 133.

72. Twilley, R. R., Chen, R. H., and Hargis, T., Carbon sinks in mangroves and their implications to carbon budget of tropical coastal ecosystems, *Water, Air, Soil Pollut.*, 64, 265, 1992.

73. Clough, B. F., Primary productivity and growth of mangrove forests, in *Tropical Mangrove Ecosystems*, Robertson, A. I. and Alongi, D. M., Eds., American Geophysical Union, Washington, D.C., chap. 8.

74. Archibold, O. W., *Ecology of World Vegetation*, Chapman & Hall, London, 1995, 510 pp.

75. Schubauer, J. P. and Hopkinson, C. S., Above- and belowground emergent macrophyte production and turnover in a coastal marsh ecosystem, Georgia, *Limnol. Oceanogr.*, 29, 1052, 1984.

76. Teal, J. M. and Howes, B. L., Interannual variability of a salt-marsh ecosystem, *Limnol. Oceanogr.*, 41, 802, 1996.

77. Day, J. W., Jr., Coronado-Molina, C., Vera-Herrera, F. R., Twilley, R. R., Rivera-Monroy, V. H., Alvarez-Guillen, H., Day, R., and Conner, W., A 7-year record of above-ground primary production in a southeastern Mexican mangrove forest, *Aquat. Bot.*, 55, 39, 1996.

78. Lugo, A. E., Fringe wetlands, in *Ecosystems of the World*. Vol. 15. *Forested Wetlands*, Lugo, A. E., Brinson, M., and Brown, S., Eds., Elsevier, Amsterdam, 1990, 143.

79. Gong, W. K., Ong, J. E., and Clough, B. F., Photosynthesis in different aged stands of a Malaysian mangrove ecosystem, in *Third ASEAN Science and Technology Week Conference Proceedings*, Vol. 6, Chou, L. M. and Wilkinson, C. R., Eds., National University Singapore, Singapore, 1992, 345.

80. Kennish, M. J., *Ecology of Estuaries*. Vol. 1. *Physical and Chemical Aspects*, CRC Press, Boca Raton, FL, 1986, 163–165.

81. Saenger, P. E., Mangroves and saltmarshes, in *Marine Biology*, Hammond, L. S. and Synnot, R. N., Eds., Longman Cheshire, Melbourne, 1994, chap. 13.

82. Clough, B. F., Ong, J. E., and Gong, G. W., Estimating leaf area index and photosynthetic production in mangrove forest canopies, *Mar. Ecol. Prog. Ser.*, in press.

83. Gong, W. K., Ong, J. E., and Wong, C. H., The light attenuation method for measuring potential primary productivity in mangrove ecosystems: an evaluation, in *Proceedings of the Regional Symposium on Living Resources in Coastal Areas*, Alcala, A. C., Ed., University of Phillippines, Manila, 1991, 399.

84. Hackney, C. T. and de la Cruz, A. A., Belowground productivity of roots and rhizomes in a giant cordgrass marsh, *Estuaries*, 9, 112, 1986.

85. Vernberg, F. J., Salt-marsh processes: a review, *Environ. Toxicol. Chem.*, 12, 2167, 1993.

86. Valiela, I. and J. M. Teal, The nitrogen budget of a salt marsh ecosystem, *Nature*, 280, 652, 1979.

87. Robertson, A. I., Alongi, D. M., and Boto, K. G., Food chains and carbon fluxes, in *Tropical Mangrove Ecosystems*, Robertson, A. I. and Alongi, D. M., Eds., American Geophysical Union, Washington, D.C., 1992, chap. 10.

88. Alongi, D. M., unpublished data.

89. Giurgevich, J. and Dunn, E., Seasonal patterns of CO_2 and water vapor exchange of the tall and short forms of *Spartina alterniflora* Loisel in a Georgia salt marsh, *Oecologia*, 43, 139,1979.

90. Andrews, Y. J., Clough, B. F., and Muller, G. J., Photosynthetic gas exchange properties and carbon isotope ratios of some mangroves in North Queensland, in *Physiology and Management of Mangroves, Tasks for Vegetation Science 9*, Teas, H. J., Ed., Junk, The Hague, 1984, 15.

91. Haines, E., Stable carbon isotope ratios in the biota, soils, and tidal water of a Georgia salt marsh, *Estuarine Coastal Mar. Sci.*, 4, 609, 1976.

92. Joshi, G. V., Karekar, M. D., Jowda, C. A., and Bhosale, L., Photosynthetic carbon metabolism and caroxylating enzymes in algae and mangroves under saline conditions, *Photosynthetica*, 8, 51, 1974.

93. Bradley, P. M. and Morris, J. T., Effect of salinity on the critical nitrogen concentration of *Spartina alterniflora* Loisel., *Aquat. Bot.*, 43, 149, 1992.

94. Ball, M. C., Ecophysiology of mangroves, *Trees*, 2, 129, 1988.

95. Passiourna, J. B., Ball, M. C., and Knight, J. H., Mangroves may salinize the soil and in so doing limit their transpiration rate, *Funct. Ecol.*, 6, 476, 1992.

96. Morris, J. T., Kjerfve, B., and Dean, J. M., Dependence of estuarine productivity on anomalites in mean sea level, *Limnol. Oceanogr.*, 35, 926, 1990.

97. de Leeuw, J., Olff, H., and Bakker, J. P., Year-to-year variation in peak aboveground biomass of six salt-marsh angiosperm communities as related to rainfall deficit and inundation frequency, *Aquat. Bot.*, 36, 139, 1990.

98. Wiegert, R. G., Chalmers, A. G., and Randerson, P. F., Productivity gradients in salt marshes: the response of *Spartina alterniflora* to experimentally manipulated soil water movement, *Oikos*, 41, 1, 1983.

99. Howes, B. L., Howarth, R. W., Teal, J. M., and Valiela, I., Oxidation-reduction potentials in a salt marsh: spatial patterns and interactions with primary production, *Limnol. Oceanogr.*, 26, 350, 1981.

100. Boto, K. G., Nutrients and mangroves, in *Pollution in Tropical Aquatic Systems*, Connell, D. W. and Hawker, O. W., Eds., CRC Press, Boca Raton, FL, 1992, 129.

101. Alongi, D. M., The dynamics of benthic nutrient pools and fluxes in tropical mangrove forests, *J. Mar. Res.*, 54, 123, 1996.

102. Andersen, F. O. and Kristensen, E., Oxygen microgradients in the rhizosphere of the mangrove *Avicennia marina*, *Mar. Ecol. Prog. Ser.*, 44, 210, 1988.

103. McKee, K. L., Mendelssohn, I. A., and Hester, M. W., Reexamination of pore water sulfide concentrations and redox potentials near the aerial roots of *Rhizophora mangle* and *Avicennia germinans*, *Am. J. Bot.*, 75, 1352, 1988.

104. McKee, K. L., Soil physicochemical patterns and mangrove species distribution—reciprocal effects?, *J. Ecol.*, 81, 477, 1993.

105. Koch, M. S., Mendelssohn, I. A., and McKee, K. L., Mechanism for the hydrogen sulfide-induced growth limitation in wetland macrophytes, *Limnol. Oceanogr.*, 35, 399, 1990.

106. Taylor, D. I. and Allanson, B. R., Impacts of dense crab populations on carbon exchanges across the surface of a salt marsh, *Mar. Ecol. Prog. Ser.*, 101, 119, 1993.

107. Bertness, M. D., Fiddler crab regulation of *Spartina alterniflora* production on a New England salt marsh, *Ecology*, 66, 1042, 1985.

108. Smith, III, T. J., Boto, K. G., Frusher, S. D., and Giddins, R. L., Keystone species and mangrove forest dynamics: the influence of burrowing by crabs on soil nutrient status and forest productivity, *Estuarine Coastal Shelf. Sci.*, 33, 419, 1991.

109. Mackin, J. E. and Aller, R. C., Ammonium adsorption in marine sediments, *Limnol. Oceanogr.*, 29, 250, 1984.

110. Howes, B. L. and Goehringer, D. D., Porewater drainage and dissolved organic carbon and nutrient losses through the intertidal creekbanks of a New England salt marsh, *Mar. Ecol. Prog. Ser.*, 114, 289, 1994.

111. Clough, B. F., Boto, K. G., and Attiwill, P. M., Mangroves and sewage: a reevaluation, in *Biology and Ecology of Mangroves, Tasks for Vegetation Science 8*, Teas, H. J., Ed., Junk, The Hague, 1983, 151.

112. Boto, K. G., Saffigna, P., and Clough. B. F., Role of nitrate in nitrogen nutrition of the mangrove *Avicennia marina*, *Mar. Ecol. Prog. Ser.*, 21, 259, 1985.

113. Naidoo, G., Effects of nitrate, ammonium and salinity on growth of the mangrove *Bruguiera gymnorrhiza* (L.) Lam., *Aquat. Bot.*, 38, 209, 1990.

114. Naidoo, G., Effects of salinity and nitrogen on the growth and water relations in the mangrove, *Avicennia marina* (Forsk.) Vierh., *New Phytol.*, 107, 317, 1987.

115. Ornes, W. H. and Kaplan, D. I., Macronutrient status of tall and short forms of *Spartina alterniflora* in a South Carolina salt marsh, *Mar. Ecol. Prog. Ser.*, 55, 63, 1989.

116. Boto, K. G. and Wellington, J. W., Soil characteristics and nutrient status in a northern Australian mangrove forest, *Estuaries*, 7, 117, 1984.

117. Osgood, D. T. and Zieman, J. C., Factors controlling aboveground *Spartina alterniflora* (smooth cordgrass) tissue element composition and production in different-age barrier island marshes, *Estuaries*, 16, 815,1993.

118. Boto, K. G. and Wellington, J. T., Phosphorus and nitrogen nutritional status of a northern Australian mangrove forest, *Mar. Ecol. Prog. Ser.*, 11, 63, 1983.

119. Pfeiffer, W. J. and Wiegert, R. G., Grazers on *Spartina* and their predators, in *The Ecology of a Salt Marsh*, Pomeroy, L. R. and Wiegert, R. G., Eds, Springer-Verlag, New York, 1981, 87.

120. Robertson, A. I., Plant-animal interactions and the structure and function of mangrove forest ecosystem, *Aust. J. Ecol.*, 16, 433, 1991.

121. Page, H. M., Dugan, J. E., and Hubbard, D. M., Comparative effects of infaunal bivalves on an epibenthic microalgal community, *J. Exp. Mar. Biol. Ecol.*, 157, 247, 1992.

122. Peterson, B. and Howarth, R. W., Sulfur, carbon, and nitrogen isotopes used to trace organic matter flow in the salt-marsh estuaries of Sapelo Island, Georgia, *Limnol. Oceanogr.*, 32, 1195, 1987.

123. Peterson, B., Howarth, R. W., and Garritt, R. H., Multiple stable isotopes used to trace the flow of organic matter in estuarine food webs, *Science*, 227, 1361, 1985.

124. Sullivan, M. J. and Montcreiff, C. A., Edaphic algae are an important component of salt marsh food webs: evidence from multiple stable isotope analyses, *Mar. Ecol.Prog. Ser.*, 62, 149, 1990.

125. Rodelli, M. R., Gearing, J. N., Gearing, P. J., Marshall, N., and Sasekumar, A., Stable isotope ratio as a tracer of mangrove carbon in Malaysian ecosystems, *Oecologia*, 61, 326, 1984.

126. Newell, R. I. E., Marshall, N., Sasekumar, A., and Chong, V. C., Relative importance of benthic microalgae, phytoplankton, and mangroves as sources of nutrition for penaeid prawns and other coastal invertebrates from Malaysia, *Mar. Biol.*, 123, 595, 1995.

127. Ambler, J. W., Alcala-Herrera, J., and Burke, R., Trophic roles of particle feeders and detritus in a mangrove island prop root ecosystem, *Hydrobiologia*, 292/293, 437, 1994.

128. Primavera, J. H., Stable carbon and nitrogen isotope ratios of penaeid juveniles and primary producers in a riverine mangrove in Guimaras, Philippines, *Bull. Mar. Sci.*, 58, 675, 1996.

129. Odum, W. E. and Heald, E. J., The detritus-based food web of an estuarine mangrove community, in *Estuarine Research*, Cronin, L. E., Ed., Academic Press, New York, 1975, 265.

130. Teal, J. M., Energy flow in the salt marsh ecosystem in Georgia, *Ecology*, 43, 614, 1962.

131. Newell, S. Y. and Barlocher, F., Removal of fungal and total organic matter from decaying cordgrass leaves by shredder snails, *J. Exp. Mar. Biol. Ecol.*, 171, 39, 1993.

132. Takeda, S. and Kurihara, Y., The effects of burrowing of *Helice tridens* (De Haan) on the soil of a salt-marsh habitat, *J. Exp. Mar. Biol. Ecol.*, 113, 79, 1987.

133. Newell, S. Y., Fallon, R. D., and Miller, J. D., Decomposition and microbial dynamics for standing, naturally positioned leaves of the salt-marsh grass *Spartina alterniflora*, *Mar. Biol.*, 101, 471, 1989.

134. Currin, C. A., Newell, S. Y., and Paerl, H. W., The role of standing dead *Spartina alterniflora* and benthic microalgae in salt marsh food webs: considerations based on multiple stable isotope analysis, *Mar. Ecol. Prog. Ser.*, 121, 99, 1995.

135. Micheli, F., Feeding ecology of mangrove crabs in north eastern Australia: mangrove litter consumption by *Sesarma messa* and *Sesarma smithii*, *J. Exp. Mar. Biol. Ecol.*, 171, 165, 1993.

136. Micheli, F., Effect of mangrove litter species and availability on survival, moulting, and reproduction of the mangrove crab *Sesarma messa*, *J. Exp. Mar. Biol. Ecol.*, 171, 149, 1993.

137. Enriquez, S., Duarte, C. M., and Sand-Jensen, K., Patterns in decomposition ratés among photosynthetic organisms: the importance of detritus C:N:P content, *Oecologia*, 94, 457, 1993.

138. Zieman, J. C., Macko, S. A., and Mills, A. L., Role of seagrasses and mangroves in estuarine food webs: temporal and spatial changes in stable isotope composition and amino acid content during decomposition, *Bull. Mar. Sci.*, 35, 380, 1984.

139. Lee, K. Y., Moran, M. A., Benner, R., and Hodson, R. E., Influence of soluble components of red mangrove *(Rhizophora mangle)* leaves on microbial decomposition of structural (lignocellulosic) leaf components in seawater, *Bull. Mar. Sci.*, 46, 374, 1990.

140. Benner, R., Weliky, K., and Hedges, J. L., Early diagenesis of mangrove leaves in a tropical estuary: molecular-level analyses of neutral sugars and lignin-derived phenols, *Geochim. Cosmochim. Acta*, 54, 1991, 1990.

141. Marinucci, A. C., Hobbie, J. E., and Helfrich, J. V. K., Effect of litter nitrogen on decomposition and microbial biomass in *Spartina alterniflora*, *Micro. Ecol.*, 9, 23, 1983.

142. White, D. S. and Howes, B. L., Nitrogen incorporation into decomposing litter of *Spartina alterniflora*, *Limnol. Oceanogr.*, 39, 133, 1994.

143. Tenore, K. R., Nitrogen in benthic food chains, in *Nitrogen Cycling in Coastal Marine Environments*, Blackburn, T. H. and Sorensen, J., Eds., John Wiley & Sons, Chichester, 1988, chap. 9.

144. Newell, S. Y., Fallon, R. D., Cal Rodriguez, R. M., and Groene, L. C., Influence of rain, tidal wetting and relative humidity on release of carbon dioxide by standing-dead salt-marsh plants, *Oecologia*, 68, 73, 1985.

145. Hemminga, M. A., Kok, C. J., and de Munck, W., Decomposition of *Spartina anglica* roots and rhizomes in a salt marsh of the Westerschelde Estuary, *Mar. Ecol.Prog. Ser.*, 48, 175, 1988.

146. Benner, R., Fogel, M. L., and Sprague, E. K., Diagenesis of belowground biomass of *Spartina alterniflora* in salt-marsh sediments, *Limnol. Oceanogr.*, 36, 1358, 1991.

147. Blum, L. K., *Spartina alterniflora* root dynamics in a Virginia marsh, *Mar. Ecol. Prog. Ser.*, 102, 169, 1993.

148. Howarth, R. W. and Hobbie, J. E., The regulation of decomposition and heterotrophic microbial activity in salt marsh soils: a review, in *Estuarine Comparisons*, Kennedy, V. S., Ed., Academic Press, New York, 1982, 183.

149. Howarth, R. W., Microbial processes in salt-marsh sediments, in *Aquatic Microbiology: An Ecological Approach*, Ford, T. E., Ed., Blackwell, Boston, 1993, chap. 10.

150. Fogel, M. L., Sprague, E. K., Gize, A. P., and Frey, R. W., Diagenesis of organic matter in Georgia salt marshes, *Estuarine Coastal Shelf Sci.*, 28, 211, 1989.

151. Alongi, D. M., unpublished data.

152. Kristensen, E., Holmer, M., and Bussarawit, N., Benthic metabolism and sulfate reduction in a southeast Asian mangrove swamp, *Mar. Ecol. Prog. Ser.*, 73, 93, 1991.

153. Kristensen, E., Devol, A. H., Ahmed, S. I., and Saleem, M., Preliminary study of benthic metabolism and sulfate reduction in a mangrove swamp of the Indus Delta, Pakistan, *Mar. Ecol. Prog. Ser.*, 90, 287, 1992.

154. Nedwell, D. B., Blackburn, T. H., and Wiebe, W. J., Dynamic nature of the turnover of organic carbon, nitrogen and sulphur in the sediments of a Jamaican mangrove forest, *Mar. Ecol. Prog. Ser.*, 110, 223, 1994.

155. King, G. M., Patterns of sulfate reduction and the sulfur cycle in a South Carolina salt marsh, *Limnol. Oceanogr.*, 33, 376, 1988.

156. Hines, M. E., Knollmeyer, S. L., and Tugel, J. B., Sulfate reduction and other sedimentary biogeochemistry in a northern New England salt marsh, *Limnol. Oceanogr.*, 34, 578, 1989.

157. Sotomayor, D., Corredor, J. E., and Morell, J. M., Methane flux from mangrove sediments along the southwestern coast of Puerto Rico, *Estuaries*, 17, 140, 1994.

158. Spratt, Jr., H. G. and Hodson, R. E., The effect of changing water chemistry on rates of manganese oxidation in surface sediments of a temperate saltmarsh and a tropical mangrove estuary, *Estuarine Coastal Shelf Sci.*, 38, 119, 1994.

159. Howes, B. L., Dacey, J. W. H., and Teal, J. M., Annual carbon mineralization and belowground production of *Spartina alterniflora* in a New England salt marsh, *Ecology*, 66, 595, 1985.

160. Sharma, P., Gardner, L. R., Moore, W. S., and Bollinger, M. S., Sedimentation and bioturbation in a salt marsh as revealed by ^{210}Pb, ^{137}Cs, and ^{7}Be studies, *Limnol. Oceanogr.*, 32, 313, 1987.

161. Lynch, J. C., Meriwether, J. R., McKee, B. A., Vera-Herrera, F., and Twilley, R. R., Recent accretion in mangrove ecosystems based on ^{137}Cs and ^{210}Pb, *Estuaries*, 12, 284, 1989.

162. Robertson, A. I. and Blaber, S., Plankton, epibenthos and fish communities, in *Tropical Mangrove Ecosystems*, Robertson, A. I. and Alongi, D. M., Eds., American Geophysical Union, Washington, D.C., 1992, chap. 7.

163. Ducklow, H. W. and Shiah, F.-K., Bacterial production in estuaries, in *Aquatic Microbiology: An Ecological Approach*, Ford, T. E., Ed., Blackwell, Boston, 1993, chap. 11.

164. Healey, M. J., Moll, R. A., and D'allo, C. O., Abundance and distribution of bacterioplankton in the Gambia River, West Africa, *Micro. Ecol.*, 16, 291, 1988.

165. Newell, S. Y., Fallon, R. D., Sherr, B. F., and Sherr, E. B., Mesoscale temporal variation in bacterial standing crop, percent active cells, productivity and output in a salt-marsh tidal river, *Verh. Int. Verein. Limnol.*, 23, 1839, 1988.

166. Shiah, F.-W. and Ducklow, H. W., Multiscale variability in bacterioplankton abundance, production, and specific growth rate in a temperate salt-marsh tidal creek, *Limnol. Oceanogr.*, 40, 55, 1993.

167. Fallon, R. D., Newell, S. Y., Sherr, B. F., and Sherr, E. B., Factors affecting bacterial biomass and growth in the Duplin River estuary and coastal Atlantic Ocean, *Proc. Int. Colloq. Mar. Bacteriol.*, 2, 137, 1987.

168. Valiela, I., Nitrogen in salt marsh ecosystems, in *Nitrogen in the Marine Environment*, Carpenter, E. J. and Capone, D. G., Eds., Academic Press, New York, 1983, chap. 17.

169. Alongi, D. M., Boto, K. G., and Robertson, A. I., Nitrogen and phosphorus cycles, in *Tropical Mangrove Ecosystems*, Robertson, A. I. and Alongi, D. M., Eds., American Geophysical Union, Washington, D.C., 1992, chap. 9.

170. Rivera-Monroy, V. H., Twilley, R. R., Boustany, R. G., Day, J. W., Vera-Herrera, F., and del Carmen Ramirez, M., Direct denitrification in mangrove sediments in Terminos Lagoon, Mexico, *Mar. Ecol. Prog. Ser.*, 126, 97, 1995.

171. Rivera-Monroy, V. H. and Twilley, R. R., The relative role of denitrification and immobilization in the fate of inorganic nitrogen in mangrove sediments (Terminos Lagoon, Mexico), *Limnol. Oceanogr.*, 41, 284, 1996.

172. Blackburn, T. H., Christensen, D., Fanger, A. M., Henriksen, K., Iizimi, H., Iversen, N., and Limpsaichol, P., Mineralization processes in mangrove and seagrass sediments, in *Ao Yan-A Mangrove in the Andaman Sea*, Hylleberg, J., Ed., Institute of Ecology and Genetics, University Aarhus, Aarhus, 1987, 22.

173. Azni b Abd. Aziz, S. and Nedwell, D. B., The nitrogen cycle of an east coast, U.K. saltmarsh. II. Nitrogen fixation, nitrification, denitrification, tidal exchange, *Estuarine Coastal Shelf Sci.*, 22, 689, 1986.

174. White, D. S. and Howes, B. L., Long-term ^{15}N-nitrogen retention in the vegetated sediments of a New England salt marsh, *Limnol. Oceanogr.*, 39, 1878, 1994.

175. Rivera-Monroy, V. H., Day, J. W. Twilley, R. R., Vera-Herrera, F., and Coronado-Molina, C., Flux of nitrogen and sediment in a fringe mangrove forest in Terminos Lagoon, Mexico, *Estuarine Coastal Shelf Sci.*, 40, 139, 1995.

176. Whiting, G. J., McKellar, Jr., H. N., Spurrier, J. D., and Wolaver, T. G., Nitrogen exchange between a portion of vegetated salt marsh and the adjoining creek, *Limnol. Oceanogr.*, 34, 463, 1989.

177. Twilley, R. R., Coupling of mangroves to the productivity of estuarine and coastal waters, in *Coastal-Offshore Ecosystem Interactions*, Jansson, B.-O., Ed., Springer-Verlag, Berlin, 1988, 155.

178. Hopkinson, C. S., Patterns of organic carbon exchange between coastal ecosystems: the mass balance approach in salt marsh ecosystems, in *Coastal-Offshore Ecosystem Interactions*, Jansson, B.-O., Ed., Springer-Verlag, Berlin, 1988, 122.

179. Taylor, D. I. and Allanson, B. R., Organic carbon fluxes between a high marsh and estuary, and the inapplicability of the outwelling hypothesis, *Mar. Ecol. Prog. Ser.*, 120, 263, 1995.

180. Rezende, C. E., Lacerda, L. D., Ovalle, A. R. C., Silva, C. A. R., and Martinelli, L. A., Nature of POC transport in a mangrove ecosystem: a carbon stable isotope study, *Estuarine Coastal Shelf Sci.*, 30, 641, 1990.

181. Roman, C. T. and Daiber, F. C., Organic carbon flux through a Delaware Bay salt marsh: tidal exchange, particle size distribution, and storms, *Mar. Ecol. Prog. Ser.*, 54, 149, 1989.

182. Wolanski, E., Mazda, Y., and Ridd, P., Mangrove hydrodynamics, in *Tropical Mangrove Ecosystems*, Robertson, A. I. and Alongi, D. M., Eds., American Geophysical Union, Washington, D.C., 1992, chap. 3.

183. Leonard, L. A. and Luther, M. E., Flow hydrodynamics in tidal marsh canopies, *Limnol. Oceanogr.*, 40, 1474, 1995.

184. Williams, G. A., The relationship between shade and molluscan grazing in structuring communities on a moderately-exposed tropical rocky shore, *J. Exp. Mar. Biol. Ecol.*, 178, 79, 1994.

185. Charpy-Robaud, C. and Sournia, A., The comparative estimation of phytoplankton, microphytobenthic, and macrophytobenthic primary production in the oceans, *Mar. Micro. Food Webs*, 4, 31,1990.

186. Smith, S. V., Marine macrophytes as a global carbon sink, *Science*, 211, 838, 1981.

187. Short, F. T. and Wyllie-Echeverra, S., Natural and human-induced disturbance of seagrasses, *Environ. Conserv.*, 23, 17, 1996.

188. Hillman, K., Walker, D. I., Larkum, A. W. D., and McComb, A. J., Productivity and nutrient limitation, in *Biology of Seagrasses*, Larkum, A. W. D., McComb, A. J., and Shepard, S. A., Eds., Elsevier, Amsterdam, 1989, chap. 19.

189. Vermaat, J. E., Agawin, N. S. R., Duarte, C. M., Fortes, M. D., Marba, N., and Uri, J. S., Meadow maintenence, growth and productivity of a mixed Philippine seagrass bed, *Mar. Ecol. Prog. Ser.*, 124, 215, 1995.

190. van Tussenbroek, B. I., Techniques of rapid assessment of seagrass production yield, applied to *Thalassia testudinum* in a Mexican tropical reef lagoon, in *Seagrass Biology: Proceedings of an International Workshop*, Kuo, J., Phillips, R. C., Walker, D. I., and Kirkman, H., Eds., Western Australian Museum, Perth, 1996, 131.

191. Duarte, C. M., Temporal biomass variability and production/biomass relationships of seagrass communities, *Mar. Ecol. Prog. Ser.*, 51, 269, 1989.

192. Neushul, M., Benson, J., Harger, B. W. W., and Charters, A. C., Macroalgal farming in the sea: water motion and nitrate uptake, *J. Appl. Phycol.*, 4, 255, 1992.

193. Mann, K. H., *Ecology of Coastal Waters. A Systems Approach.* University of California Press, Berkeley, 1982, 322 pp.

194. Heip, C. H. R., Goosen, N. K., Herman, P. M. J., Kromkamp, J., Middelburg, J. J., and Soetaert, K., Production and consumption of biological particles in temperate tidal estuaries, *Oceanogr. Mar. Biol. Ann. Rev.*, 33, 1, 1995.

195. Branch, G. M. and Griffiths, C. L., The Benguela ecosystem. Part V. The coastal zone, *Oceanogr. Mar. Biol. Ann. Rev.*, 26, 395, 1988.

196. Borowitzka, M. A. and Lethbridge, R. C., Seagrass epiphytes, in *Biology of Seagrasses*, Larkum, A. W. D., McComb, A. J., and Shepard, S. A., Eds., Elsevier, Amsterdam, 1989, chap. 14.

197. Duarte, C. M., Submerged aquatic vegetation in relation to different nutrient regimes, *Ophelia*, 41, 87, 1995.

198. Lobban, C. S., Harrison, P. J., and Duncan, M. J., *Seaweed Ecology and Physiology*, Cambridge University Press, London, 1993.

199. Fourqurean, J. W. and Zieman, J. C., Photosynthesis, respiration and whole plant carbon budget of the seagrass *Thalassia testudinum*, *Mar. Ecol. Prog. Ser.*, 69, 161, 1991.

200. Kraemer, G. P. and Alberte, R. S., Age-related patterns of metabolism and biomass in subterranean tissues of *Zostera marina* (eelgrass), *Mar. Ecol. Prog. Ser.*, 95, 193, 1993.

201. Hatcher, B., G., Chapman, A. R. O., and Mann, K. H., An annual carbon budget for the kelp *Laminaria longicruris*, *Mar. Biol.*, 44, 85, 1977.

202. Dunton, K. H. and Schell, D. M., Seasonal carbon budget and growth of *Lamineria solidungula* in the Alaskan High Arctic, *Mar. Ecol. Prog. Ser.*, 31, 57, 1986.

203. Maberly, S. C. and Madsen, T. V., Contribution of air and water to the carbon balance of *Fucus spiralis*, *Mar. Ecol. Prog. Ser.*, 62, 175, 1990.

204. Luning, K., Environmental and internal control of seasonal growth in seaweeds, *Hydrobiologia*, 260/261, 1, 1993.

205. Larkum, A. W. D. and James, P. L., Towards a model for inorganic carbon uptake in seagrasses involving carbonic anhydrase, in *Seagrass Biology: Proceedings of an International Workshop*, Kuo, J., Phillips, R. C., Walker, D. I., and Kirkman, H., Eds., Western Australian Museum, Perth, 1996, 191.

206. Mann, K. H. and Lazier, J. R. N., *Dynamics of Marine Ecosystems*, Blackwell Scientific, Boston, 1991, chap. 2.

207. Ackerman, J. D. and Okubo, A., Reduced mixing in a marine macrophyte canopy, *Funct. Ecol.*, 7, 305, 1993.

208. Koch, E. W., Hydrodynamics, diffusion-boundary layers and photosynthesis of the seagrasses *Thalassia testudinum* and *Cymodocea nodosa*, *Mar. Biol.*, 118, 767, 1994.

209. Gambi, M. C., Nowell, A. R. M., and Jumars, P. A., Flume observations on flow dynamics in *Zostera marina* (eelgrass) beds, *Mar. Ecol. Prog. Ser.*, 61, 159, 1990.

210. Worcester, S. E., Effects of eelgrass beds on advection and turbulent mixing in low current and low shoot density environments, *Mar. Ecol. Prog. Ser.*, 126, 223, 1995.

211. Hurd, C. L., Harrison, P. J., and Druehl, L. D., Effect of seawater velocity on inorganic nitrogen uptake by morphologically distinct forms of *Macrocystis integrifolia* from wave-sheltered and exposed sites, *Mar. Biol.*, 126, 205, 1996.

212. Eckman, J. E., Duggins, D. O., and Sewell, A. T., Ecology of understory kelp environments. I. Effects of kelps on flow and particle transport near the bottom, *J. Exp. Mar. Biol. Ecol.*, 129, 173, 1989.

213. Koehl, M. A. R. and Alberte, R. S., Flow, flapping, and photosynthesis of *Nereocystis leutkeana*: a functional comparison of undulate and flat blade morphologies, *Mar. Biol.*, 99, 435, 1988.

214. Short, F. T., Effects of sediment nutrients on seagrasses: literature review and mesocosm experiment, *Aquat. Bot.*, 27, 41, 1987.

215. Short, F. T., Dennison, W. C., and Capone, D. G., Phosphorus-limited growth of the tropical seagrass *Syringodium filiforme* in carbonate sediments, *Mar. Ecol. Prog. Ser.*, 62, 169, 1990.

216. Perez, M., Romero, J., Duarte, C. M., and Sand-Jensen, K., Phosphorus limitation of *Cymodocea nodosa* growth, *Mar. Biol.*, 109, 129, 1991.

217. Fourqurean, J. W., Zieman, J. C., and Powell, G. V. N., Phosphorus limitation of primary production in Florida Bay: evidence from C:N:P ratios of the dominant seagrass *Thalassia testudinum*, *Limnol Oceanogr.*, 37, 162, 1992.

218. Fourqurean, J. W., Zieman, J. C., and Powell, G. V. N., Relationships between porewater nutrients and seagrasses in a subtropical carbonate environment, *Mar. Biol.*, 114, 57, 1992.

219. Erftemeijer, P. L. A. and Middelburg, J. J., Sediment-nutrient interactions in tropical seagrass beds: a comparison between a terrigenous and a carbonate sedimentary environment in south Sulawesi (Indonesia), *Mar. Ecol. Prog. Ser.*, 102, 187, 1993.

220. Erftemeijer, P. L. A., Differences in nutrient concentrations and resources between seagrass communities on carbonate and terrigenous sediments in South Sulawesi, Indonesia, *Bull. Mar. Sci.*, 54, 403, 1994.

221. Erftemeijer, P. L. A., Stapel, J., Smekens, M. J. E., and Drossaert, W. M. E., The limited effect of *in situ* phosphorus and nitrogen additions to seagrass beds on carbonate and terrigenous sediments in South Sulawesi, Indonesia, *J. Exp. Mar. Biol. Ecol.*, 182, 123, 1994.

222. Duarte, C. M., Merino, M., and Gallegos, M., Evidence of iron deficiency in seagrasses growing above carbonate sediments, *Limnol. Oceanogr.*, 40, 1153, 1995.

223. Lapointe, B. E., Macroalgal production and nutrient relations in oligotrophic areas of Florida Bay, *Bull. Mar. Sci.*, 44, 312, 1989.

224. Lapointe, B. E., Littler, M. M., and Littler, D. E. , A comparison of nutrient-limited productivity in macroalgae from a Caribbean barrier reef and from a mangrove ecosystem, *Aquat. Bot.*, 28, 243, 1987.

225. Zimmerman, R. C. and Kremer, J. N., *In situ* growth and chemical composition of the giant kelp, *Macrocystis pyrifera*: response to temporal changes in ambient nutrient availability, *Mar. Ecol. Prog. Ser.*, 27, 277, 1986.

226. Chapman, A. O. and Craigie, J. S., Seasonal growth in *Lamineria longicruris*: relations with dissolved inorganic nutrients and internal reserves of nitrogen, *Mar. Biol.*, 40, 197, 1977.

227. Hemminga, M. A., Harrison, P. G., and van Lent, F., The balance of nutrient losses and gains in seagrass meadows, *Mar. Ecol. Prog. Ser.*, 71, 85, 1991.

228. Hemminga, M. A., Koutstaal, B. P., van Soelen, J., and Merks, A. G. A., The nitrogen supply to intertidal eelgrass (*Zostera marina*), *Mar. Biol.*, 118, 223, 1994.

229. Perez-Llorens, J. L., de Visscher, P., Nienhuis, P. H., and Niell, F. X., Light-dependent uptake, translocation and foliar release of phosphorus by the intertidal seagrass *Zostera noltii* Hornem., *J. Exp. Mar. Biol. Ecol.*, 166, 165, 1993.

230. Short, F. T., Burdick, D. M., and Kaldy, III, J. E., Mesocosm experiments quantify the effects of eutrophication on eelgrass, *Zostera marina*, *Limnol. Oceanogr.*, 40, 740, 1995.

231. Neckles, H. A., Wetzel, R. L., and Orth, R. J., Relative effects of nutrient enrichment and grazing on epiphyte-macrophyte (*Zostera marina* L.) dynamics, *Oecologia*, 93, 285, 1993.

232. Williams, S. L. and Ruckelshaus, M. H., Effects of nitrogen availability and herbivory on eelgrass (*Zostera marina*) and epiphytes, *Ecology*, 74, 904, 1993.

233. Peterson, C. J. G., The sea bottom and its production of fish food. A survey of work done in connection with valuation of the Danish waters from 1883–1917, *Rep. Danish Biol. Sta.*, 25, 1, 1918.

234. Dayton, P. K., Tegner, M. J., Parnell, P. E., and Edwards, P. B., Temporal and spatial patterns of disturbance and recovery in a kelp forest community, *Ecol. Monogr.*, 62, 421, 1992.

235. Neighbors, M. A. and Horn, M. H., Nutritional quality of macrophytes eaten and not eaten by two temperate-zone herbivorous fishes: a multivariate comparison, *Mar. Biol.*, 108, 471, 1991.

236. de Iongh, H. H., Wenno, B. J., and Meelis, E., Seagrass distribution and seasonal biomass changes in relation to dugong grazing in the Moluccas, East Indonesia, *Aquat. Bot.*, 50, 1, 1995.

237. Lanyon, J., Limpus, C. J., and Marsh, H., Dugongs and turtles: grazers in the seagrass system, in *Biology of Seagrasses*, Larkum, A. W. D., McComb, A. J., and Shepard, S. A., Eds., Elsevier, Amsterdam, 1989, chap. 18.

238. Nienhuis, P. H. and Groenendijk, A. M., Consumption of eelgrass (*Zostera marina*) by birds and invertebrates: an annual budget, *Mar. Ecol. Prog. Ser.*, 29, 29, 1986.

239. Schmitt, T. M., Hay, M. E., and Lindquist, N., Constraints on chemically mediated coevolution: multiple functions for seaweed secondary metabolites, *Ecology*, 76, 107, 1995.

240. Thayer, G. W., Bjorndal, K. A., Ogden, J. C., Williams, S. L., and Zieman, J. C., Role of larger herbivores in seagrass communities, *Estuaries*, 7, 351, 1984.

241. Klumpp, D. W., Howard, R. K., and Pollard, D. A., Trophodynamics and nutritional ecology of seagrass communities, in *Biology of Seagrasses*, Larkum, A. W. D., McComb, A. J., and Shepard, S. A., Eds., Elsevier, Amsterdam, 1989, chap. 13.

242. Klumpp, D. W., Salita-Espinosa, J. S., and Fortes, M. D., The role of epiphytic periphyton and macroinvertebrate grazers in the trophic flux of a tropical seagrass community, *Aquat. Bot.*, 43, 327, 1992.

243. Klumpp, D. W., Salita-Espinosa, J. T., and Fortes, M. D., Feeding ecology and trophic role of sea urchins in a tropical seagrass community, *Aquat. Bot.*, 45, 205, 1993.

244. Duarte, C. M., Nutrient concentration of aquatic plants: patterns across species, *Limnol. Oceanogr.*, 37, 882, 1992.

245. Romero, J., Pergent, G., Pergent-Martini, C., Mateo, M.-A., and Regnier, C., The detritic compartment in a *Posidonia oceanica* meadow: litter features, decomposition rates, and mineral stocks, *P.S.Z.N.I. Mar. Ecol.*, 13, 69, 1992.

246. Buchsbaum, R., Valiela, I., Swain, T., Dzierzeski, M., and Allen, S., Available and refractory nitrogen in detritus of coastal vascular plants and macroalgae, *Mar. Ecol. Prog. Ser.*, 72, 131, 1991.

247. Harrison, P. G., Detrital processing in seagrass systems: a review of factors affecting decay rates, remineralization and detritivory, *Aquat. Bot.*, 23, 263, 1989.

248. Opsahl, S. and Benner, R., Decomposition of senescent blades of the seagrass *Halodule wrightii* in a subtropical lagoon, *Mar. Ecol. Prog. Ser.*, 94, 191, 1993.

249. Kristensen, E., Decomposition of macroalgae, vascular plants and sediment detritus on seawater: use of stepwise thermogravimetry, *Biogeochemistry*, 26, 1, 1994.

250. Blum, L. K. and Mills, A. L., Microbial growth and activity during the initial stages of seagrass decomposition, *Mar. Ecol. Prog. Ser.*, 70, 73, 1991.

251. Kenworthy, W. J. and Thayer, G. W., Production and decomposition of the roots and rhizomes of seagrasses, *Zostera marina* and *Thalassia testudinum*, in temperate and subtropical marine ecosystems, *Bull. Mar. Sci.*, 35, 364, 1984.

252. Peduzzi, P. and Herndl, G. J., Decomposition and significance of seagrass leaf litter (*Cymodocea nodosa*) for the microbial food web in coastal waters (Gulf of Trieste, Northen Adriatic Sea), *Mar. Ecol. Prog. Ser.*, 71, 163, 1991.

253. Newell, R. C., The energetics of detritus utilisation in coastal lagoons and nearshore waters, *Oceanol. Acta*, SP, 347, 1982.

254. Robinson, J. D., Mann, K. H., and Novitsky, J. A., Conversion of the particulate fraction of seaweed detritus to bacterial biomass, *Limnol. Oceanogr.*, 27, 1072, 1982.

255. Moriarty, D. J. W. and Boon, P. I., Interactions of seagrasses with sediment and water, in *Biology of Seagrasses*, Larkum, A. W. D., McComb, A. J., and Shepard, S. A., Eds., Elsevier, Amsterdam, 1989, chap. 15.

256. Romero, J., Perez, M., Mateo, M. A., and Sala, E., The belowground organs of the Mediterranean seagrass *Posidonia oceanica* as a biogeochemical sink, *Aquat. Bot.*, 47, 13, 1994.

257. Pollard, P. C. and Kogure, K., Bacterial decomposition of detritus in a tropical seagrass (*Syringodium isoetifolium*) ecosystem, measured with [methyl-^{3}H] thymidine, *Aust. J. Mar. Freshw. Res.*, 44, 155, 1993.

258. Pollard, P. C. and Moriarty, D. J. W., Organic carbon decomposition, primary and bacterial productivity, and sulphate reduction, in tropical seagrass beds of the Gulf of Carpentaria, Australia, *Mar. Ecol. Prog. Ser.*, 69, 149, 1991.

259. Hines, M. E., The role of certain infauna and vascular plants in the mediation of redox reactions in marine sediments, in *Diversity of Environmental Biogeochemistry*, Berthelin, J., Ed., Elsevier, Amsterdam, 1991, 275.

260. Pedersen, M. F. and Borum, J., Phosphorus recycling in the seagrass *Zostera marina* L., in *Biology and Ecology of Shallow Coastal Waters*, Eleftheriou, A., Ansell, A. D., and Smith, C. J., Eds., Olsen & Olsen, Fredensborg, 1995, 45.

261. Blackburn, T. H., Nedwell, D. B., and Wiebe, W. J., Active mineral cycling in a Jamaican seagrass sediment, *Mar. Ecol. Prog. Ser.*, 110, 233, 1994.

262. Welsh, D. T., Bourgues, S., deWit, R., and Herbert, R. A., Seasonal variations in nitrogen-fixation (acetylene reduction) and sulfate-reduction rates in the rhizosphere of *Zostera noltii*: nitrogen fixation by sulfate-reducing bacteria, *Mar. Biol.*, 125, 619, 1996.

263. Pedersen, M. F. and Borum, J., An annual nitrogen budget for a seagrass *Zostera marina* population, *Mar. Ecol. Prog. Ser.*, 101, 169, 1993.

264. Caffrey, J. M. and Kemp, W. M., Nitrogen cycling in sediments with estuarine populations of *Potamogeton perfoliatus* and *Zostera marina*, *Mar. Ecol. Prog. Ser.*, 66, 147, 1990.

265. Caffrey, J. M. and Kemp, W. M., Influence of the submersed plant, *Potamogeton perfoliatus*, on nitrogen cycling in estuarine sediments, *Limnol. Oceanogr.*, 37, 1483, 1992.

266. Smith, B. D., Cabot, E. L., and Foreman, R. E., Seaweed detritus versus benthic diatoms as important food resources for two dominant subtidal gastropods, *J. Exp. Mar. Biol. Ecol.*, 92, 143, 1985.

267. Duggins, D. O., Simenstad, C. A., and Estes, J. A., Magnification of secondary production by kelp detritus in coastal marine ecosystems, *Science*, 245, 170, 1989.

268. Velimirov, B. and Walenta-Simon, M., Seasonal changes in specific growth rates, production and biomass of a bacterial community in the water column above a Mediterranean seagrass system, *Mar. Ecol. Prog. Ser.*, 80, 237, 1992.

269. Velimirov, B. and Walenta-Simon, M., Bacterial growth rates and productivity within a seagrass system: seasonal variations in a *Posidonia oceanica* bed, *Mar. Ecol. Prog. Ser.*, 96, 101, 1993.

270. Thybo-Christensen, M., Rasmussen, M. B., and Blackburn, T. H., Nutrient fluxes and growth of *Cladophora sericea* in a shallow Danish Bay, *Mar. Ecol. Prog. Ser.*, 100, 273, 1993.

271. Duggins, D. O. and Eckman, J. E., The role of kelp detritus in the growth of benthic suspension feeders in an understory kelp forest, *J. Exp. Mar. Biol. Ecol.*, 176, 53, 1994.

272. Velimirov, B., Organic matter derived from a seagrass meadow: origin, properties, and quality of particles, *P.S.Z.N.I. Mar. Ecol.*, 8, 143, 1987.

273. Dauby, P., Bale, A. J., Bloomer, N., Canon, C., Ling, R. D., Norro, A., Robertson, J. E., Simon, A., Theate, J.-M., Watson, A. J., and Frankignoulle, M., Particle fluxes over a Mediterranean seagrass bed: a one year case study, *Mar. Ecol. Prog. Ser.*, 126, 233, 1995.

274. Thayer, G. W., Adams, S. M., and La Croix, W. W., Structural and functional aspects of a recently established *Zostera marina* community, *in Estuarine Research*, Vol. 1, Cronin, L. E., Ed., Academic Press, New York, 1975, 518.

275. Newell, R. C. and Field, J. G., The contribution of bacteria and detritus to carbon and nitrogen flow in a benthic community, *Mar. Biol. Lett.*, 4, 23, 1983.

276. Newell, R. C. and Field, J. G., Relative flux of carbon and nitrogen in a kelp-dominated system, *Mar. Biol. Lett.*, 4, 249, 1983.

277. Field, J. G., Flow patterns of energy and matter, in *Marine Ecology*, Vol. 5, Pt. 2, Kinne, O., Ed., John Wiley & Sons, Chichester, 1983, 758.

278. Hawkins, S. J., Hartnoll, R. G., Kain, J. M., and Norton, T. A., Plant-animal interactions on hard substrata in the North-East Atlantic, in *Plant-Animal Interactions in the Marine Benthos*, John, D. M., Hawkins, S. J., and Price, J. H., Eds., Clarendon Press, Oxford, 1992, 1.

279. Peduzzi, P. and Herndl, G. J., Mucus trails in the rocky intertidal: a highly active microenvironment, *Mar. Ecol. Prog. Ser.*, 75, 267, 1991.

280. Erftemeijer, P. L. A. and Middelburg, J. J., Mass balance constraints on nutrient cycling in tropical seagrass beds, *Aquat. Bot.*, 50, 21, 1995.

281. Frankignoulle, M. and Bouquegneau, J. M., Seasonal variation of the diel carbon budget of a marine macrophyte ecosystem, *Mar. Ecol. Prog. Ser.*, 38, 197, 1987.

282. Thresher, R. E., Nichols, P. D., Gunn, J. S., Bruce, B. D., and Furlani, D. M., Seagrass detritus as the basis of a coastal planktonic food chain, *Limnol. Oceanogr.*, 37, 1754, 1992.

283. Fortes, M. D., Philippine seagrasses: status and perspectives, in *Proceedings of the Third ASEAN-Australia Symposium on Living Coastal Resources*, Vol. 1, Sudara, S., Wilkinson, C. R., and Ming, C. L., Eds., Chulalongkorn University, Bangkok, 1994, 291.

284. Hemminga, M. A., Slim, Kazungu, J., Ganssen, G. M., Nieuwenhuize, J., and Kruyt, N. M., Carbon outwelling from a mangrove forest with adjacent seagrass beds and coral reefs (Gazi Bay, Kenya), *Mar. Ecol. Prog. Ser.*, 106, 291, 1994.

285. Kitheka, J. U., Water circulation and coastal trapping of brackish water in a tropical mangrove-dominated bay in Kenya, *Limnol. Oceanogr.*, 41, 169, 1996.

286. Schiel, D. R., Kelp communities, in *Marine Biology*, Hammond, L. S. and Synnot, R. N., Eds., Longman Cheshire, Melbourne, 1994, chap. 19.

287. Niell, F. X., Rocky intertidal benthic systems in temperate seas: a synthesis of their functional performances, *Helgo. Wiss. Meersunters.*, 30, 315, 1977.

288. Stevenson, J. C., Comparative ecology of submerged grass beds in freshwater, estuarine, and marine environments, *Limnol. Oceanogr.*, 33, 867, 1988.
289. Fortes, M. D., Comparative study of structure and productivity of seagrass communities in the ASEAN region, in *Third ASEAN Science and Technology Conference Proceedings*, Vol. 6., *Marine Science: Living Coastal Resources*, Department of Zoology, National University of Singapore and National Science Technology Board, Singpore, 1992, 223.
290. Lindeboom, H. J. and Sandee, A. J. J., Production and consumption of tropical seagrass fields in eastern Indonesia measured with bell jars and microelectrodes, *Neth. J. Sea Res.*, 23, 181, 1989.
291. Pollard, P. C. and Kogure, K., The role of epiphytic and epibenthic algal productivity in a tropical seagrass, *Syringodium isoetifolium* (Aschers.) Dandy, community, *Aust. J. Mar. Freshwater Res.*, 44, 141, 1993.
292. Zieman, J. C., Thayer, G. W., Robblee, M. B., and Zieman, R. T., Production and export of seagrasses from a tropical bay, in *Ecological Processes in Coastal and Marine Systems*, Livingston, R. J., Ed., Plenum Press, New York, 1979, 21.
293. Nojima, S. and Mukai, H., The rate and fate of production of seagrass debris in cages over a *Syringodium isoetifolium* (Aschers.) Dandy meadow in Fiji, in *Seagrass Biology: Proceedings of an International Workshop*, Kuo, J., Phillips, R. C., Walker, D. I., and Kirkman, H., Eds., Western Australian Museum, Perth, 1996, 149.
294. Bach, S. D., Thayer, G. W., and LaCroix, M. W., Export of detritus from eelgrass (*Zostera marina*) beds near Beaufort, North Carolina, U.S., *Mar. Ecol. Prog. Ser.*, 28, 265, 1986.
295. Fry, B. and Virnstein, R. W., Leaf production and export of the seagrass *Syringodium filiforme* Kutz. in Indian River lagoon, Florida, *Aquat. Bot.*, 30, 261, 1988.
296. Sorokin, Y. I., *Coral Reef Ecology*, Springer-Verlag, Berlin, 1995, 465 pp.
297. Smith, S. V., Coral reef area and the contributions of reefs to processes and resources of the world's oceans, *Nature*, 273, 225, 1978.
298. Hatcher, B. G., Coral reef primary productivity: a beggar's banquet, *Trends in Ecol. Evol.*, 3, 106, 1988.
299. Shreeve, J., Are algae — not coral — reefs' master builders?, *Science*, 271, 597, 1996.
300. Barnes, D. J. and Chalker, B. E., Calcification and photosynthesis in reef-building corals and algae, in *Ecosystems of the World*. Vol. 25. *Coral Reefs*, Dubinsky, Z., Ed., Elsevier, Amsterdam, 1990, chap. 6.
301. Smith, S. V., The Houtman Abrolhas Islands: carbon metabolism of coral reefs at high latitude, *Limnol. Oceanogr.*, 26, 612, 1981.
302. Kinsey, D. W., Metabolism, calcification and carbon production. I. System-level studies, *Proc. Fifth Coral Reef Congr.*, 4, 505, 1985.
303. Muscatine, L., The role of symbiotic algae in carbon and energy flux in reef corals, in *Ecosystems of the World*. Vol. 25. *Coral Reefs*, Dubinsky, Z., Ed., Elsevier, Amsterdam, 1990, chap. 4.
304. Jokiel, P. L. and Morrissey, J. I., Influence of size on primary production in the reef coral *Pocillopora damicornis* and the macroalga *Acanthophora spicifera*, *Mar. Biol.*, 91, 15, 1986.
305. Dubinsky, Z. and Jokiel, P. L., Ratio of energy and nutrient fluxes regulates symbiosis between zooxanthellae and corals, *Pac. Sci.*, 48, 313, 1994.
306. Snidvongs, A. and Kinzie, III, R. A., Effects of nitrogen and phosphorus enrichment on *in vivo* symbiotic zooxanthellae of *Pocillopora damicornis*, *Mar. Biol.*, 118, 705, 1994.

307. Lesser, M. P., Weis, V. M., Patterson, M. R., and Jokiel, P. L., Effects of morphology and water motion on carbon delivery and productivity in the reef coral, *Pocillopora damicornis* (Linnaeus): diffusion barriers, inorganic carbon limitation, and biochemical plasticity, *J. Exp. Mar. Biol. Ecol.*, 178, 153, 1994.

308. Kuhl, M., Cohen, Y., Dalsgaard, T., Jorgensen, B. B., and Revsbech, N. P., Microenvironment and photosynthesis of zooxanthellae in scleractian corals studied with microsensors for O_2, pH and light, *Mar. Ecol. Prog. Ser.*, 117, 159, 1995.

309. Gladfelter, E., Michel, G., and Sanfelici, A., Metabolic gradients along a branch of the reef coral *Acropora palmata*, *Bull. Mar. Sci.*, 44, 1166, 1989.

310. Marshall, A. T., Calcification in hermatypic and ahermatypic corals, *Science*, 271, 637, 1996.

311. Klumpp, D. W. and McKinnon, A. D., Temporal and spatial patterns in primary production of a coral-reef epilithic algal community, *J. Exp. Mar. Biol. Ecol.*, 131, 1, 1989.

312. Williams, S. L. and Carpenter, R. C., Photosynthesis/photon flux density relationships among components of coral reef algal turfs, *J. Phycol.*, 26, 40, 1990.

313. Alongi, D. M., The role of soft-bottom benthic communities in tropical mangrove and coral reef ecosystems, *Rev. Aquat. Sci.*, 1, 243, 1989.

314. Furnas, M. J., Phytoplankton biomass and primary production in semi-enclosed reef lagoons of the central Great Barrier Reef, Australia, *Coral Reefs*, 9, 1, 1990.

315. Delesalle, B. and Sournia, A., Residence time of water and phytoplankton biomass in coral reef lagoons, *Cont. Shelf Res.*, 12, 939, 1992.

316. Delesalle, B., Pichon, M., Frankignoulle, M., and Gattuso, J. P., Effects of a cyclone on coral reef phytoplankton biomass, primary production and composition (Moorea Island, French Polynesia), *J. Plank. Res.*, 15, 1413, 1993.

317. Steneck, R. S., Herbivory on coral reefs: a synthesis, *Proc. Sixth Coral Reef Symp.*, 1, 37, 1988.

318. Hay, M. E., Patterns of fish and urchin grazing on Caribbean coral reefs: are previous results typical?, *Ecology*, 65, 446, 1984.

319. Carpenter, R. C., Partitioning herbivory and its effects on coral reef algal communities, *Ecol. Monogr.*, 56, 345, 1986.

320. Hatcher, B. G., Grazing in coral reef ecosystems, in *Perspectives on Coral Reefs*, Barnes, D. J., Ed., Australian Institute of Marine Science, Townsville, 1983, chap. 10.

321. McClanahan, T. R., Resource utilization, competition, and predation: a model and example from coral reef grazers, *Ecol. Model.*, 61, 195, 1992.

322. Klumpp, D. W. and Polunin, N. V. C., Partitioning among grazers of food resources within damselfish territories on a coral reef, *J. Exp. Mar. Biol. Ecol.*, 125, 145, 1989.

323. Klumpp, D. W. and Polunin, N. V. C., Algal production, grazers and habitat partitioning on a coral reef: positive correlation between grazing rate and food availability, in *Trophic Relationships in the Marine Environment*, Barnes, M. and Gibson, R. N., Eds., Aberdeen University, Aberdeen, 1990, 372.

324. Polunin, N. V. C. and Klumpp, D. W., A trophodynamic model of fish production on a windward reef tract, in *Plant-Animal Interactions in the Marine Benthos*, John, D. M., Hawkins, S. J., and Price, J. H., Eds., Clarendon Press, Oxford, 1992, 213.

325. Polunin, N. V. C. and Klumpp, D. W., Algal food supply and grazer demand in a very productive coral-reef zone, *J. Exp. Mar. Biol. Ecol.*, 164, 1, 1992.

326. Alongi, D. M., Population structure and trophic composition of the free-living nematodes inhabiting carbonate sands of Davies Reef, Great Barrier Reef, *Aust. J. Mar. Freshwater Res.*, 37, 609, 1986.
327. Fabricius, K. E., Benayahu, Y., and Genin, A., Herbivory in asymbiotic soft corals, *Science*, 268, 90, 1995.
328. Sorokin, Y. I., Aspects of trophic relations, productivity and energy balance in coral reef ecosystems, in *Ecosystems of the World*. Vol. 25. *Coral Reefs*, Dubinsky, Z., Ed., Elsevier, Amsterdam, 1990, chap. 14.
329. Fabricius, K. E. and Klumpp, D. W., Widespread mixotrophy in reef-inhabiting soft corals: the influence of depth, and colony expansion and contraction on photosynthesis, *Mar. Ecol. Prog. Ser.*, 125, 195, 1995.
330. Glynn, P. W., Feeding ecology of selected coral-reef macroconsumers: patterns and effects on coral community structure, in *Ecosystems of the World*. Vol. 25. *Coral Reefs*, Dubinsky, Z., Ed., Elsevier, Amsterdam, 1990, chap. 13.
331. Johnson, C., Klumpp, D., Field, J., and Bradbury, R. Carbon flux on coral reefs: effects of large shifts in community structure, *Mar. Ecol. Prog. Ser.*, 126, 123, 1995.
332. Hamner, W. M., Jones, M. S., Carleton, J. H., Hauri, I. R., and Williams, D. McB., Zooplankton, planktivorous fish, and water currents on a windward reef face: Great Barrier Reef, Australia, *Bull. Mar. Sci.*, 42, 459, 1988.
333. Hatcher, B. G., The role of detritus in the metabolism and secondary production of coral reef ecosystems, in *Proceedings of the Great Barrier Reef Conference*, Baker, J. T., Carter, R. M., Sammarco, P. W., and Stark, K. P., Eds., James Cook University and Australian Institute of Marine Science, Townsville, 1983, 317.
334. Alongi, D. M., Detritus in coral reef ecosystems: fluxes and fates, *Proc. Sixth Int. Coral Reef Symp.*, 1, 29, 1988.
335. Wolanski, E., Some evidence for boundary mixing near coral reefs, *Limnol. Oceanogr.*, 32, 735, 1987.
336. Koop, K. and Larkum, A. W. D., Deposition of organic material in a coral reef lagoon, One Tree Island, Great Barrier Reef, *Estuarine Coastal Shelf Sci.*, 25, 1, 1987.
337. Charpy, L. and Charpy-Roubaud, C. J., Particulate organic matter fluxes in a Tuamotu atoll lagoon (French Polynesia), *Mar. Ecol. Prog. Ser.*, 71, 53, 1991.
338. Chardy, P. and Clavier, J., An attempt to estimate the carbon budget for the south west lagoon of New Caledonia, *Proc. Sixth Int. Coral Reef Symp.*, 2, 521, 1988.
339. Hansen, J. A., Klumpp, D. W., Alongi, D. M., Dayton, P. K., and Riddle, M. J., Detrital pathways in a coral reef lagoon. II. Detritus deposition, benthic microbial biomass and production, *Mar. Biol.*, 113, 363, 1992.
340. Coffroth, M. A., Mucous sheet formation on poritid corals: an evaluation of coral mucus as a nutrient source on reefs, *Mar. Biol.*, 105, 39, 1990.
341. Riddle, M. J., Alongi, D. M., Dayton, P. K., Hansen, J. A., and Klump, D. W., Detrital pathways in a coral reef lagoon. I. Macrofaunal biomass and estimates of production, *Mar. Biol.*, 104, 109, 190.
342. Hansen, J. A., Alongi, D. M., Moriarty, D. J. W., and Pollard, P. C., The dynamics of benthic microbial communities at Davies Reef, Great Barrier Reef, *Coral Reefs*, 6, 63, 1987.
343. Boucher, G. and Clavier, J., Contribution of benthic biomass to overall metabolism in New Caledonia lagoon sediments, *Mar. Ecol. Prog. Ser.*, 64, 271, 1990.
344. Villiers, L., Christien, D., and Severe, A., Investigations sur l'ecologie des sables lagonaires biogenes de l'atoll de Mururoa (Tuamotu, Polynesie Francaise), *ORSTOM Oceanographie Notes Documents No. 36*, 1987, 98 pp.

345. Skyring. G. W., Anaerobic microbial processes in coral reef sediments, *Proc. Fifth Int. Coral Reef Cong.*, 3, 421, 1985.

346. Nedwell, D. B. and Blackburn, T. H., Anaerobic metabolism in lagoon sediments from Davies Reef, Great Barrier Reef, *Estuarine Coastal Shelf Sci.*, 25, 347, 1987.

347. Boucher, G., Clavier, J., and Garrigue, C., Oxygen and carbon dioxide fluxes at the water-sediment interface of a tropical lagoon, *Mar. Ecol. Prog. Ser.*, 107, 185, 1994.

348. Johnstone, R. W., Koop, K., and Larkum, A. W. D., Physical aspects of coral reef lagoon sediments in relation to detritus processing and primary production, *Mar. Ecol. Prog. Ser.*, 66, 273, 1990.

349. Hines, M. E., Microbial biogeochemistry in shallow water sediments of Bermuda, *Proc. Fifth Coral Reef Congr.*, 3, 427, 1985.

350. Moriarty, D. J. W., Pollard, P. C., Hunt, W. G., Moriarty, C. M., and Wassenberg, T. J., Productivity of bacteria and microalgae and the effect of grazing by holothurians in sediments on a coral reef flat, *Mar. Biol.*, 85, 293, 1985.

351. Klinger, T. S., Johnson, C. R., and Jell, J., Sediment utilization, feeding-niche breadth, and feed-niche overlap of Aspidochirotida (Echinodermata: Holothuroidea) at Heron Island, Great Barrier Reef, in *Echinoderms through Space and Time*, David, B., Guille, A., Feral, J. P., and Roux, M., Eds., Balkema, Rotterdam, 1994, 523.

352. Hansen, J. A. and Skilleter, G. A., Effects of the gastropod *Rhinoclavis aspera* (Linnaeus, 1758) on microbial biomass and productivity in coral-reef sediments, *Aust. J. Mar. Freshwater Res.*, 45, 569, 1994.

353. Ducklow, H. W., The biomass, production and fate of bacteria in coral reefs, in *Ecosystems of the World.* Vol. 25. *Coral Reefs*, Dubinsky, Z., Ed., Elsevier, Amsterdam, 1990, chap. 10.

354. Ayukai, T., Retention of phytoplankton and planktonic microbes on coral reefs within the Great Barrier Reef, Australia, *Coral Reefs*, 14, 141, 1995.

355. D'Elia, C. F., The cycling of essential elements in coral reefs, in *Concepts of Ecosystem Ecology*, Pomeroy, L. R. and Alberts, J. J., Eds., Springer-Verlag, New York, 1988, chap. 10.

356. D'Elia, C. F. and Wiebe, W. J., Biogeochemical nutrient cycles in coral-reef ecosystems, in *Ecosystems of the World.* Vol. 25. *Coral Reefs*, Dubinsky, Z., Ed., Elsevier, Amsterdam, 1990, chap. 3.

357. Smith, S. V., Phosphorus versus nitrogen limitation in the marine environment, *Limnol. Oceanogr.*, 29, 1149, 1984.

358. Capone, D. G., Dunham, S. E., Horrigan, S. G., and Duguay, L. E., Microbial nitrogen transformations in unconsolidated coral reef sediments, *Mar. Ecol. Prog. Ser.*, 80, 75, 1992.

359. Shashar, N., Cohen, Y., Loya, Y., and Sar, N., Nitrogen fixation (acetylene reduction) in stony corals: evidence for coral-bacteria interactions, *Mar. Ecol. Prog. Ser.*, 111, 259, 1994.

360. Wafar, M., Wafar, S., and David, J. J., Nitrification in reef corals, *Limnol. Oceanogr.*, 35, 725, 1990.

361. Schiller, C. and Herndl, G. J., Evidence of enhanced microbial activity in the interstitial space of branched corals: possible implications for coral metabolism, *Coral Reefs*, 7, 179, 1989.

362. Corredor, J. E., Wilkinson, C. R., Vicente, V. P., Morell, J. M., and Otero, E., Nitrate release by Caribbean reef sponges, *Limnol. Oceanogr.*, 33, 114, 1988.

363. Andrews, J. C. and Muller, H., Space-time variability of nutrients in a lagoonal patch reef, *Limnol. Oceanogr.*, 28, 215, 1983.

364. Hatcher, A. I. and Frith, C. A., The control of nitrate and ammonium concentrations in a coral reef lagoon, *Coral Reefs*, 4, 101, 1985.

365. Boucher, G., Clavier, J., and Garrigue, C., Estimation of bottom ammonium affinity in the New Caledonia lagoon, *Coral Reefs*, 13, 13, 1994.

366. Hopkinson, Jr., C. S., Sherr, B. F., and Ducklow, H. W., Microbial regeneration of ammonium in the water-column of Davies Reef, Australia, *Mar. Ecol. Prog. Ser.*, 41, 147, 1987.

367. Bishop, J. W. and Greenwood, J. G., Nitrogen excretion by some demersal macrozooplankton in Heron and One Tree Reefs, Great Barrier Reef, Australia, *Mar. Biol.*, 120, 447, 1994.

368. Sorokin, Y. I., Phosphorus metabolism in coral reef communities: dynamics in the water column, *Aust. J. Mar. Freshwater Res.*, 41, 775, 1990.

369. Sorokin, Y. I., Phosphorus metabolism in coral reef communities: exchange between the water column and bottom biotopes, *Hydrobiologia*, 242, 105, 1992.

370. Atkinson, M. J., Productivity of Enewetak Atoll reef flats predicted from mass transfer relationships, *Cont. Shelf Res.*, 12, 799, 1992.

371. Atkinson, M. J., Are coral reefs nutrient-limited?, *Proc. Sixth Int. Coral Reef Symp.*, 1, 157, 1988.

372. Bilger, R. W. and Atkinson, M. J., Anamalous mass transfer of phosphate on coral reef flats, *Limnol. Oceanogr.*, 37, 261, 1992.

373. Atkinson, M. J. and Bilger, R. W., Effects of water velocity on phosphate uptake in coral reef-flat communities, *Limnol. Oceanogr.*, 37, 273, 1992.

374. Charpy, L. and Charpy-Roubaud, C. J., Phosphorus budget in an atoll lagoon, *Proc. Sixth Int. Coral Reef Symp.*, 2, 547, 1988.

375. Rougerie, F. and Wauthy, B., The endo-upwelling concept: a new paradigm for solving an old paradox, *Proc. Sixth Int. Coral Reef Symp.*, 3, 21, 1988.

376. Ferrier, M. D., Net uptake of dissolved free amino acids by four scleractinian corals, *Coral Reefs*, 10, 183, 1991.

377. Muscatine, L., Daily budgets of photosynthetically-fixed carbon in light and shade-adapted corals, in *Proceedings of the Great Barrier Reef Conference*, Baker, J. T., Carter, R. M., Sammarco, P. W., and Stark, K. P., Eds., James Cook University and Australian Institute of Marine Science, Townsville, 1983, 341.

378. Edmunds, P. J. and Spencer Davies, P., An energy budget for *Porites porites* (Scleractina), *Mar. Biol.*, 92, 339, 1986.

379. Davies, P. S., Effect of daylight variations on the energy budgets of shallow-water corals, *Mar. Biol.*, 108, 137, 1991.

380. Gattuso, J.-P. and Jaubert, J., Effect of light on oxygen and carbon dioxide fluxes and on metabolic quotients measured *in situ* in a zooxanthellate coral, *Limnol. Oceanogr.*, 1990.

381. Patterson, M. R., Sebens, K. P., and Olson, R. R., *In situ* measurements of flow effects on primary production and dark respiration in reef corals, *Limnol. Oceanogr.*, 36, 936, 1991.

382. Atkinson, M. J., Kotler, E., and Newton, P., Effects of water velocity on respiration, calcification, and ammonium uptake of a *Porites compressa* community, *Pac. Sci.*, 48, 296, 1994.

383. Bythell, J. C., A total nitrogen and carbon budget for the elkhorn coral *Acropora palmata* (Lamarck), *Proc. Sixth Coral Reef Symp.*, 2, 535, 1988.

384. Fabricius, K. E., Genin, A., and Benayahu, Y., Flow-dependent herbivory and growth in zooxanthellate-free soft corals, *Limnol. Oceanogr.*, 40, 1290, 1995.

385. Odum, H. T. and Odum, E. P., Trophic structure and productivity of a windward coral reef community on Eniwetak Atoll, *Ecol. Monogr.*, 25, 291, 1955.

386. Sorokin, Y. I., Parameters of productivity and metabolism of coral reef ecosystems off central Vietnam, *Estuarine Coastal Shelf Sci.*, 33, 259, 1991.

387. Wiebe, W. J., Coral reef energetics, in *Concepts of Ecosystem Ecology*, Pomeroy, L. R. and Alberts, J. J., Eds., Springer-Verlag, New York, 1988, chap. 11.

388. Polovina, J. J., Model of a coral reef ecosystem. I. The ECOPATH model and its application to French Frigate Shoals, *Coral Reefs*, 3, 1, 1984.

389. Atkinson, M. J. and Grigg, R. W., Model of a coral reef ecosystem. II. Gross and net benthic primary production at French Frigate Shoals, Hawaii, *Coral Reefs*, 3, 13, 1984.

390. Grigg, R. W., Polovina, J. J., and Atkinson, M. J., Model of a coral reef ecosystem. III. Resource limitation, community regulation, fisheries yield and resource management, *Coral Reefs*, 3, 23, 1984.

391. Arias-Gonzalez, J. E., Delesalle, B., Salvat, B., and Galzin, R., Trophic functioning of the Tiahura reef sector, Moorea Island, French Polynesia, *Coral Reefs*, in review.

392. Yap, H. T., Montebon, A. R. F., and Dizon, R. M., Energy flow and seasonality in a tropical coral reef flat, *Mar. Ecol. Prog. Ser.*, 103, 35, 1994.

393. Nakamori, T., Suzuki, A., and Iryu, Y., Water circulation and carbon flux on Shiraho coral reef of the Ryukyu Islands, Japan, *Cont. Shelf Res.*, 12, 951, 1992.

394. Smith, S. V., Mass balance in coral reef-dominated areas, in *Coastal-Offshore Ecosystem Interactions*, Jansson, B.-O., Ed., Springer-Verlag, Berlin, 1988, 209.

395. Crossland, C. J., Hatcher, B. G., and Smith, S. V., Role of coral reefs in global ocean production, *Coral Reefs*, 10, 55, 1991.

396. Kinsey, D. W., The coral reef: an owner-built, high-density, fully-serviced, self-sufficient housing estate in the desert — or is it?, *Symbiosis*, 10, 1, 1991.

397. Kinsey, D. W. and Hopley, D., The significance of coral reefs as global carbon sinks — response to Greenhouse, *Paleogeogr. Paleoclimatol. Paleoecol.*, 89, 363, 1991.

398. Ware, J. R., Smith, S. V., and Reaka-Kudla, M. L., Coral reefs: sources or sinks of atmospheric CO_2?, *Coral Reefs*, 11, 127, 1991.

399. Frankignoulle, M., Canon, C., and Gattuso, J.-P., Marine calcification as a source of carbon dioxide: positive feedback of increasing atmospheric CO_2, *Limnol. Oceanogr.*, 39, 458, 1994.

400. Kayanne, H., Suzuki, A., and Saito, H., Diurnal changes in the partial pressure of carbon dioxide in coral reef water, *Science*, 269, 214, 1995.

401. Gattuso, J.-P., Pichon, M., Delesalle, B., and Frankignoulle, M., Community metabolism and air-sea CO_2 fluxes in a coral reef ecosystem (Moorea, French Polynesia), *Mar. Ecol. Prog. Ser.*, 96, 259, 1993.

402. Gattuso, J.-P., Pichon, M., and Frangignoulle, M., Biological control of air-sea CO_2 fluxes: effect of photosynthetic and calcifying marine organisms and ecosystems, *Mar. Ecol. Prog. Ser.*, 129, 307, 1995.

403. Munro, J. L., The scope of tropical reef fisheries and their management, in *Reef Fisheries*, Polunin, N. V. C. and Roberts, C. M., Eds., Chapman & Hall, London, 1996, chap. 1.

404. Isdale, P. J., Fluorescent bands in massive corals record centuries of coastal rainfall, *Nature*, 310, 578, 1984.

405. Lough, J. M. and Barnes, D. J., Possible relationships between environmental variables and skeletal density in a coral colony from the central Great Barrier Reef, *J. Exp. Mar. Biol. Ecol.*, 134, 221, 1990.

406. Ogden, J., The influence of adjacent systems on the structure and function of coral reefs, *Proc. Sixth Int. Coral Reef Symp.*, 1, 123, 1988.

407. Holligan, P. M. and Reiners, W. A., Predicting the responses of the coastal zone to global change, *Adv. Ecol. Res.*, 22, 211, 1992.

408. Mantoura, R. F. C., Matrin, J. M., and Wollast, R., Introduction, in *Ocean Margin Processes in Global Change*, Mantoura, R. F. C., Martin, J. M., and Wollast, R., Eds., John Wiley & Sons, Chichester, 1991, 1.

409. Hekstra, G. P., Global warming and rising sea levels: the policy implications, *Ecologist*, 19, 4, 1989.

410. Pritchard, D. W., What is an estuary: physical standpoint, in *Estuaries*, Lauff, G. H., Ed., American Association for the Advancement of Science, Publication 83, Washington, D.C., 1967, 3.

411. Kjerfve, B., Estuarine geomorphology and physical oceanography, in *Estuarine Ecology*, Day, Jr., J. W., Hall, C. A. S., Kemp, W. M., and Yanez-Arancibia, A., John Wiley & Sons, New York, 1989, chap. 2.

412. Dronkers, J., Inshore/offshore water exchange in shallow coastal systems, in *Coastal-Offshore Ecosystem Interactions*, Jansson, B.-O., Ed., Springer-Verlag, Berlin, 1988, 3.

413. Burrell, D. C., Carbon flow in fjords, *Oceanogr. Mar. Biol. Ann. Rev.*, 26, 143, 1988.

414. Day, Jr., J. W., Hall, C. A. S., Kemp, W. M., and Yanez-Arancibia, A., *Estuarine Ecology*, John Wiley & Sons, New York, 1989, 558 pp.

415. Nienhuis, P. H., Nutrient cycling and foodwebs in Dutch estuaries, *Hydrobioligia*, 265, 15, 1993.

416. Cloern, J. E., Phytoplankton bloom dynamics in coastal ecosystems: a review with some general lessons from sustained investigation of San Francisco Bay, California, *Rev. Geophys.*, 34, 127, 1996.

417. Qasim, S. Z. and Wafar, M. V. M., Marine resources in the tropics, in *Tropical Resources. Ecology and Development*, Furtado, J. I., Morgan, W. B., Pfafflin, J. R., and Ruddle, K., Eds., Harwood Academic, London, 1990, 141.

418. Eyre, B., Nutrient biogeochemistry in the tropical Moresby River estuary system, north Queensland, Australia, *Estuarine Coastal Shelf Sci.*, 39, 15, 1994.

419. Soulemane, B., Production primaire dans une baie tropicale eutrophe: influence de la destratification, *Vie Milieu*, 40, 273, 1990.

420. Cole, B. E. and Cloern, J. E., An empirical model for estimating phytoplankton productivity in estuaries, *Mar. Ecol. Prog. Ser.*, 36, 299, 1987.

421. Heip, C. H. R., Goosen, N. K., Herman, P. M. J., Kromkamp, J., Middelburg, J. J., and Soetaert, K., Production and consumption of biological particles in temperate tidal estuaries, *Oceanogr. Mar. Biol. Ann. Rev.*, 33, 1, 1995.

422. Mallin, M. A. and Paerl, H. W., Effects of variable irradiance on phytoplankton productivity in shallow estuaries, *Limnol. Oceanogr.*, 37, 54, 1992.

423. Pennock, J. R. and Sharp, J. H., Temporal alternation between light and nutrient limitation of phytoplankton production in a coastal plain estuary, *Mar. Ecol. Prog. Ser.*, 111, 275, 1994.

424. Gilbert, P. M., Primary productivity and pelagic nitrogen cycling, in *Nitrogen Cycling in Coastal Marine Environments*, Blackburn, T. H. and Sorensen, J., Eds, John Wiley & Sons, Chichester, 1988, chap. 1.

425. Harrison, W. G., Regeneration of nutrients, in *Primary Productivity and Biogeochemical Cycles in the Sea*, Falkowski, P. G. and Woodhead, A. D., Eds., Plenum Press, New York, 1992, 385.

426. Horrigan, S. G., Montoya, J. P., Nevins, J. L., McCarthy, J. J., Ducklow, H., Goericke, R., and Malone, T., Nitrogenous nutrient transformations in the spring and fall in the Chesapeake Bay, *Estuarine Coastal Shelf Sci.*, 30, 369, 1990.

427. Ducklow, H. W. and Carlson, C. A., Oceanic bacterial production, *Adv. Microb. Ecol.*, 12, 113, 1992.

428. Shiah, F.-K. and Ducklow, H. W., Temperature regulation of heterotrophic bacterioplankton abundance, production, and specific growth rates in Chesapeake Bay, *Limnol. Oceanogr.*, 39, 1243, 1994.

429. Eldridge, P. M. and Sieracki, M. E., Biological and hydrodynamic regulation of the microbial food web in a periodically mixed estuary, *Limnol. Oceanogr.*, 38, 1666, 1993.

430. Fuhrman, J. A. and Noble, R. T., Viruses and protists cause bacterial mortality in coastal seawater, *Limnol. Oceanogr.*, 40, 1236, 1995.
431. Sanders, R. W. and Wickham, S. A., Planktonic protozoa and metazoa: predation, food quality and population control, *Mar. Microb. Food Web.*, 7, 197, 1993.
432. Dagg, M. J., Ingestion of phytoplankton by the micro- and mesoplankton communities in a productive subtropical estuary, *J. Plank. Res.*, 17, 845, 1995.
433. Hill, A. E., A mechanism for horizontal zooplankton transport by vertical migration in tidal currents, *Mar. Biol.*, 111, 485, 1991.
434. Tyson, P., *Sedimentary Organic Matter*, Chapman & Hall, London, 1995, 615 pp.
435. Cifuentes, L. A., Sharp, J. H., and Fogel, M. L., Stable carbon and nitrogen isotope biogeochemistry in the Delaware estuary, *Limnol. Oceanogr.*, 33, 1102, 1988.
436. Hedges, J. I., Clark, W. A., and Cowie, G. L., Organic matter sources to the water column and surficial sediments of a marine bay, *Limnol. Oceanogr.*, 33, 1116, 1988.
437. Ostrom, N. E. and Macko, S. A., Sources, cycling, and distribution of water column particulate and sedimentary organic matter in northern Newfoundland fjords and bays: a stable isotope study, in *Organic Matter: Productivity, Accumulation, and Preservation in Recent and Ancient Sediments*, Whelan, J. K. and Farrington, J. W., Eds., Columbia University Press, New York, 1992, chap. 4.
438. Alber, M. and Valiela, I., Organic aggregates in detrital food webs: incorporation by bay scallops *Argopecten irradians*, *Mar. Ecol. Prog. Ser.*, 121, 117, 1995.
439. Herman, P. M. J., A set of models to investigate the role of benthic suspension feeders in estuarine ecosystems, in *Bivalve Filter Feeders in Estuarine and Coastal Ecosystem Processes*, Dame, R., Ed., NATO ASI Series, Series G: Ecological Sciences, 1993, 421.
440. Herman, P. M. J. and Scholten, H., Can suspension feeders stabilise estuarine ecosystems?, in *Trophic Relationships in the Marine Environment*, Barnes, M. and Gibson, R. N., Eds., Aberdeen University Press, Aberdeen, 1990, 104.
441. Lopez, G., Taghon, G., and Levinton, J., Eds., *Ecology of Marine Deposit Feeders*, Springer-Verlag, New York, 1989, 322 pp.
442. Tenore, K. R., Some ecological perspectives in the study of the nutrition of deposit feeders, in *Ecology of Marine Deposit Feeders*, Lopez, G., Taghon, G., and Levinton, J., Eds., Springer-Verlag, New York, 1989, chap. 14.
443. Aller, R. C., The sedimentary Mn cycle in Long Island Sound: its role as intermediate oxidant and the influence of bioturbation, O_2, and Corg flux on diagenetic reaction balances, *J. Mar. Res.*, 52, 259, 1994.
444. Mackin, J. E. and Swider, K. T., Organic matter decomposition pathways and oxygen consumption in coastal marine sediments, *J. Mar. Res.*, 47, 681, 1989.
445. Kristensen, E. and Blackburn, T. H., The fate of organic carbon and nitrogen in experimental marine sediment systems: influence of bioturbation and anoxia, *J. Mar. Res.*, 45, 231, 1987.
446. Hansen, L. S. and Blackburn, T. H., Aerobic and anaerobic mineralization of organic material in marine sediment microcosms, *Mar. Ecol. Prog. Ser.*, 75, 283, 1991.
447. Blackburn, T. H., Accumulation and regeneration: processes at the benthic boundary layer, in *Ocean Margin Processes in Global Change*, Mantoura, R. F. C., Martin, J.-M., and Wollast, R., Eds., John Wiley & Sons, Chichester, 1991, 181.
448. Kristensen, E., Benthic fauna and biogeochemical processes in marine sediments: microbial activities and fluxes, in *Nitrogen Cycling in Coastal Marine Environments*, Blackburn, T. H. and Sorensen, J., Eds., John Wiley & Sons, Chichester, 1988, chap. 12.

449. Aller, R. C., Benthic fauna and biogeochemical processes in marine sediments: the role of burrow structures, in *Nitrogen Cycling in Coastal Marine Environments*, Blackburn, T. H. and Sorensen, J., Eds., John Wiley & Sons, Chichester, 1988, chap. 13.

450. Aller, R. C., Bioturbation and remineralization of sedimentary organic matter: effects of redox oscillation, *Chem. Geol.*, 114, 331, 1994.

451. Alongi , D. M., Effect of monsoonal climate on sulfate reduction in coastal sediments of the central Great Barrier Reef lagoon, *Mar. Biol.*, 122, 497, 1995.

452. Alperin, M. J., Reeburgh, W. S., and Devol, A. H., Organic carbon remineralization and preservation in sediments of Skan Bay, Alaska, in *Organic Matter: Productivity, Accumulation, and Preservation in Recent and Ancient Sediments*, Whelan, J. and Farrington, J. W., Eds., Columbia University Press, New York, 1992, chap. 6.

453. Martens, C. S., Haddad, R. I., and Chanton, J. P., Organic matter accumulation, remineralization, and burial in an anoxic coastal sediment, in *Organic Matter: Productivity, Accumulation, and Preservation in Recent and Ancient Sediments*, Whelan, J. K. and Farrington, J. W., Eds, Columbia University Press, New York, 1992, chap. 5.

454. Billen, C. and Lancelot, C., Modelling benthic nitrogen cycling in temperate coastal ecosystems, in *Nitrogen Cycling in Coastal Marine Environments*, Blackburn, T. H. and Sorensen, J., Eds., John Wiley & Sons, Chichester, 1988, chap. 14.

455. Kemp, W. M., Sampou, P., Caffrey, J., Mayer, M., Henriksen, K., and Boynton, W., R., Ammonium recycling versus denitrification in Chesapeake Bay sediments, *Limnol. Oceanogr.*, 35, 1545, 1990.

456. Seitzinger, S. P., Denitrification in freshwater and coastal marine ecosystems: ecological and geochemical significance, *Limnol. Oceanogr.*, 33, 702, 1988.

457. Zimmerman, A. R. and Benner, R., Denitrification, nutrient regeneration and carbon mineralization in sediments of Galveston Bay, Texas, U.S., *Mar. Ecol. Prog. Ser.*, 114, 275, 1994.

458. Graf, G., Benthic-pelagic coupling: a benthic view, *Oceanogr. Mar. Biol. Ann. Rev.*, 30, 149, 1992.

459. Lucotte, M., Hillaire-Marcel, C., and Louchouarn, P., First-order organic carbon budget in the St. Lawrence lower estuary from ^{13}C data, *Estuarine Coastal Shelf Sci.*, 32, 297, 1991.

460. Roden, E. E., Tuttle, J. H., Boynton, W. R., and Krmp, W. M., Carbon cycling in mesohaline Chesapeake Bay sediments. 1. POC deposition rates and mineralization pathways, *J. Mar. Res.*, 53, 799, 1995.

461. Roden, E. E. and Tuttle, J. H., Carbon cycling in mesohaline Chesapeake Bay sediments. 2. Kinetics of particulate and dissolved organic carbon turnover, *J. Mar. Res.*, 54, 343, 1996.

462. Kemp, W. M., Sampou, P. A., Garber, J., Tuttle, J., and Boynton, W. R., Seasonal depletion of oxygen from bottom waters of Chesapeake Bay: roles of benthic and panktonic respiration and physical exchange processes, *Mar. Ecol. Prog. Ser.*, 85, 137, 1992.

463. Dollar, S. J., Smith, S. V., Vink, S. M., Obrebski, S., and Hollibaugh, J. T., Annual cycle of benthic nutrient fluxes in Tomales Bay, California, and contribution of the benthos to total ecosystem metabolism, *Mar. Ecol. Prog. Ser.*, 79, 115, 1991.

464. Baird, D. and Ulanowitcz, R. E., Comparative study on the trophic structure, cycling and ecosystem properties of four tidal estuaries, *Mar. Ecol. Prog. Ser.*, 99, 221, 1993.

465. Baird, D. and Ulanowitcz, R. E., The seasonal dynamics of the Chesapeake Bay ecosystem, *Ecol. Monogr.*, 59, 329, 1989.

466. Baird, D., Ulanowicz, R. E., and Boynton, W. R., Seasonal nitrogen dynamics in Chesapeake Bay: a network approach, *Estuarine Coastal Shelf Sci.*, 41, 137, 1995.

467. Mantoura, R. F. C., Owens, N. J. P., and Burkill, P. H., Nitrogen biogeochemistry and modelling of Carmarthen Bay, in *Nitrogen Cycling in Coastal Marine Environments*, Blackburn, T. H. and Sorensen, J., Eds., John Wiley & Sons, Chichester, 1988, chap. 16.

468. Smith, S. V., Hollibaugh, J. T., Dollar, S. J., and Vink, S., Tomales Bay metabolism: C-N-P stoichiometry and ecosystem heterotrophy at the land-sea interface, *Estuarine Coastal Shelf Sci.*, 33, 223, 1991.

469. Kimmerer, W. J., Smith, S. V., and Hollibaugh, J. T., A simple heuristic model of nutrient cycling in an estuary, *Estuarine Coastal Shelf Sci.*, 37, 145, 1993.

470. Nixon, S. W., Granger, S. L., and Nowicki, B. L., An assessment of the annual mass balance of carbon, nitrogen, and phosphorus in Narragansett Bay, *Biogeochemistry*, 31, 15, 1995.

471. Hargrave, B. T., Harding, G. C., Drinkwater, K. F., Lambert, T. C., and Harrison, W. G., Dynamics of the pelagic food web in St. Georges Bay, southern Gulf of St. Lawrence, *Mar. Ecol. Prog. Ser.*, 20, 221, 1985.

472. Kjerfve, B., in *Coastal Lagoon Processes*, Kjerfve, B., Ed., Elsevier, Amsterdam, 1994, 577 pp.

473. Knoppers, B., Aquatic primary production in coastal lagoons, in *Coastal Lagoon Processes*, Kjerfve, B., Ed., Elsevier, Amsterdam, 1994, chap. 9.

474. Arfi, R. and Bouvy, M., Size, composition and distribution of particles related to wind-induced resuspension in a shallow tropical lagoon, *J. Plank. Res.*, 17, 557, 1995.

475. Guiral, D., L'instabilite physique, facteur d'organisation et de structuration d'un ecosysteme tropical saumatre peu profond: La lagune Ebrie, *Vie Milieu*, 42, 73, 1992.

476. Yanez-Arancibia, A. and Day, Jr., J. W., *Ecology of Coastal Ecosystems in the Southern Gulf of Mexico — The Terminos Lagoon Region*, Universidad Nacional Autonoma de Mexico, 1988, 637 pp.

477. Herrera-Silveira, J. A., Phytoplankton productivity and submerged macrophyte biomass variation in a tropical coastal lagoon with groundwater discharge, *Vie Milieu*, 44, 257, 1994.

478. Herrera-Silveira, J. A., Salinity and nutrients in a tropical coastal lagoon with groundwater discharges to the Gulf of Mexico, *Hydrobiologia*, 321, 165, 1996.

479. Torreton, J.-P., Guiral, D., and Arfi, R., Bacterioplankton biomass and production during destratification in a monomictic eutrophic bay of a tropical lagoon, *Mar. Ecol. Prog. Ser.*, 57, 53, 1989.

480. Abreu, P. C., Biddanda, B. B,. and Odebrecht, C., Bacterial dynamics of the Patos Lagoon estuary, southern Brazil (32°S, 52°W): relationship with phytoplankton production and suspended material, *Estuarine Coastal Shelf Sci.*, 35, 621, 1992.

481. Wikner, J. and Hagstrom, A., Annual study of bacterioplankton community dynamics, *Limnol. Oceanogr.*, 36, 1313,1991.

482. Garcia, J. R. and Lopez, J. M., Seasonal patterns of phytoplankton productivity, zooplankton abundance and hydrological conditions in Laguna Joyuda, Puerto Rico, *Sci. Mar.*, 53, 625, 1989.

483. Alvarez-Borrego, S., Secondary productivity in coastal lagoons, in *Coastal Lagoon Processes*, Kjerfve, B., Ed., Elsevier, Amsterdam, 1994, chap. 10.

484. Kristenesen, E., Seasonal variations in benthic community metabolism and nitrogen dynamics in a shallow, organic-poor Danish lagoon, *Estuarine Coastal Shelf Sci.*, 367, 565, 1993.

485. Kristenesen, E., Andersen, F. O., and Blackburn, T. H., Effects of benthic macrofauna and temperature on degradation of macroalgal detritus: the fate of organic carbon, *Limnol. Oceanogr.*, 37, 1404, 1992.

486. Blackburn, T. H. and Henriksen, K., Nitrogen cycling in different types of sediments from Danish waters, *Limnol. Oceanogr.*, 28, 477, 1983.

487. Jensen, H. S., Mortensen, P. B., Andersen, F. O., Rasmussen, E., and Jensen, A., Phosphorus cycling in a coastal marine sediment, Aarhus Bay, Denmark, *Limnol. Oceanogr.*, 40, 908, 1995.

488. Fores, E., Christian, R. R., Comin, F. A., and Menendez, M., Network analysis on nitrogen cycling in a coastal lagoon, *Mar. Ecol. Prog. Ser.*, 106, 283, 1994.

489. Herrera-Silveira, J. A. and Comin, F. A., Nutrient fluxes in a tropical coastal lagoon, *Ophelia*, 42, 127, 1995.

490. Espino, G. de la lanza and Medina, M. A. R., Nutrient exchange between subtropical lagoons and the marine environment, *Estuaries*, 16, 273, 1993.

491. Nienhuis, P. H., Ecology of coastal lagoons in the Netherlands (Veerse Meer and Grevelingen), *Vie Milieu*, 42, 59, 1992.

492. Kautsky, U. and Kautsky, H., Coastal productivity in the Baltic Sea, in *Biology and Ecology of Shallow Coastal Waters*, Eleftheriou, A., Ansell, A. D., and Smith, C. J., Eds., Olsen & Olsen, Fredensborg, 1995, 31.

493. Kapetsky, J. M., Coastal lagoon fisheries around the world: some perspectives on fishery yields and other comparative fishery characteristics, in *Management of Coastal Lagoon Fisheries*, Kapetsky, J. M. and Lasserre, G., Eds., FAO Stud. Rev. GFCM No. 61, Vol. 1, Rome, 1984, 97.

494. Milliman, J. D., Flux and fate of fluvial sediment and water in coastal seas, in *Ocean Margin Processes in Global Change*, Mantoura, R. F. C., Martin, J.-M., and Wollast, R., Eds., John Wiley & Sons, Chicester, 1991, 69.

495. Milliman, J. D. and Syvitski, J. P. M., Geomorphic/tectonic control of sediment discharge to the ocean: the importance of small mountainous rivers, *J. Geol.*, 100, 525, 1992.

496. Meybeck, M., C, N, P and S in rivers: from sources to global inputs, in *Interactions of C, N, P and S Biogeochemical Cycles and Global Change*, Wollast, R., Mackenzie, F. T., and Chou, L., Eds., Springer-Verlag, Berlin, 1993, 163.

497. Ittekkot, V., Global trends in the nature of organic matter in river suspensions, *Nature*, 332, 436, 1988.

498. Lohrenz, S. E., Dagg, M. J., and Whitledge, T. E., Enhanced primary production at the plume/oceanic interface of the Mississippi River, *Cont. Shelf Res.*, 10, 639, 1990.

499. Bierman, Jr., V. J., Hinz, S. C., Zhu, D.-W., Wiseman, Jr., W. J., Rabalais, N. N., and Turner, R. E., A preliminary mass balance model of primary productivity and dissolved oxygen in the Mississippi River plume/inner gulf shelf region, *Estuaries*, 17, 886, 1994.

500. Chin-Leo, G. and Benner, R., Enhanced bacterioplankton production and respiration at intermediate salinities in the Mississippi River plume, *Mar. Ecol. Prog. Ser.*, 87, 87, 1992.

501. Pakulski, J. D., Benner, R., Amon, R., Eadie, B., and Whitledge, T., Community metabolism and nutrient cycling in the Mississippi River plume: evidence for intense nitrification at intermediate salinities, *Mar. Ecol. Prog. Ser.*, 117, 207, 1995.

502. Dagg, M. J., Copepod grazing and the fate of phytoplankton in the northern Gulf of Mexico, *Cont. Shelf Res.*, 15, 1303, 1995.

503. Strom, S. L. and Strom, M. W., Microplankton growth, grazing, and community structure in the northern Gulf of Mexico, *Mar. Ecol. Prog. Ser.*, 130, 229, 1996.
504. Biddanda, B., Opsahl, S., and Benner, R., Plankton respiration and carbon flux through bacterioplankton on the Louisiana Shelf, *Limnol. Oceanogr.*, 39, 1259, 1994.
505. Rabalais, N. N., Turner, R. E., Wiseman, Jr., W. J., and Boesch, D. F., A brief summary of hypoxia on the northern Gulf of Mexico continental shelf: 1985–1988, in *Modern and Ancient Continental Shelf Anoxia*, Tyson, R. V. and Pearson, T. H., Eds., Geological Society Special Publ. No. 58, London, 1991, 35.
506. Turner, R. E. and Rabalais, N. N., Coastal eutrophication near the Mississippi River delta, *Nature*, 368, 619, 1994.
507. Eadie, B. J., McKee, B. A., Lansing, M. B., Robbins, J. A., Metz, S., and Trefry, J. H., Records of nutrient-enhanced coastal ocean productivity in sediments from the Louisiana continental shelf, *Estuaries*, 17, 754, 1994.
508. Lin, S. and Morse, J. W., Sulfate reduction and iron sulfide mineral formation in Gulf of Mexico anoxic sediments, *Am. J. Sci.*, 291, 55, 1991.
509. Soto, Y., Bianchi, M., Martinez, J., and Rego, J. V., Seasonal evolution of microplanktonic communities in the estuarine front ecosystem of the Rhone River plume (north-western Mediterranean Sea), *Estuarine Coastal Shelf Sci.*, 37, 1, 1993.
510. Bianchi, M., Bonin, P., and Feliatra, Bacterial nitrification and denitrification rates in the Rhone River plume (northwestern Mediterranean Sea), *Mar. Ecol. Prog. Ser.*, 103, 197, 1994.
511. Bianchi, T. S., Findlay, S. E. G., and Dawson, R., Organic matter sources in the water column and sediments of the Hudson River estuary: the use of plant pigments as tracers, *Estuarine Coastal Shelf Sci.*, 36, 359, 1993.
512. Revelante, N. and Gilmartin, M., The lateral advection of particulate organic matter from the Po delta region during sumer stratification, and its implications for the Northern Adriatic, *Estuarine Coastal Shelf Sci.*, 35, 191, 1992.
513. Zoppini, A., Pettine, M., Totti, C., Puddu, A., Artegiani, A., and Pagnotta, R., Nutrients, standing crop and primary production in western coastal waters of the Adriatic Sea, *Estuarine Coastal Shelf Sci.*, 41, 493, 1995.
514. Vollenweider, R. A., Rinaldi, A., and Montanari, G., Eutrophication, structure and dynamics of a marine coastal system: results of ten-year monitoring along the Emilia-Romagna coast (northwest Adriatic Sea), *Sci. Tot. Environ. (Suppl.)*, 63, 1992.
515. Justic, D., Hypoxic conditions in the northern Adriatic Sea: historical development and ecological significance, in *Modern and Ancient Continental Shelf Anoxia*, Tyson, R. V. and Pearson, T. H., Eds., Geological Society Special Publ. No. 58, 1991, 95.
516. Giordani, P., Hammond, D. E., Berelson, W. M., Montanari, G., Poletti, R., Milandri, A., Frignani, M., Langone, L., Ravaioli, M., Rovatti, G., and Rabbi, E., Benthic fluxes and nutrient budgets for sediments in the northern Adriatic Sea: burial and recycling efficiencies, *Sci. Tot. Environ. (Suppl.)*, 251, 1992.
517. Stanley, D. J. and Warne, A. G., Nile delta: recent geological evolution and human impact, *Science*, 260, 628, 1993.
518. Justic, D., Rabalais, N. N., Turner, R. E., and Dortch, Q., Changes in nutrient structure of river-dominated coastal waters: stoichiometric nutrient balance and its consequences, *Estuarine Coastal Shelf Sci.*, 40, 339, 1995.
519. Milliman, J. D., Sediment discharge to the ocean from small mountainous rivers: the New Guinea example, *Geo-Mar. Lett.*, 15, 127, 1995.
520. Nittrouer, C. A., Brunskill, G. J., and Figueiredo, A. G., Importance of tropical coastal environments, *Geo-Mar. Lett.*, 15, 121, 1995.

521. Milliman, J. D. and Qingming, J., Eds., Sediment dynamics of the Changjiang estuary and the adjacent East China Sea, *Cont. Shelf Res.*, 4(1/2), 1985 (special issue).

522. Wolanski, E., Norro, A., and King, B., Water circulation in the Gulf of Papua, *Cont. Shelf Res.*, 15, 185, 1995.

523. Nittrouer, C. A. and DeMaster, D. J., Eds., Oceanography of the Amazon continental shelf, *Cont. Shelf Res.*, 16(5/6), 551, 1996 (special issue).

524. Alongi, D. M. and Robertson, A. I., Factors regulating benthic food chains in tropical river deltas and adjacent shelf areas, *Geo-Mar. Lett.*, 15, 145, 1995.

525. Robertson, A. I., Daniel, P. A., Dixon, P., and Alongi, D. M., Pelagic biological processes along a salinity gradient in the Fly delta and adjacent river plume (Papua New Guinea), *Cont. Shelf Res.*, 13, 205, 1993.

526. Robertson, A. I., Dixon, P., and Alongi, D. M., The influence of fluvial discharge on pelagic production in the Gulf of Papua, northern Coral Sea, *Estuarine Coastal Shelf Sci.*, in press.

527. van Bennekom, A. J., Berger, G. W., Helder, W., and DeVries, R. T. P., Nutrient distributions in the Zaire estuary and river plume, *Neth. J. Sea Res.*, 12, 296, 1978.

528. DeMaster, D. J., Smith, Jr., W. O., Nelson, D. M., and Aller, J. Y., Biogeochemical processes in Amazon shelf waters: chemical distributions and uptake rates of silicon, carbon and nitrogen, *Cont. Shelf Res.*, 16, 617, 1996.

529. DeMaster, D. J. and Pope, R. H., Nutrient dynamics in Amazon shelf waters: results from AMASSEDS, *Cont. Shelf Res.*, 16, 263, 1996.

530. Edmond, J. M., Boyle, E. A., Grant, G., and Stallard, R. F., The chemical mass balance in the Amazon plume. I. The nutrients, *Deep-Sea Res.*, 28, 1339, 1981.

531. Turner, R. E., Rabalais, N. N., and Zhang, Z. N., Phytoplankton biomass, production and growth limitation on the Huanghe (Yellow River) continental shelf, *Cont. Shelf Sci.*, 10, 545, 1990.

532. Dortch, Q. and Whitledge, T. E., Does nitrogen or silicon limit phytoplankton production in the Mississippi River plume and nearby regions?, *Cont. Shelf Res.*, 12, 1293, 1992.

533. Courties, C., Shi, J., Ning, X., Chen, Z., and Lasserre, P., Respiration rates in the Changjiang River mouth and the adjacent East China Sea: relations with bacteria and phytoplankton, *Sci. Mar.*, 53, 167, 1989.

534. Aller, J. Y. and Stupakoff, I., The distribution and seasonal characteristics of benthic communities on the Amazon shelf as indicators of physical processes, *Cont. Shelf Res.*, 16, 717, 1996.

535. Aller, J. Y., Molluscan death assemblages on the Amazon shelf: implication for physical and biological controls on benthic populations, *Palaeogeogr. Palaeoclimatol. Palaeoecol.*, 118, 181, 1995.

536. Rhoads, D. C., Boesch, D. F., Tang, Z.-c., Xu, F.-s., Huang, L.-q., and Nilsen, K. J., Macrobenthos and sedimentary facies on the Changjiang delta platform and adjacent continental shelf, East China Sea, *Cont. Shelf Sci.*, 4, 189, 1985.

537. Dalzell, P. and Pauly, D., Assessment of the fish resources of Southeast Asia, with emphasis on the Banda and Arafura Seas, *Neth. J. Sea Res.*, 24, 641, 1989.

538. Villegas, L. and Dragovich, A., The Guianas-Brazil shrimp fishery, its problems and management aspects, in *Penaeid Shrimps — Their Biology and Management*, Gullard, J. A. and Rothschild, B. J., Eds., Fishing News Books, Farnham, 1982, 60.

539. Sharp, G. D., Fish populations and fisheries, in *Ecosystems of the World*. Vol. 27. *Continental Shelves*, Postma, H. and Zijlstra, J. J., Eds., Elsevier, Amsterdam, 1988, 155.

540. Alongi, D. M., Christoffersen, P., Tirendi, F., and Robertson, A. I., The influence of freshwater and material export on sedimentary facies and benthic processes within the Fly Delta and adjacent Gulf of Papua (Papua New Guinea), *Cont. Shelf Res.*, 12, 287, 1992.

541. Alongi, D. M., Decomposition and recycling of organic matter in muds of the Gulf of Papua, northern Coral Sea, *Cont. Shelf Res.*, 15, 1319, 1995.

542. Alongi, D. M., Boyle, S. G., Tirendi, F., and Payn, C., Composition and behavior of trace metals in post-oxic sediments of the Gulf of Papua, Papua New Guinea, *Estuarine Coastal Shelf Sci.*, 42, 197, 1996.

543. Aller, R. C., Mackin, J. E., and Cox, Jr., R. T., Diagenesis of Fe and S in Amazon inner shelf muds: apparent dominance of Fe reduction and implications for the genesis of ironstones, *Cont. Shelf Res.*, 6, 263, 1986.

544. Aller, R. C., Blair, N. E., Xia, Q., and Rude, P. D., Remineralization rates, recycling, and storage of carbon in Amazon shelf sediments, *Cont. Shelf Res.*, 16, 753, 1996.

545. Mackin, J. E., Aller, R. C., and Ullman, W. J., The effects of iron reduction and nonsteady-state diagenesis on iodine, ammonium, and boron distributions in sediments from the Amazon continental shelf, *Cont. Shelf Res.*, 8, 363, 1988.

546. Blair, N. E. and Aller, R. C., Anaerobic methane oxidation on the Amazon shelf, *Geochim. Cosmochim. Acta*, 59, 3707, 1995.

547. Aller, R. C., Mackin, J. E., Ullman, W. J., Wang, C.-H., Tsai, S.-H., Jin, J.-C., Sui, Y.-N., and Hong, J.-Z., Early chemical diagenesis, sediment-water solute exchange, and storage of reactive organic matter near the mouth of the Changjiang, East China Sea, *Cont. Shelf Res.*, 4, 227, 1985.

548. Manjunatha, B. R. and Shankar, R., Signature of non-steady-state diagensis in continental shelf sediments, *Estuarine Coastal Shelf Sci.*, 42, 361, 1996.

549. Michalopoulos, P. and Aller, R. C., Rapid clay mineral formation in Amazon delta sediments: reverse weathering and oceanic elemental cycles, *Science*, 270, 614, 1995.

550. Walsh, J. J., *On the Nature of Continental Shelves*, Academic Press, New York, 1988, 520 pp.

551. Postma, H. and Zijlstra, J. J., Eds., *Ecosystems of the World. Vol. 27. Continental Shelves*, Elsevier, Amsterdam, 1988, chaps. 4, 6, and 7.

552. Loder, J. W. and Platt, T., Physical controls on phytoplankton production at tidal fronts, in *Proceedings of the Nineteenth European Marine Biology Symposium*, Gibbs, P. E., Ed., Cambridge University Press, London, 1985, 3.

553. Fogg, G. E., Biological activities at a front in the western Irish Sea, in *Proceedings of the Nineteenth European Marine Biology Symposium*, Gibbs, P. E., Ed., Cambridge University Press, London, 1985, 87.

554. Lochte, K. and Turley, C. M., Heterotrophic activity and carbon flow via bacteria in waters associated with a tidal mixing front, *Proceedings of the Nineteenth European Marine Biology Symposium*, Gibbs, P. E., Ed., Cambridge University Press, London, 1985, 73.

555. Holligan, P. M., Williams, P. J., Le, B., Purdie, D., and Harris, R. P., Photosynthesis, respiration and nitrogen supply of plankton populations in stratified, frontal and tidally mixed shelf waters, *Mar. Ecol. Prog. Ser.*, 17, 201, 1984.

556. Linley, E. A. S., Newell, R. C., and Lucus, M. I., Quantitative relationships between phytoplankton, bacteria and heterotrophic microflagellates in shelf waters, *Mar. Ecol. Prog. Ser.*, 12, 77, 1983.

557. Hesse, K.-J., Liu, Z. L., and Schaumann, K., Phytoplankton and fronts in the German Bight, *Sci. Mar.*, 53, 187, 1989.

558. Zijlistra, J. J., The North Sea ecosystem, in *Ecosystems of the World*. Vol. 27. *Continental Shelves*, Postma, H. and Zijlistra, J. J., Eds., Elsevier, Amsterdam, 1988, chap. 8.

559. Radach, G., Ecosystem functioning in the German Bight under continental nutrient inputs by rivers, *Estuaries*, 15, 477, 1992.

560. Reid, P. C., Lancelot, C., Gieskes, W. W. C., Hagmeier, E., and Weichart, G., Phytoplankton of the North Sea and its dynamics: a review, *Neth. J. Sea Res.*, 26, 295, 1990.

561. Aebischer, N. J., Coulson, J. C., and Colebrook, J. M., Parallel long-term trends across four marine trophic levels and weather, *Nature*, 347, 753, 1990.

562. Billen, G., Joiris, C., Meyer-Reil, L., and Lindeboom, H., Role of bacteria in the North Sea ecosystem, *Neth. J. Sea Res.*, 26, 265, 1990.

563. van Duyl, F. C., Bak, R. P. M., Kop, A. J., Nieuwland, G., Berghuis, E. M., and Kok, A., Mesocosm experiments: mimicking seasonal developments of microbial variables in North Sea sediments, *Hydrobiologia*, 235/236, 267, 1992.

564. Laane, R. W. P. M., Turkstra, E., and Mook, W. G., Stable carbon isotope composition of pelagic and benthic organic matter in the North Sea and adjacent estuaries, in *Facets of Modern Biogeochemistry*, Ittekot, V., Kemp, S., Michaelis, W., and Spitzy, A., Eds., Springer-Verlag, Berlin, 1990, 214.

565. Bak, R. P. M., van Duyl, F. C., and Nieuwland, G., Organic sedimentation and macrofauna as forcing factors in marine benthic nanoflagellate communities, *Micro. Ecol.*, 29, 173, 1995.

566. Upton, A. C., Nedwell, D. B., Parkes, R. J., and Harvey, S. M., Seasonal benthic microbial activity in the southern North Sea; oxygen uptake and sulfate reduction, *Mar. Ecol. Prog. Ser.*, 101, 273, 1993.

567. Cramer, A., Seasonal variation in benthic metabolic activity in a frontal system in the North Sea, in *Trophic Relationships in the Marine Environment*, Barnes, M. and Gibson, R. N., Eds., Aberdeen University Press, Aberdeen, 1990, 54.

568. Lohse, L., Malschaert, J. F. P., Slomp, C. P., Helder, W., and van Raaphorst, W., Nitrogen cycling in North Sea sediments: interaction of denitrification and nitrification in offshore and coastal areas, *Mar. Ecol. Prog. Ser.*, 101, 283, 1993.

569. Ray, G. C., Sustainable use of the global ocean, in *Changing the Global Environment: Perspectives on Human Involvement*, Botkin, D. B., Ed., Academic Press, New York, 1989, chap. 5.

570. Mills, E. L. and Fournier, R. O., Fish production and the marine ecosystems of the Scotian shelf, eastern Canada, *Mar. Biol.*, 54, 101, 1979.

571. Mills, E. L., Pittman, K. and Tan, F. C., Food-web structure on the Scotian shelf, eastern Canada: a study using ^{13}C as a food-chain tracer, *Rapp. P.-v. Reun. Cons. Int. Explor. Mer.*, 183, 111, 1984.

572. Sherman, K., Grosslein, M., Mountain, D., Busch, D., O'Reilly, J., and Theroux, R., The continental shelf ecosystem off the northeast coast of the United States, in *Ecosystems of the World*. Vol. 27. *Continental Shelves*, Postma, H. and Zijlistra, J. J., Eds., Elsevier, Amsterdam, 1988, chap. 9.

573. Rivkin, R. B., Legendre, L., Deibel, D., Tremblay, J.-E., Klein, B., Crocker, K., Roy, S., Silverberg, N., Lovejoy, C., Mespie, F., Romero, N., Anderson, M. R., Matthews, P., Savenkoff, C., Vezina, A., Therriault, J.-C., Wesson, J., Berube, C., and Ingram, R. G., Vertical flux of biogenic carbon in the ocean: is there food web control?, *Science*, 272, 1163, 1996.

574. Cochlan, W. P., Seasonal study of uptake and regeneration of nitrogen on the Scotian shelf, *Cont. Shelf Res.*, 5, 555, 1986.

575. Christensen, J. P., Townsend, D. W., and Montoya, J. P., Water column nutrients and sedimentary denitrification in the Gulf of Maine, *Cont. Shelf Res.*, 16, 489, 1996.

576. Townsend, D. W., Influences of oceanographic processes on the biological productivity of the Gulf of Maine, *Rev. Aquat. Sci.*, 5, 211, 1991.
577. Pocklington, R., Leonard, J. D., and Crewe, N. F., Sources of organic matter to surficial sediments from the Scotian shelf and slope, Canada, *Cont. Shelf Res.*, 11, 1069, 1991.
578. Grant, J., Emerson, C. W., Hargrave, B. T., and Shortle, J. L., Benthic oxygen consumption on continental shelves off Eastern Canada, *Cont. Shelf Res.*, 11, 1083, 1991.
579. Hines, M. E., Bazylinski, D. A., Tugel, J. B., and Lyons, W. B., Anaerobic microbial biogeochemistry in sediments from two basins in the Gulf of Maine: evidence for iron and manganese reduction, *Estuarine Coastal Shelf Sci.*, 32, 313, 1991.
580. Walsh, J. J., Rowe, G. T., Iverson, R. L., and McRoy, C. P., Biological export of shelf carbon is a neglected sink of the global CO_2 cycle, *Nature*, 291, 196, 1981.
581. Walsh, J. J., Biscaye, P. E., and Csanady, G. T., The 1983–1984 Shelf Edge Exchange Processes (SEEP)-I experiment: hypotheses and highlights, *Cont. Shelf Res.*, 8, 435, 1988.
582. Boesch, D. F., Benthic ecological studies: macrobenthos, Special Report 194, *Appl. Mar. Sci. Ocean Eng.*, VIMS, Gloucester Pt, 301, 1979.
583. Rowe, G. T., Theroux, R., Phoel, W., Quinby, H., Wilke, R., Koschoreck, D., Whitledge, T. E., Falkowski, P. G., and Fray, C., Benthic carbon budgets for the continental shelf south of New England, *Cont. Shelf Res.*, 8, 511, 1988.
584. Biscayne, P. E., Flagg, C. N., and Falkowski, P. G., The Shelf Edge Exchange Processes experiment, SEEP-II: an introduction to hypotheses, results and conclusions, *Deep-Sea Res. II*, 41, 231, 1994.
585. Wirick, C. D., Exchange of phytoplankton across the continental shelf-slope boundary of the Middle Atlantic Bight during spring 1988, *Deep-Sea Res. II*, 41, 391, 1994.
586. Flagg, C. N., Wirick, C. D., and Smith, S. L., The interaction of phytoplankton, zooplankton and currents from 15 months of continuous data in the mid-Atlantic Bight, *Deep-Sea Res. II*, 41, 411, 1994.
587. Verity, P. G., Paffenhofer, G.-A., Wallace, D., Sherr, E., and Sheer, B., Composition and biomass of plankton in spring on the Cape Hatteras shelf, with implications for carbon flux, *Cont. Shelf Res.*, 16, 1087, 1996.
588. Kemp, P. F., Microbial carbon utilization on the continental shelf and slope during the SEEP-II experiment, *Deep-Sea Res. II*, 41, 563, 1994.
589. Hanson, R. B., Tenore, K. R., Bishop, S., Chamberlain, C., Pamatmat, M. M., and Tietjen, J. H., Benthic enrichment in the Georgia Bight related to Gulf Stream intrusions and estuarine outwelling, *J. Mar. Res.*, 39, 417, 1981.
590. Tenore, K. R., Seasonal changes in soft bottom macroinfauna of the U.S. South Atlantic Bight, in *Oceanography of the Southeastern U.S. Continental Shelf*, Coastal and Estuarine Sciences 2 series, Atkinson, L. P., Menzel, D. W., and Bush, K. A., Eds, American Geophysical Union, Washington, D.C., 1985, 130.
591. Hopkinson, C. S., Shallow-water benthic and pelagic metabolism: evidence of heterotrophy in the nearshore Georgia Bight, *Mar. Biol.*, 87, 19, 1985.
592. Moran, M. A., Pomeroy, L. R., Sheppard, E. S., Atkinson, L. P., and Hodson, R. E., Distribution of terrestrially derived dissolved organic matter on the southeastern U.S. continental shelf, *Limnol. Oceanogr.*, 36, 1134, 1991.
593. Moran, M. A. and Hodson, R. E., Dissolved humic substances of vascular plant origin in a coastal marine environment, *Limnol. Oceanogr.*, 39, 762, 1994.
594. Verity, P. G., Yoder, J. A., Bishop, S. S., Nelson, J. R., Craven, D. B., Blanton, J. O., Robertson, C. Y., and Tronzo, C. R., Composition, productivity and nutrient chemistry of a coastal ocean planktonic food web, *Cont. Shelf Res.*, 13, 741, 1993.

595. Coachman, L. K. and Hansell, D. A., Eds., ISHTAR-Inner Shelf Transfer and Recycling in the Bering and Chukchi Seas, *Cont. Shelf Res.*, 13(5/6), 1993 (special issue).

596. Springer, A. M. and McRoy, C. P., The paradox of pelagic food webs in the northern Bering Sea. III. Patterns of primary production, *Cont. Shelf Res.*, 13, 575, 1993.

597. Springer, A. M., McRoy, P. C., and Flint, M. V., The Bering Sea Green Belt: shelf-edge processes and ecosystem production, *Fish. Oceanogr.*, 5, 205, 1996.

598. Henriksen, K., Blackburn, T. H., Lomstein, B. A., and McRoy, C. P., Rates of nitrification, distribution of nitrifying bacteria and inorganic N fluxes in northern Bering-Chukchi shelf sediments, *Cont. Shelf Res.*, 13, 629, 1993.

599. Grebmeier, J. M., Studies of pelagic-benthic coupling extended onto the Soviet continental shelf in the northern Bering and Chukchi seas, *Cont. Shelf Res.*, 13, 653, 1993.

600. Walsh, J. J. and Dieterle, D. A., CO_2 cycling in the coastal ocean. I. A numerical analysis of the southeastern Bering Sea with applications to the Chukchi Sea and the northern Gulf of Mexico, *Prog. Oceanogr.*, 34, 335, 1994.

601. Naidu, A. S., Scalan, R. S., Feder, H. M., Goering, J. J., Hameedi, M. J., Parker, P. L., Behrens, E. W., Caughey, M. E and Jewett, S. C., Stable organic carbon isotopes in sediments of the north Bering-south Chukchi seas, Alaska-Soviet Arctic Shelf, *Cont. Shelf Res.*, 13, 669, 1993.

602. Wolanski, E., *Physical Oceanographic Processes of the Great Barrier Reef*, CRC Press, Boca Raton, FL, 1994, 194 pp.

603. Belperio, A. P. and Searle, D. E., Terrigenous and carbonate sedimentation in the Great Barrier Reef Province, in *Carbonate-Clastic Transitions, Developments in Sedimentology 42*, Doyle, L. J. and Roberts, H. H., Eds., Elsevier, Amsterdam, 1988, chap. 5.

604. Furnas, M. J. and Mitchell, A. W., Shelf-scale estimates of phytoplankton primary production in the Great Barrier Reef, *Proc. Sixth Int. Coral Reef Symp.*, 2, 557, 1988.

605. McKinnon, A. D. and Thorrold, S. R., Zooplankton community structure and copepod egg production in coastal waters of the central Great Barrier Reef lagoon, *J. Plank. Res.*, 15, 1387, 1993.

606. Thorrold, S. R. and McKinnon, A. D., Response of larval fish assemblages to a riverine plume in coastal waters of the central Great Barrier Reef lagoon, *Limnol. Oceanogr.*, 40, 177, 1995.

607. Daniel, P. A. and Robertson, A. I., Epibenthos of mangrove waterways and open embayments: community structure and the relationship between exported mangrove detritus and epifaunal standing stocks, *Estuarine Coastal Shelf Sci.*, 31, 599, 1990.

608. Alongi, D. M., Benthic processes across mixed terrigenous-carbonate sedimentary facies on the central Great Barrier Reef continental shelf, *Cont. Shelf Res.*, 9, 629, 1989.

609. Alongi, D. M., The role of bacteria in nutrient recycling in tropical mangrove and other coastal benthic ecosystems, *Hydrobiologia*, 285, 19, 1994.

610. Alongi, D. M., Vertical profiles of bacterial abundance, productivity and growth rates in coastal sediments of the central Great Barrier Reef lagoon, *Mar. Biol.*, 112, 657, 1992.

611. Alongi, D. M., Effect of mangrove detrital outwelling on nutrient regeneration and oxygen fluxes in coastal sediments of the central Great Barrier Reef lagoon, *Estuarine Coastal Shelf Sci.*, 31, 581, 1990.

612. Alongi, D. M., Abundance of benthic microfauna in relation to outwelling of mangrove detritus in a tropical coastal region, *Mar. Ecol. Prog. Ser.*, 63, 53, 1990.

613. Alongi, D. M. and Christoffersen, P., Benthic infauna and organism-sediment relations in a shallow, tropical coastal area: influence of outwelled mangrove detritus and physical disturbance, *Mar. Ecol. Prog. Ser.*, 81, 229, 1992.

614. Alongi, D. M., Boto, K. G., and Tirendi, F., Effect of exported mangrove litter on bacterial productivity and dissolved organic carbon fluxes in adjacent tropical nearshore sediments, *Mar. Ecol. Prog. Ser.*, 56, 133, 1989.

615. Alongi, D. M., Tirendi, F., and Christoffersen, P., Sedimentary profiles and sediment-water solute exchange of iron and manganese in reef- and river-dominated shelf regions of the Coral Sea, *Cont. Shelf Res.*, 13, 287, 1993.

616. Torgersen, T. and Chivas, A. R., Terrestrial organic carbon in marine sediment: a preliminary balance for a mangrove environment derived from ^{13}C, *Chem. Geol.*, 52, 379, 1985.

617. Alongi, D.M., unpublished data.

618. Furnas, M. J., Mitchell, A. W., and Skuza, M., *Nitrogen and Phosphorus Budgets for the Central Great Barrier Reef Shelf*, Great Barrier Reef Marine Park Authority Research Publication 36, 1995, 194 pp.

619. Longhurst, A. R. and Pauly, D., *Ecology of Tropical Oceans*, Academic Press, San Diego, 1987, 407 pp.

620. Longhurst, A., Benthic-pelagic coupling and export of organic carbon from a tropical Atlantic continental shelf-Sierra Leone, *Estuarine Coastal Shelf Sci.*, 17, 261, 1983.

621. Pauly, D. and Christensen, V., Stratified models of large marine ecosystems: a general approach and an application to the South China Sea, in *Large Marine Ecosystems: Stress, Mitigation and Sustainability*, Sherman, K., Alexander, L. M., and Gold, B. D., Eds., American Association for the Advancement of Science Press, Washington, D.C., 1993, chap. 15.

622. Zijlstra, J. and Baars, M. A., Productivity and fisheries potential of the Banda Sea ecosystem, in *Large Marine Ecosystems: Patterns, Processes and Yields*, Sherman, K., Alexander, L. M., and Gold, B. D., Eds., American Association for the Advancement of Science Press, Washington, D.C., 1990, chap. 5.

623. Summerhayes, C. P., Prell, W. L., and Emeis, K. C., Eds., *Upwelling Systems: Evolution Since the Early Miocene*, Geological Society Special Publ. No. 64, London, 1992, 459 pp.

624. Summerhayes, C. P., Emeis, K.-C., Angel, M. V., Smith, R. L., and Zeitzschel, B., Eds., *Upwelling in the Ocean: Modern Processes and Ancient Records*, Environmental Sciences Research Report 18, John Wiley & Sons, Chichester, 1995, 422 pp.

625. Chavez, F. P., Barber, R. T., Kosro, P. M., Huyer, A., Ramp, S. R., Stanton, T. P., and Rojas de Mendiola, B., Horizontal transport and the distribution of nutrients in the coastal transition zone off Northern California: effects on primary production, phytoplankton biomass and species composition, *J. Geophys. Res.*, 96, 14833, 1991.

626. Smith, R. L., Huyer, A., Godfrey, J. S., and Church, J. A., The Leeuwin Current off Western Australia, *J. Phys. Oceanogr.*, 21, 323, 1991.

627. Baird, D., McGlade, J. M., and Ulanowicz, R. E., The comparative ecology of six marine ecosystems, *Phil. Trans. Roy. Soc. Lond. B*, 333, 15, 1991.

628. Peterson, W. T., Arcos, D. F., McManus, G. B., Dam, H., Bellantoni, D., Johnson, T., and Tiselius, P., The nearshore zone during coastal upwelling: daily variability and coupling between primary and secondary production off Central Chile, *Prog. Oceanogr.*, 20, 1, 1988.

629. Wolff, W. J., van der Land, J., Nienhuis, P. H., and de Wilde, P. A. W. J., The functioning of the ecosystem of the Banc d'Arguin, Mauritania: a review, *Hydrobiologia*, 258, 211, 1993.

630. Binet, D. and Marchal, E., The large marine ecosystem of shelf areas in the Gulf of Guinea: long-term variability induced by climatic changes, in *Large Marine Ecosystem: Stress, Mitigation and Sustainability*, Sherman, K., Alexander, L. M., and Gold, B. D., Eds., American Society for the Advancement of Science Press, Washington, D.C., 1993, chap. 12.

631. Binet, D., Marchal, E., and Pezennec, O., Sardinella aurita de Cote d'Ivoire et du Ghana: fluctuations halieutiques et changements climatiques, in *Pecheries Ouest-Afrcaines: Variabilite, Instabilite et Changement*, Cury, P. and Roy, C., Eds., ORSTOM, Paris, 1991, 320.

632. Banse, K., Overview of the hydrography and associated biological phenomena in the Arabian Sea off Pakistan, in *Marine Geology and Oceanography of Arabian Sea and Coastal Pakistan*, Haq, B. U. and Milliman, J. D., Eds., Van Nostrand Reinhold Co., New York, 1984.

633. Sreekumaran Nair, S. R., Devassy, V. P., and Madhupratap, M., Blooms of phytoplankton along the west coast of India associated with nutrient enrichment and the response of zooplankton, *Sci. Tot. Environ.*, Suppl. 1, 819, 1992.

634. Smith, S. S., Banse, K., Cochran, J. K., Codispoti, L. A., Ducklow, H. W., Luther, M. E., Olson, D. B., Peterson, W. T., Prell, W.L., Surgi, N., Swallow, J. C., and Wishner, K., U.S. JGOFS: Arabian Sea Process Study, *U.S. JGOFS Planning Report No. 13*, Woods Hole Oceanographic Institute, Woods Hole, 1991, 164.

635. Brock, J. C. and McClain, C. R., Interannual variability in phytoplankton blooms observed in the Northwestern Arabian Sea during the southwest monsoon, *J. Geophys. Res.*, 97, 733, 1992.

636. Tenore, K. R., Boyer, L. F., Cal, R. M., Corral, J., Garcia-Fernandez, C., Gonzalez, N., Gonzalez-Gurriaran, Hanson, R. B., Iglesias, J., Krom, M., Lopez-Jamar, E., McClain, J., Pamatmat, M. M., Perez, A., Rhoads, D. C., de Santiago, G., Tietjen, J. H., Westrich, J., and Windom, H. L., Coastal upwelling in the Rias Bajas, northwest Spain: contrasting the benthic regimes of the Rias de Arosa and de Muros, *J. Mar. Res.*, 40, 701, 1982.

637. Lopez-Jamar, E., Cal, R. M., Gonzalez, G., Hanson, R. B., Rey, J., Santiago, G., and Tenore, K. R., Upwelling and outwelling effects on the benthic regime of the continental shelf off Galicia, northwest Spain, *J. Mar. Res.*, 50, 465, 1992.

638. Prego, R., General aspects of carbon biogeochemistry in the Ria of Vigo, northwestern Spain, *Geochim. Cosmochim Acta*, 57, 2041, 1993.

639. Prego, R., Nitrogen interchanges generated by biogeochemical processes in a Galician ria, *Mar. Chem.*, 45, 167, 1994.

640. Prego, R., Biogeochemical pathways of phosphate in a Galician ria (northwestern Iberian Peninsula), *Estuarine Coastal Shelf Sci.*, 37, 437, 1993.

641. Walsh, J. J., Importance of continental margins in the marine biogeochemical cycling of carbon and nitrogen, *Nature*, 350, 53, 1991.

642. Kuehl, S. A., Hariu, T. M., and Moore, W. S., Shelf sedimentation off the Ganges-Brahmaputra river system: evidence for sediment bypassing to the Bengal fan, *Geology*, 17, 1132, 1989.

643. France-Lanord, C. and Derry, L. A., δ ^{13}C of organic carbon in the Bengal Fan: source evaluation and transport of C_3 and C_4 plant carbon to marine sediments, *Geochim. Cosmochim Acta,*, 58, 4809, 1994.

644. Ittekkot, V., Hakke, B., Bartsch, M., Nair, R. R., and Ramaswamy, V., Organic carbon removal in the sea: the continental connection, in *Upwelling Systems: Evolution Since the Early Miocene*, Summerhayes, C. P., Prell, W. L., and Emeis, K. C., Eds., Geological Society Special Publ. No. 64, London, 1992, 167.

645. Paropkari, A. L., Prakash Babu, C., and Mascarenhas, A., A critical evaluation of depositional parameters controlling the variability of organic carbon in Arabian Sea sediments, *Mar. Geol.*, 107, 213, 1992.

646. Ittekkot, V. and Arain, R., Nature of particulate organic matter in the river Indus, Pakistan, *Geochim. Cosmochim. Acta,* 50, 1643, 1986.

647. Mariotti, A., Gadel, F., Giresse, P., and Kinga-Mouzeo, Carbon isotope composition and geochemistry of particulate organic matter in the Congo River (Central Africa): application to the study of Quaternary sediments off the mouth of the river, *Chem. Geol.,* 86, 345, 1991.

648. Bird, M. I., Brunskill, G. J., and Chivas, A. R., Carbon-isotope composition of sediments from the Gulf of Papua, *Geo-Mar. Lett.,* 15, 153, 1995.

649. Brunskill, G. J., Woolfe, K. J., and Zagorskis, I., Distribution of riverine sediment chemistry on the shelf, slope and rise of the Gulf of Papua, *Geo-Mar. Lett.,* 15, 160, 1995.

650. Robertson, A. I. and Alongi, D. M., Role of riverine mangrove forests in organic carbon export to the tropical coastal ocean: a preliminary mass balance for the Fly delta (Papua New Guinea), *Geo-Mar. Lett.,* 15, 134, 1995.

651. Alongi, D. M., Bacterial growth rates, production and estimates of detrital carbon utilization in deep-sea sediments of the Solomon and Coral Seas, *Deep-Sea Res.,* 37, 731, 1990.

652. Wollast, R., The coastal organic carbon cycle: fluxes, sources, and sinks, in *Ocean Margin Processes in Global Change,* Mantoura, R. F. C., Martin, J.-M., and Wollast, R., Eds., John Wiley & Sons, Chichester, 1991, 365.

653. Smith, S. V. and Hollibaugh, J. T., Coastal metabolism and the oceanic organic carbon balance, *Rev. Geophys.,* 31, 75, 1993.

654. Mackenzie, F. T., Ver, L. M., Sabine, C., Lane, M., and Lerman, A., C, N, P, S global biogeochemical cycles and modeling of global change, in *Interactions of C, N, P and S Biogeochemical Cycles and Global Change,* Wollast, R., Mackenzie, F. T., and Chou, L., Eds., Springer-Verlag, Berlin, 1993, 1.

655. Wollast, R., Interactions of carbon and nitrogen cycles in the coastal zone, in *Interactions of C, N, P and S Biogeochemical Cycles and Global Change,* Wollast, R., Mackenzie, F. T. and Chou, L., Eds., Springer-Verlag, Berlin, 1993, 195.

656. Christensen, J. P., Murray, J. W., Devol, A. H., and Codispoti, L. A., Denitrification in continental shelf sediments has major impact on the oceanic nitrogen budget, *Global Biogeochem. Cycles,* 1, 97, 1987.

657. Codispoti, L. A., Phosphorus vs. nitrogen limitation of new and export production, in *Productivity of the Ocean: Present and Past,* Berger, W. H., Smetacek, V. S., and Wefer, G., Eds., John Wiley & Sons, London, 1989, 377.

658. Devol, A. H., Direct measurement of nitrogen gas fluxes from continental shelf sediments, *Nature,* 349, 319, 1991.

659. Christensen, J. P., Carbon export from continental shelves, denitrification and atmospheric carbon dioxide, *Cont. Shelf Res.,* 14, 547, 1994.

660. Berner, R. A. and Rao, J.-L., Phosphorus in sediments of the Amazon River and estuary: implications for the global flux of phosphorus to the sea, *Geochim. Cosmochim. Acta,* 58, 2333, 1994.

661. Treguer, P., Nelson, D. M., Van Bennekom, A. J., DeMaster, D. J., Leynaert, A., and Queguiner, B., The silica balance in the world ocean: a reestimate, *Science,* 268, 375, 1995.

662. Canfield, D. E., Factors influencing organic carbon preservation in marine sediments, *Chem. Geol.,* 114, 315, 1994.

663. Kristensen, E., Ahmed, S. I., and Devol, A. H., Aerobic and anaerobic decomposition of organic matter in marine sediment: which is fastest?, *Limnol. Oceanogr.,* 40, 1430, 1995.

664. Lee, C., Controls on organic carbon preservation: the use of stratified water bodies to compare intrinsic rates of decomposition in oxic and anoxic systems, *Geochim. Cosmochim. Acta.,* 56, 3323, 1992.

665. Lee, C., Controls on carbon preservation—new perspectives, *Chem. Geol.*, 114, 285, 1994.
666. Canuel, E. A. and Martens, C. S., Seasonal variation in the sources and alteration of organic matter associated with recently-deposited sediments, *Org. Geochem.*, 20, 563, 1993.
667. Macko, S. A., Engel, M. H., and Qian, Y., Early diagenesis and organic matter preservation—a molecular stable carbon isotope perspective, *Chem. Geol.*, 114, 365, 1994.
668. Heath, M., An holistic analysis of the coupling between physical and biological processes in the coastal zone, *Ophelia*, 42, 95, 1995.
669. Nixon, S. W., Physical energy inputs and the comparative ecology of lake and marine ecosystems, *Limnol. Oceanogr.*, 33, 1005, 1988.
670. Mann, K. H., Physical oceanography, food chains, and fish stocks: a review, *ICES J. Mar. Sci.*, 50, 105, 1993.
671. de Silva, A. J., River runoff and shrimp abundance in a tropical coastal ecosystem—the example of the Sofala Bank (central Mozambique), in *The Role of Freshwater Outflow in Coastal Marine Ecosystems*, Skreslet, S., Ed., Springer-Verlag, Berlin, 1986, 329.
672. Lin, C., Xu, B., and Huang, S., Long-term variations in the oceanic environment of the East China Sea and their influence on fisheries resources, in *Climate Change and Northern Fish Populations*, Beamish, R. J., Ed., National Research Council of Canada, Ottawa, No. 121, 1995, 307.
673. Dwivedi, S. N., Long-term variability in the food chains, biomass yield, and oceanography of the Bay of Bengal ecosystem, in *Large Marine Ecosystem: Stress, Mitigation and Sustainability*, Sherman, K., Alexander, L. M., and Gold, B. D., Eds., American Association for the Advancement of Science Press, Washington, D.C., 1993, chap. 7.
674. Hopner-Petersen, G., Energy flow in comparable aquatic ecosystems from different climatic zones, *Rapp. P.-v. Reun. Cons. Int. Explor. Mer.*, 183, 119, 1984.
675. Ursin, E., The tropical, the temperate and the Arctic seas as media for fish production, *Dana*, 3, 43, 1984.
676. Houde, E. D. and Rutherford, E. S., Recend trends in estuarine fisheries: predictions of fish production and yield, *Estuaries*, 16, 161, 1993.
677. FAO, *Fish and Fishery Products*, Fisheries Circular No. 821, Rev. 2, Food and Agricultural Organization, Rome, 1992, 3.
678. Goldberg, E. D., *Coastal Zone Space — Prelude to Conflict?*, UNESCO Report, Paris, 1994, 5.
679. Kelletat, D., Biosphere and man as agents in coastal geomorphology and ecology, *Geookodynamik*, 10, 215, 1989.
680. Viles, H. and Spencer, T., *Coastal Problems. Geomorphology, Ecology and Society at the Coast*, Edward Arnold, London, 1995, 350 pp.
681. GESAMP, *The State of the Marine Environment*, Blackwell Scientific, Oxford, 1990, 146 pp.
682. National Research Council, *Priorities for Coastal Ecosystem Science*, National Academy Press, Washington, D.C., 1994, 106 pp.
683. Linden, O., Human impact on tropical coastal zones, *Nature Resources*, 26, 3, 1990.
684. Taylor, P., The state of the marine environment: a critique of the work and role of the Joint Group of Experts on Scientific Aspects of Marine Pollution (GESAMP), *Mar. Poll. Bull.*, 26, 120, 1993.
685. Stocker, T. F., Climate change: the variable ocean, *Nature*, 367, 221, 1994.
686. Sindermann, C. J., *Ocean Pollution. Effects on Living Resources and Humans*, CRC Press, Boca Raton, FL, 1996, 275 pp.

687. Walker, C. H. and Livingstone, D. R., *Persistent Pollutants in Marine Ecosystems*, Pergamon Press, Oxford, 1992, 273 pp.

688. Kennish, M. J., *Ecology of Estuaries: Anthropogenic Effects*, CRC Press, Boca Raton, FL, 1992, 476 pp.

689. Gray, J. S., Eutrophication in the sea, in *Marine Eutrophication and Population Dynamics*, Colombo, G., Ferrari, I., Ugo Ceccherelli, V., and Rossi, R., Eds., Olsen & Olsen, Fredensborg, 1992, 3.

690. Nixon, S. W., Coastal marine eutrophication: a definition, social causes, and future concerns, *Ophelia*, 41, 199, 1995.

691. Paerl, H. W., Coastal eutrophication in relation to atmospheric nitrogen deposition: current perspectives, *Ophelia*, 41, 237, 1995.

692. Peierls, B. L., Caraco, N. F., Pace, M. L., and Cole, J. J., Human influence on river nitrogen, *Nature*, 350, 386, 1991.

693. Soetaert, K. and Herman, P. M. J., Nitrogen dynamics in the Westerchelde estuary (southwest Netherlands) estimated by means of the ecosystem model MOSES, *Hydrobiologia*, 311, 225, 1995.

694. Soetaert, K. and Herman, P. M. J., Carbon flows in the Westerschelde estuary (The Netherlands) evaluated by means of an ecosystem model (MOSES), *Hydrobiologia*, 311, 247, 1995.

695. Dudgeon, D., Endangered ecosystems: a review of the conservation status of tropical Asian Rivers, *Hydrobioligia*, 248, 167, 1992.

696. Ramaiah, N., Ramaiah, N., Chandramohan, D., and Nair, V. R., Autotrophic and heterotrophic characteristics in a polluted tropical estuarine complex, *Estuarine Coastal Shelf Sci.*, 40, 45, 1995.

697. Lean, G., Hinrichsen, D., and Markham, A., *Atlas of the Environment*, Arrow Books and World Wide Fund (WWF) for Nature, London, 1990, 194 pp.

698. Chou, L. M., Marine environmental issues of Southeast Asia: state and development, in *Ecology and Conservation of Southeast Asian Marine and Freshwater Environments Including Wetlands*, Sasekumar, A., Marshall, N., and Macintosh, D. J., Eds., Kluwer Academic, Dordrecht, 1994, 139.

699. Gabric, A. J. and Bell, P. R. F., Review of the effects of non-point nutrient loading on coastal ecosystems, *Aust. J. Mar. Freshwater Res.*, 44, 261, 1993.

700. Done, T. J., Ogden, J. C., Wiebe, W. J., and Rosen, B. R., Biodiversity and ecosystem function of coral reefs, in *Functional Roles of Biodiversity: Global Perspectives*, Mooney, H. A., Cushman, J. H., Medina, E., Sala, O. E., and Schulze, E.-D., Eds., John Wiley & Sons, Chichester, 1996, chap. 15.

701. Dugan, P., Ed., *Wetlands in Danger*, M. Beazley and IUCN, London, 1993, 192 pp.

702. Mitsch, W. J., Mitsch, R. H., and Turner, R. E., Wetlands of the old and new worlds: ecology and management, in *Global Wetlands: Old World and New*, Mitsch, W. J., Ed., Elsevier, Amsterdam, 1994, 3.

703. Wilkinson, C. R., Coral reefs of the world are facing widespread devastation: can we prevent this through sustainable management practices?, *Proc. Seventh Coral Reef Symp. Guam*, 1, 11, 1993.

704. Hughes, T. P., Catastrophes, phase shifts, and large-scale degradation of a Caribbean coral reef, *Science*, 265, 1547, 1994.

705. Chou, L. M., Wilkinson, C. R., Gomez, E., and Sudara, S., Status of coral reefs in the ASEAN region, in *Living Coastal Resources of Southeast Asia: Status and Management*, Wilkinson, C. R., Ed., Australian Institute of Marine Science, Townsville, 1994, 8.

706. Wilkinson, C. R. and Rahman, R. A., Causes of coral reef degradation within Southeast Asia, in *Living Coastal Resources of Southeast Asia: Status and Management*, Wilkinson, C. R., Ed., Australian Institute of Marine Science, Townsville, 1994, 18.

707. Ong, J. E., The status of mangroves in ASEAN, in *Living Coastal Resources of Southeast Asia: Status and Management*, Wilkinson, C. R., Ed., Australian Institute of Marine Science, Townsville, 1994, 52.

708. Fortes, M., Seagrass resources of ASEAN, in *Living Coastal Resources of Southeast Asia: Status and Management*, Wilkinson, C. R., Ed., Australian Institute of Marine Science, Townsville, 1994, 106.

709. Twilley, R. R., Bodero, A., and Robadue, D., Mangrove ecosystem biodiversity and conservation in Ecuador, in *Perspectives on Biodiversity: Case Studies of Genetic Resource Conservation and Development*, Potter, C. S., Cohen, J. L., and Janczewski, D., Eds., American Association for the Advancement of Science Press, Washington, D.C., 1993, chap. 9.

710. Montague, C. L., Zale, A. V., and Percival, H. F., Ecological effects of coastal marsh impoundments: a review, *Environ. Manag.*, 11, 743, 1987.

711. Wong, Y. S., Lan, C. Y., Chen, G. Z., Li, S. H., Chen, X. R. Liu, Z. P., and Tam, N. F. Y., Effect of wastewater discharge on nutrient contamination of mangrove soils and plants, *Asia-Pacific Symposium on Mangrove Ecosystems*, Wong, Y. S. and Tum, N. F. Y., Eds., Kluwer Academic, Dordrecht, 1995, 243.

712. Thorhaug, A., Restoration of mangroves and seagrasses — economic benefits for fisheries and mariculture, in *Environmental Restoration: Science and Strategies for Restoring the Earth*, Berger, J. J., Ed., Island Press, Washington, D.C., 1990, 265 pp.

713. Thorhaug, A., Miller, B., Jupp, B., and Booker, F., Effects of a variety of impacts on seagrass restoration in Jamaica, *Mar. Poll. Bull.*, 167, 355, 1985.

714. McLaughlin, P. A., Treat, S. A., Thorhaug, A., and Lematril, R., Restored seagrass and its animal communities, *Environ. Conserv.*, 10, 247, 1983.

715. Fonseca, M. S., Kenworthy, W. J., and Courtney, F. X., Development of planted seagrass beds in Tampa Bay, Florida, U.S. I. Plant components, *Mar. Ecol. Prog. Ser.*, 132, 127, 1996.

716. Field, C., *Journey Amongst Mangroves*, International Tropical Timber Organization and International Society of Mangrove Ecosystems, Hong Kong, 1995, 140 pp.

717. Saenger, P. and Siddiqi, N. A., Land from the sea: the mangrove afforestation of Bangladesh, *Ocean. Coastal Manag.*, 20, 23, 1993.

718. Rice, R. C., Environmental degradation, pollution, and the exploitation of Indonesia's fishery resources, in *Indonesia: Resources, Ecology, and Environment*, Harjono, J., Ed., Oxford University Press, Singapore, 1991, chap. 8.

719. Gibson, K. D., Zedler, J. B., and Langis, R., Limited response of cordgrass (*Spartina foliosa*) to soil amendments in a constructed marsh, *Ecol. Appl.*, 4, 757, 1994.

720. Wilsey, B. J., McKee, K. L., and Mendelssohn, I. A., Effects of increased elevation and macro- and micronutrient additions on *Spartina alterniflora* transplant success in salt-marsh dieback areas in Louisiana, *Environ. Manage.*, 16, 505, 1992.

721. Havens, K. J., Varnell, L. M., and Bradshaw, J. G., An assessment of ecological conditions in a constructed tidal marsh and two natural reference tidal marshes in coastal Virginia, *Ecol. Eng.*, 4, 117, 1995.

722. Hilborn, R., Walters, C. J., and Ludwig, D., Sustainable exploitation of renewable resources, *Ann. Rev. Ecol. Syst.*, 26, 45, 1995.

723. Haron, Hj. Abu Hassan, *A Working Plan for the Second 30-Year Rotation of the Matang Mangrove Forest Reserve, Perak*, State Forestry Dept., Perak, 1981, 115 pp.

724. Gong, W. K., Ong, C. H., Wong, C. H., and Dhanarajan, G., Productivity of mangrove trees and its significance in a managed mangrove ecosystem in Malaysia, in *Proceedings of the UNESCO Asian Symposium on Mangrove Environment — Research and Management*, Soepadmo, E., Rao, A. N., and Macintosh, D. J., Eds., Universiti Malaya, Kuala Lumpur, 1984, 216.

725. Larsson, J., Folke, C., and Kautsky, N., Ecological limitations and appropriation of ecosystem support by shrimp farming in Colombia, *Environ. Manage.*, 18, 663, 1994.

726. Vallega, A., The coastal use structure within the coastal system: a sustainable development-consistent approach, *J. Mar. Syst.*, 7, 95, 1996.

727. Robertson, A. I. and Phillips, M. J., Mangroves as filters of shrimp pond effluent: predictions and biogeochemical research needs, in *Asia-Pacific Symposium on Mangrove Ecosystems*, Wong, Y. S. and Tam, N. F. Y., Eds., Kluwer Academic, Dordrecht, 1995, 311.

728. Sherman, K., Sustainability, biomass yields, and health of coastal ecosystems: an ecological perspective, *Mar. Ecol. Prog. Ser.*, 112, 277, 1994.

729. Price, A. R. G. and Humphrey, S. L., Eds., *Application of the Biosphere Concept to Coastal Marine Areas: Papers Presented at the UNESCO/IUCN San Francisco Workshop of 14–20 August, 1989*, A Marine Conservation and Development Report, IUCN, Gland, Switzerland, 1993, 114 pp.

730. Kelleher, G., Bleakley, C., and Wells, S., Eds., *A Global Representative System of Marine Protected Areas*, Vol. 1, The Great Barrier Reef Marine Park Authority, the World Bank, and IUCN, Washington, D.C., 1995, 219 pp.

731. Polunin, N. V. C., Marine regulated areas: an expanded approach for the tropics, in *Tropical Resources: Ecology and Development*, Furtado, J. I., Morgan, W. B., Pfafflin, J. R., and Ruddle, K., Eds, Harwood Academic, London, 1990, 283.

732. Norse, E. A., Ed., *Global Marine Biological Diversity. A Strategy for Building Conservation into Decision Making*, Island Press, Washington, D.C., 1993, 383 pp.

733. Ruitenbeck, H. J., Modelling economy-ecology linkages in mangroves: economic evidence for promoting conservation in Bintuni Bay, Indonesia, *Ecol. Econ.*, 10, 233, 1994.

734. Pernetta, J. C., Leemans, R., Elder, D., and Humphrey, S., Eds., *Impacts of Climate Change on Ecosystems and Species: Marine and Coastal Ecosystems*, A Marine Conservation and Development Report, IUCN, Gland, Switzerland, 1994, 108 pp.

735. Bardach, J. E., Global warming and the coastal zone, *Clim. Change*, 15, 117, 1989.

736. Pirazzoli, P. A., Global sea-level changes and their measurement, *Global Planet. Change*, 8, 135, 1993.

737. Bakun, A., Global climate change and intensification of coastal ocean upwelling, *Science*, 247, 198, 1990.

738. Hsieh, W. W. and Boer, G. J., Global climate change and ocean upwelling, *Fish. Oceanogr.*, 1, 333, 1992.

739. UNEP, *Assessment and Monitoring of Climatic Change Impacts on Mangrove Ecosystems*, UNEP Regional Seas Reports and Studies No. 154, United Nationals Environmental Programme, 1994, 61 pp.

740. Wilkinson, C. R. and Buddemeier, R. W., *Global Climate Change and Coral Reefs: Implications for People and Reefs*, Report of the UNEP-IOC-IUCN Global Task Team on the Implications of Climate Change on Coral Reefs, IUCN, Gland, Switzerland, 1994, 124 pp.

INDEX